Lecture Notes in Computer Science 1928

Edited by G. Goos, J. Hartmanis and J. van Leeuwen

T0259790

Springer
Berlin
Heidelberg
New York
Barcelona
Hong Kong
London
Milan
Paris
Singapore
Tokyo

Ulrik Brandes   Dorothea Wagner (Eds.)

Graph-Theoretic Concepts in Computer Science

26th International Workshop, WG 2000
Konstanz, Germany, June 15-17, 2000
Proceedings

 Springer

Series Editors

Gerhard Goos, Karlsruhe University, Germany
Juris Hartmanis, Cornell University, NY, USA
Jan van Leeuwen, Utrecht University, The Netherlands

Volume Editors

Ulrik Brandes
Dorothea Wagner
University of Konstanz
Department of Computer and Information Science
Box D 188, 78457 Konstanz, Germany
E-mail: {ulrik.brandes,dorothea.wagner}@uni-konstanz.de

Cataloging-in-Publication Data applied for

Die Deutsche Bibliothek - CIP-Einheitsaufnahme

Graph theoretic concepts in computer science : 26th international
workshop ; proceedings / WG 2000, Konstanz, Germany, June 15 - 17,
2000. Ulrik Brandes ; Dorothea Wagner (ed.). - Berlin ; Heidelberg ;
New York ; Barcelona ; Hong Kong ; London ; Milan ; Paris ; Singapore ;
Tokyo : Springer, 2000
  (Lecture notes in computer science ; Vol. 1928)
  ISBN 3-540-41183-6

CR Subject Classification (1998): F.2, G.1.2, G.1.6, G.2, G.3, E.1, I.3.5

ISSN 0302-9743
ISBN 3-540-41183-6 Springer-Verlag Berlin Heidelberg New York

Springer-Verlag Berlin Heidelberg New York
a member of BertelsmannSpringer Science+Business Media GmbH
© Springer-Verlag Berlin Heidelberg 2000
Printed in Germany

Typesetting: Camera-ready by author, data conversion by PTP-Berlin, Stefan Sossna
Printed on acid-free paper   SPIN: 10722890   06/3142   5 4 3 2 1 0

Preface

The 26th International Workshop on Graph-Theoretic Concepts in Computer Science (WG 2000) was held at Waldhaus Jakob, in Konstanz, Germany, on 15–17 June 2000. It was organized by the Algorithms and Data Structures Group of the Department of Computer and Information Science, University of Konstanz, and sponsored by Deutsche Forschungsgemeinschaft (DFG) and Universitätsgesellschaft Konstanz.

The workshop aims at uniting theory and practice by demonstrating how graph-theoretic concepts can be applied to various areas in computer science, or by extracting new problems from applications. The goal is to present recent research results and to identify and explore directions for future research. The workshop looks back on a remarkable tradition of more than a quarter of a century. Previous Workshops have been organized in various places in Europe, and submissions come from all over the world.

This year, 57 attendees from 13 different countries gathered in the relaxing atmosphere of Lake Constance, also known as the Bodensee. Out of 51 submissions, the program committee carefully selected 26 papers for presentation at the workshop. This selection reflects current research directions, among them graph and network algorithms and their complexity, algorithms for special graph classes, communication networks, and distributed algorithms. The present volume contains these papers together with the survey presented in an invited lecture by Ingo Wegener (University of Dortmund) and an extended abstract of the invited lecture given by Emo Welzl (ETH Zürich).

It is my pleasure to thank all the people whose contributions made WG 2000 a successful event: the authors of papers for submitting their work and presenting accepted papers at the workshop; the invited speakers for their remarkable lectures; all reviewers for their careful reports; the members of the program committee for their timely evaluations and intensive discussion during the selection process, and in particular Peter Widmayer for coming to Konstanz for the final discussion; the DFG, the Universitätsgesellschaft Konstanz, and Springer-Verlag for financial support; the Friedrich-Naumann-Stiftung for funding the organization of the workshop at Waldhaus Jakob and Andrea Stern for supporting us during the workshop.

Special thanks go to Sabine Cornelsen, Dagmar Handke, Annegret Liebers, Barbara Lüthke, and Thomas Willhalm. They were responsible for the excellent preparation and perfect organization of the workshop. Furthermore, I would like to express my gratitude to Ulrik Brandes. While I had only to make decisions about insignificant details, he had "a finger in every pie," including the organization of the workshop, the electronic submission and discussion, and the preparation of this volume.

Konstanz, August 2000   Dorothea Wagner

The Tradition of WG

Hosts – Location

1975 U. Pape – Berlin
1976 H. Noltemeier – Göttingen
1977 J. Mühlbacher – Linz
1978 M. Nagl, H.J. Schneider – Schloß Feuerstein, near Erlangen
1979 U. Pape – Berlin
1980 H. Noltemeier – Bad Honnef
1981 J. Mühlbacher – Linz
1982 H.J. Schneider, H. Göttler – Neunkirchen, near Erlangen
1983 M. Nagl, J. Perl – Haus Ohrbeck, near Osnabrück
1984 U. Pape – Berlin
1985 H. Noltemeier – Schloß Schwanenberg, near Würzburg
1986 G. Tinhofer, G. Schmidt – Stift Bernried, near München
1987 H. Göttler, H.J. Schneider – Schloß Banz, near Bamberg
1988 J. van Leeuwen – Amsterdam
1989 M. Nagl – Schloß Rolduc, near Aachen
1990 R.H. Möhring – Johannesstift Berlin
1991 G. Schmidt, R. Berghammer – Richterheim Fischbachau, München
1992 E.W. Mayr – Wilhelm-Kempf-Haus, Wiesbaden-Naurod
1993 J. van Leeuwen – Sports Center Papendal, near Utrecht
1994 G. Tinhofer, E.W. Mayr, G. Schmidt – Herrsching, near München
1995 M. Nagl – Haus Eich, Aachen
1996 G. Ausiello, A. Marchetti-Spaccamela – Cadenabbia
1997 R.H. Möhring – Bildungszentrum am Müggelsee, Berlin
1998 J. Hromkovič, O. Sýkora – Smolenice-Castle, near Bratislava
1999 P. Widmayer – Centro Stefano Franscini, Monte Verità, Ascona
2000 D. Wagner – Waldhaus Jakob, Konstanz

Program Committee

Hans Bodlaender	Utrecht University, The Netherlands
Andreas Brandstädt	University of Rostock, Germany
Michel Habib	LIRMM Montpellier, France
Juraj Hromkovič	RWTH Aachen, Germany
Michael Kaufmann	University of Tübingen, Germany
Luděk Kučera	Charles University, Prague, Czech Republic
Alberto Marchetti-Spaccamela	Università di Roma "La Sapienzia," Italy
Ernst W. Mayr	TU München, Germany
Rolf H. Möhring	TU Berlin, Germany
Manfred Nagl	RWTH Aachen, Germany
Hartmut Noltemeier	University of Würzburg, Germany
Ondrej Sýkora	Loughborough University, United Kingdom
Gottfried Tinhofer	TU München, Germany
Dorothea Wagner	University of Konstanz, Germany (chair)
Peter Widmayer	ETH Zürich, Switzerland
Christos Zaroliagis	University of Patras, Greece

Additional Reviewers

Luca Becchetti	Yubao Guo	Thomas Schickinger
Sergei Bezrukov	Dagmar Handke	Konrad Schlude
Hans-J. Boeckenhauer	Arne Hoffmann	Sebastian Seibert
Jean-Paul Bordat	Klaus Jansen	Jop Sibeyn
Vincent Bouchitté	Ekkehard Köhler	Martin Skutella
Ulrik Brandes	Ralf Klasing	Ladislav Stacho
Bogdan Chlebus	Ton Kloks	Yannis C. Stamatiou
Mark Cieliebak	Dieter Kratsch	Elias C. Stavropoulos
Andrea Clementi	Van Bang Le	László A. Székely
Sabine Cornelsen	Annegret Liebers	Wolfgang Thomas
Peter Damaschke	Zsuzsanna Lipták	Eberhard Triesch
Miriam Di Ianni	Haiko Müller	Marc Uetz
Feodor F. Dragan	Michael Naatz	Walter Unger
Jürgen Ebert	Gabriele Neyer	Marián Vajteršic
Pavlos Efraimidis	Ulrich Quernheim	Imrich Vrťo
Stephan Eidenbenz	Dieter Rautenbach	Karsten Weihe
Jens Ernst	Mark Scharbrodt	Thomas Willhalm

Sponsoring Institutions

Deutsche Forschungsgemeinschaft (DFG)
Universitätsgesellschaft Konstanz

Table of Contents

On the Expected Runtime and the Success Probability of Evolutionary Algorithms [*]
(Invited Presentation)

Ingo Wegener

FB Informatik, LS2, Universität Dortmund, 44221 Dortmund, Germany.
wegener@ls2.cs.uni-dortmund.de

Abstract. Evolutionary algorithms are randomized search heuristics whose general variants have been successfully applied in black box optimization. In this scenario the function f to be optimized is not known in advance and knowledge on f can be obtained only by sampling search points a revealing the value of $f(a)$. In order to analyze the behavior of different variants of evolutionary algorithms on certain functions f, the expected runtime until some optimal search point is sampled and the success probability, i.e., the probability that an optimal search point is among the first sampled points, are of particular interest. Here a simple method for the analysis is discussed and applied to several functions. For specific situations more involved techniques are necessary. Two such results are presented. First, it is shown that the most simple evolutionary algorithm optimizes each pseudo-boolean linear function in an expected time of $O(n \log n)$. Second, an example is shown where crossover decreases the expected runtime from superpolynomial to polynomial.

1   Introduction

Evolutionary algorithm is the generic term for a class of randomized search heuristics. These search heuristics are investigated and applied since the mid-sixties and more intensively since the mid-eighties (Fogel (1985), Goldberg (1989), Holland (1975), Rechenberg (1994), or Schwefel (1995)). Despite many successes in applications people working in the area of efficient algorithms (including graph theoretic methods) have ignored evolutionary algorithms. The main reasons for this are the following ones. Many papers on evolutionary algorithms are quite imprecise and claim too general results which then are not proved rigorously. In particular, general evolutionary algorithms are mixed with hybrid algorithms for special problems which use ideas from efficient algorithms for the considered problem. Finally, an analysis of search heuristics for the most often considered problems is often too difficult. One of the few successful approaches is the analysis of the Metropolis algorithm for the graph bisection problem by Jerrum and Sorkin (1998). In order to understand the complicated stochastic process

[*] This work was supported by the Deutsche Forschungsgemeinschaft (DFG) as part of the Collaborative Research Center "Computational Intelligence" (531).

U. Brandes and D. Wagner (Eds.): WG 2000, LNCS 1928, pp. 1–10, 2000.

behind evolutionary algorithms and related search heuristics like simulated annealing, Rabani, Rabinovich, and Sinclair (1998) and Rabinovich, Sinclair, and Wigderson (1992) have investigated isolated features of these algorithms. These results are fundamental for a well-founded theory but they do not contain any results on the behavior of evolutionary algorithms when applied to certain specific functions. These papers have been ignored in the community working on evolutionary algorithms.

We strongly believe that the analysis of randomized search heuristics is an interesting area which should be part of the area of efficient algorithms.

In order to distinguish general evolutionary algorithms from specialized ones, we introduce and discuss in Section 2 the scenario of black box optimization. In Section 3, we present those variants of evolutionary algorithms which are investigated in this paper. In Section 4, we discuss a general and simple method to analyze the behavior of evolutionary algorithms on certain functions. This method leads in surprisingly many situations to quite precise results. In order to prove that a variant of evolutionary algorithms may have a specific property, it is useful to look for example functions where this method of analysis works. However, this method has a limited power. For specific problems one has to develop specific tools which hopefully will turn out to be useful in further situations. In Section 5, we prove that the simplest evolutionary algorithm is efficient on all pseudo-boolean linear functions and, in Section 6, it is proved that crossover, although unnecessary for many problems, can decrease the expected runtime of an evolutionary algorithm from superpolynomial to polynomial.

2   Randomized Search Heuristics in Black Box Optimization

The scenario in the area of efficient algorithms is the following one. We are concerned with a specific and well-defined optimization problem. The structure behind this problem is investigated and this leads to a specialized algorithm which then is analyzed and implemented.

However, there are situations where this scenario is not adequate. If a problem does not belong to the class of problems investigated in textbooks or scientific papers, people may not have the time, skills or other resources to develop a specialized algorithm. Moreover, the functions f to be optimized may not be given in a clearly structured way. In order to optimize a technical system it may be the only possibility to perform or simulate an experiment to evaluate f on a specific input a. In these situations one needs randomized search heuristics with a good behavior on typical problems.

Such situations are captured by the scenario of black box optimization. The problem is to maximize an unknown function $f : S \to \mathbb{R}$. Here we only consider pseudo-boolean functions $f : \{0,1\}^n \to \mathbb{R}$. The algorithm can choose the first or more generally the t-th search point based only on the current knowledge namely the $t - 1$ first search points together with their f-values.

This scenario has some unusual features. The randomized search strategy is not developed for a specific function. Hence, there will be no strategy which is superior to all the other ones. The aim is to develop a strategy with a good behavior on many problems, in particular, those problems expected to be considered in applications.

Moreover, it should be obvious that algorithms designed for a special class of functions should be better than a search heuristic for black box optimization when applied to the same class of functions. Since black box optimization is a typical scenario in applications, the design and analysis of randomized search heuristics is well motivated. The results reveal some knowledge on the type of problems where the given heuristic has good properties.

In black box optimization, the algorithm does not "know" if the best point ever seen is optimal. Hence, a stopping criterion is applied. After the algorithm has been stopped, the best of the sampled search points is presented as a solution. We abstract from this problem and consider search heuristics as infinite stochastic processes. We are interested in the random variable $T_{A,f}$ describing for the algorithm A and the function f the first point of time when an optimal input is sampled. The expected runtime $R_{A,f}$ of A for f is the expected value of $T_{A,f}$. Another important parameter is the success probability $S_{A,f,t}$ that A applied to f has sampled an optimal input within the first t steps. It is possible that $R_{A,f}$ increases exponentially while the success probability for a polynomial number of steps is not too small (constant or $1/poly(n)$). Then a multistart variant of A is efficient for f.

3   Some Variants of Evolutionary Algorithms

The most simple variant of an evolutionary algorithm is the so-called (1+1)EA working with a population size of 1 and, therefore, without crossover.

The first search point x is chosen randomly in $\{0,1\}^n$. This search point x is also the actual one. Later, a new search point x' is obtained from the actual one by mutation, i.e., the bits x_i' are considered independently and x_i' differs from x_i with the mutation probability $p_m(n)$. The default value of $p_m(n)$ equals $1/n$ ensuring that the expected number of flipping bits equals 1. The new point x' becomes the actual one iff $f(x') \geq f(x)$. Otherwise, x remains actual. The notation (1+1) describes that we select among one parent and one child. It will turn out that $p_m(n) = 1/n$ often is a good choice but sometimes other values are better. There are methods to find a good value of $p_m(n)$ during the search process (Bäck (1993, 1998)). However, a quite simple dynamic version called dynamic (1+1)EA turns out to have some good features (Jansen and Wegener (2000)). The first mutation step uses the parameter $p_m(n) = 1/n$. Before each new step $p_m(n)$ gets doubled and a value larger than $1/2$ is replaced immediately by $1/n$. This leads to phases of approximately $\log n$ steps where for each meaningful mutation probability p one step uses a mutation probability which is close to p.

The analysis of evolutionary algorithms without crossover is much easier than the analysis in the presence of the crossover operator. Although many people have claimed that crossover is a useful operator, Jansen and Wegener (1999) were the first to prove this for some well-chosen function (see Section 6). They work with a population size of n, i.e., the algorithm stores n search points and their f-values. The first n search points are chosen randomly and independently. With probability $p_c(n)$, the algorithm chooses randomly and independently two parents x' and x'' from the current population. Uniform crossover leads to a random child z where the bits z_i are chosen independently and $z_i = x'_i$ if $x'_i = x''_i$. Otherwise, $z_i = x'_i$ with probability $1/2$ and $z_i = x''_i$ otherwise. Then y is the result of mutation with parameter $1/n$ applied to z. With probability $1 - p_c(n)$, y is the result of mutation with parameter $1/n$ applied to a random parent from the current population. Let x be a random element of those elements of the current population which have the smallest f-value. The new element y replaces x iff $f(y) \geq f(x)$ and y is not a replication of (one of) its parent(s).

4   A Simple Method for the Analysis of Mutation Based Evolutionary Algorithms and Some Applications

Let $A, B \subseteq \{0,1\}^n$. Then $A <_f B$ if $f(x) <_f (y)$ for all $x \in A$ and $y \in B$. Let $A_1, \ldots, A_p$ be a partition of the search space $\{0,1\}^n$ such that $A_1 <_f A_2 <_f \ldots <_f A_p$ and A_p is the set of all optimal search points.

For $x \in A_i$ let $s(x)$ be the probability that mutation changes x into some $y \in A_j$ where $j > i$. and let $s(i) = \min\{s(x)|x \in A_i\}$. The expected time to leave A_i has an upper bound of $r(i) = 1/s(i)$. Hence, the expected runtime of the (1+1)EA on f is bounded above by the sum of all $r(i)$, $1 \leq i \leq p-1$. The bound can be improved by taking into account the probabilities $\pi(i)$ that the first search point lies in A_i. It is essential to choose a "good" partition of $\{0,1\}^n$. The number of sets should not be too large but, nevertheless, the probabilities $s(i)$ should not be too small. More involved applications of this technique consider also the probabilities that the search may omit some sets of the partition by jumping from A_i to A_{i+d} for some $d \geq 2$.

We look for an upper bound on the success probability $S_{f,t}$ of the (1+1)EA which implies a lower bound of $(1 - S_{f,t}) \cdot t$ on the expected runtime. We have no success within t steps if

- we start with a search point in $A_1 \cup \cdots \cup A_i$,
- no jump from some A_j to $A_{j+d} \cup \cdots \cup A_p$ takes place,
- and less than $(p - i)/(d - 1)$ times a change of the A-set takes place.

Hence, we have to prove that long jumps have a very small probability and that $s^*(i) = \max\{s(x)|x \in A_i\}$ is small. (It is easy to generalize the upper and lower bound technique to other mutation based evolutionary algorithms.)

In the following, we analyze the (1+1)EA on several functions. A function $f : \{0,1\}^n \to \mathbb{R}$ is called unimodal if each non-optimal search point x has a neighbor y ($H(x,y) = 1$ for the Hamming distance H) such that $f(y) > f(x)$.

Theorem 1. *Let $f : \{0,1\}^n \to \mathbb{R}$ be unimodal and let $I(f)$ be the size of the set $image(f) = \{f(x)|x \in \{0,1\}^n\}$. The expected runtime of the (1+1)EA is bounded above by $e \cdot n \cdot (I(f) - 1)$.*

Proof. We partition $\{0,1\}^n$ into the sets of inputs with the same f-value. Because of the unimodality of f there is for each non-optimal x a mutation changing one bit of x and leading to an improvement. Hence, $s(x) \geq \frac{1}{n}(1 - \frac{1}{n})^{n-1} \geq \frac{1}{e \cdot n}$ and $r(i) \leq e \cdot n$ for all non-optimal A-sets. $\qquad\square$

The function LO_n (LEADING ONES) counts the length of the longest prefix of x consisting of ones only. LO_n is unimodal and $I(LO_n) = n + 1$. The upper bound of Theorem 1 for LO_n equals $e \cdot n^2$.

Theorem 2. *The expected runtime of the (1+1)EA on LO_n equals $\Theta(n^2)$.*

Sketch of Proof. (Droste, Jansen, and Wegener, 2000). With a probability of at least $1 - \frac{1}{n}$ the initial search point has at most $\log n$ leading ones. The probability that the f-value is increased during one step equals $1/n$, since one specific bit has to flip. The suffix of the actual point x behind the prefix $1^i 0$ is a random string. Hence, the f-value increases in a successful step with probability $(1/2)^j$ by $j + 1$. The result follows by an appropriate application of Chernoff bounds (Hagerup and Rüb (1989)). $\qquad\square$

A function $f : \{0,1\}^n \to \mathbb{R}$ is called linear if $f(x) = w_1 x_1 + \cdots + w_n x_n$ for some weights. The linear function $ONEMAX_n$ is defined by $w_1 = \cdots = w_n = 1$ and the linear function BV_n (BINARY VALUE) by $w_i = 2^{n-i}$. All linear functions are unimodal. The following results have been proved by Mühlenbein (1992), Rudolph (1997b), and Droste, Jansen, and Wegener (1998).

Theorem 3. *The expected runtime of the (1+1)EA on a linear function f where $w_i \neq 0$ for all i is bounded below by $\Omega(n \log n)$. It is bounded above for all linear functions by $O(n^2)$ and by $O(n \log n)$ for $ONEMAX_n$.*

Sketch of Proof. The lower bound follows by an application of the coupon collector's theorem (see Motwani and Raghaven (1995)). For the general upper bound we assume w.l.o.g. $w_1 \geq \cdots \geq w_n \geq 0$. Let $A_i = \{x|w_1 + \cdots + w_{i-1} \leq f(x) < w_1 + \cdots + w_i\}$, $1 \leq i \leq n + 1$. Then $s(i) \geq \frac{1}{e \cdot n}$ for $i \leq n$. For $ONEMAX_n$ even $s(i) \geq \frac{n+1-i}{e \cdot n}$, since inputs from A_i have $n+1-i$ neighbors with a larger function value. $\qquad\square$

In Section 5, an upper bound of $O(n \log n)$ on the expected runtime of the (1+1)EA for an arbitrary linear function is established. The simple method used in this section seems to be too weak for this purpose. Unimodality is a natural property of many functions. Hence, people have looked for unimodal functions where the (1+1)EA needs exponential time. Such a function has to take exponentially many different function values and it has to be unlikely that one jumps over many possible function values. Long path functions (Horn, Goldberg, and Deb (1994)) have the property that there is a path (Hamming distance between a point and its successor equals one) of exponential length where the value of the function increases along the path. The values of points not on the path lead the (1+1)EA to the source of the path. Rudolph (1997a) has described explicitly paths which additionally have the property that for each $i \leq n^{1/2}$ and each point x on the path there is at most one successor on the path whose Hamming distance to x equals i. Rudolph (1997a) has applied Theorem 1 to this function. Droste, Jansen, and Wegener (2000) have obtained an asymptotically matching lower bound. They have proved that the (1+1)EA meets the path with probability at least $1/2$ in the first half of the path and that with high probability it needs $\Omega(d)$ improvement steps if its distance to the terminal of the path equals d.

Theorem 4. *There are explicitly defined unimodal functions where the expected runtime of the (1+1)EA grows exponentially. The success probability for some exponentially increasing time bound is exponentially small.*

The method also works for the dynamic (1+1)EA. If the mutation probability $1/n$ is good, we expect not to loose more than a factor of $O(\log n)$.

Theorem 5. *The expected runtime of the dynamic (1+1)EA is bounded by*

- $O(n(\log n)I(f))$ *for unimodal functions,*
- $\Theta(n^2 \log n)$ *for LO_n,*
- $O(n^2 \log n)$ *for all linear functions,*
- $O(n \log^2 n)$ *for $ONEMAX_n$.*

We omit the proofs due to Jansen and Wegener (2000). They also have shown that the dynamic (1+1)EA can be very useful. We explain their results not in detail. They have defined a function PJ_n (path and jump) with the following features. There is an island I of optimal points (all vectors with $\log n$ ones and only zeros at the first $2 \log n$ positions), a path P of length $n/4$ from $1^{n/4}0^{3n/4}$ to 0^n consisting of all $1^i 0^{n-i}$, and a bridge B of all x with $n/4$ ones. Outside $I \cup P \cup B$ the function PJ_n is defined such that one efficiently finds the bridge but not the island or the path. The PJ-value on the bridge increases with the number of ones among the first $n/4$ positions. Hence, we look for $1^{n/4}0^{3n/4}$, the source of P. For this purpose, mutation probabilities of size $1/n$ are good. The same holds on the path from $1^{n/4}0^{3n/4}$ to 0^n. It is very unlikely to find the island before having reached 0^n. Then a jump to a point in Hamming distance $\log n$ is necessary and mutation probabilities of size $(\log n)/n$ are good.

Theorem 6. *Let $p_m(n) = \frac{\alpha(n) \ln n}{n}$. If $\alpha(n) \to \infty$ or $\alpha(n) \to 0$ for $n \to \infty$, the probability that the $(1+1)EA$ on PJ_n needs a superpolynomial number of steps converges to 1. If $\alpha(n) = c$, the expected runtime is $O(n^{2+c} \ln^{-1} n + n^{c-\log c - \log \ln 2})$ which is $O(n^{2.361})$ for $c = 1/(4 \ln 2)$. The expected runtime of the dynamic $(1+1)EA$ on PJ_n is $O(n^2 \log n)$ and the success probability is $1 - e^{-\Omega(n)}$ if $t \geq dn^3 \log n$ for some constant d.*

Until now we have seen examples where steps with bad values of $p_m(n)$ have neither positive nor negative effects. However, there are examples where bad values of $p_m(n)$ lead to a disaster. There is a function PWT_n (path with a trap) with a path P of polynomial length and a trap T. PWT_n takes its optimal value at the end of P, the value is increasing along the path but the values in T are second best. All values outside $P \cup T$ are smaller and are defined in such a way that it is very likely that $P \cup T$ is reached at the beginning of P. For $p_m(n) = 1/n$ it is very likely that we follow the path to the optimum without reaching the trap which is too far away (distance approximately $\log n$). Using the dynamic $(1 + 1)$ EA there are steps with mutation probabilities close to $(\log n)/n$. Since the trap is large enough it is likely to reach the trap. The probability to return to the path is reduced to a jump to the unique optimum which is quite unlikely.

5   The $(1 + 1)$EA on Linear Functions

Theorem 7. *The expected runtime of the $(1 + 1)EA$ on a linear function is* $O(n \log n)$.

Sketch of Proof. (Droste, Jansen, and Wegener (1998, 2000)).
We have to improve the simple $O(n^2)$ bound of Theorem 3. W.l.o.g. $w_1 \geq w_2 \geq \ldots \geq w_n > 0$. It is possible that the Hamming distance from the unique optimum 1^n increases during the application of the $(1 + 1)$EA. E.g., for BV_n 10^{n-1} is better than 01^{n-1}. The following technical trick is well-known from the analysis of efficient algorithms. The $(1 + 1)$EA works on some linear function f but we investigate its behavior w.r.t. to some other function, a so-called potential function.

It has turned out that one linear function can serve as a potential function for all linear functions, namely the linear function val_n with the weights $w_1 = \cdots = w_{n/2} = 2$ and $w_{n/2+1} = \cdots = w_n = 1$. Obviously, $I(\mathrm{val}_n) = (3/2)n + 1$. we are investigating the $(1 + 1)$EA working on f. A step is called successful if it changes the actual point. Let $t(x)$ be the random number of successful steps until the $(1+1)$EA on f starting at x samples for the first time a point x' where $\mathrm{val}_n(x') > \mathrm{val}_n(x)$. The main step is the proof that $E(t(x)) \leq c$ for a constant c which does not depend on x. Since val_n is close enough to ONEMAX_n, this result implies the upper bound of $O(n \log n)$ on the expected runtime by bounding the expected number of unsuccessful steps similarly as in the proof of Theorem 3 for ONEMAX_n.

The claim $E(t(x)) \leq c$ is hard to prove. Let $x \in \{0, 1\}^n$ and let x' be the random point produced by the $(1 + 1)$EA on f by a successful step. Let S be the random set of all i where $x_i = 0$ and $x'_i = 1$. Then $S \neq \emptyset$ and $D_S(x) = \mathrm{val}(x') - \mathrm{val}(x)$ describes for some fixed set S the random gain w.r.t. val. The random variable $D_S(x)$ is difficult to analyze. Therefore, a random variable $D_S^*(x)$ taking integer values is defined such that $D_S^*(x) \leq 1$ and $\mathrm{Prob}(D_S^*(x) \leq r) \geq \mathrm{Prob}(D_S(x) \leq r)$ for all r. Then the existence of a constant $d > 0$ (which neither depends on x nor on S) such that $E(D_S^*(x)) > d$ is proved. These steps need an involved case inspection. Finally, a slight generalization of Wald's identity is proved to obtain the result.   □

This result indicates the difficulty of obtaining tight results on the expected runtime of simple variants of evolutionary algorithms on simple functions. However, typical tools used for the analysis of randomized algorithms have to be applied also for the analysis of evolutionary algorithms.

6   The Hard Case – Evolutionary Algorithms with Crossover

Evolutionary algorithms without crossover are much easier to analyze than those with crossover. Crossover is defined for two points. The success of crossover depends in the case of uniform crossover on the number and the position of those bits where the parents differ. For many functions, among them many of those discussed in this paper, crossover seems to be unable to improve evolutionary algorithms. Jansen and Wegener (1999) were the first to prove for a specific function the usefulness of crossover. Let $\text{JUMP}_{m,n}(x) = ||x||_1 = x_1 + \cdots + x_n$ if $0 \leq ||x||_1 \leq n - m$ or $||x||_1 = n$ and let $\text{JUMP}_{m,n}(x) = -||x||_1$ otherwise. The set of x where $\text{JUMP}_{m,n}(x) < 0$ is called the hole. We are only interested in the case $m = O(\log n)$. A random population of polynomial size contains with overwhelming probability no optimal point and no point in the hole . If the selection procedure rejects points in the hole and if we only use mutation, an evolutionary algorithm needs with overwhelming probability $\Omega(n^{m-\epsilon})$ steps for the optimization of $\text{JUMP}_{m,n}$ ($\epsilon > 0$ constant).

Theorem 8. *Let A be the evolutionary algorithm with crossover described in Section 3 and let $p_c(n) = 1/(n \log n)$.*

a) *Let m be a constant and $f = \text{JUMP}_{m,n}$. Then*
   - $S_{A,f,t} = 1 - \Omega(n^{-k})$ *if $t \geq c_k n^2 \log n$ for some constant c_k,*
   - $S_{A,f,t} = 1 - e^{-\Omega(n)}$ *if $t \geq cn^3$ for some constant c, and*
   - $R_{A,f} = O(n^2 \log n)$.

b) *Let $m = O(\log n)$ and $f = \text{JUMP}_{m,n}$. Then*
   - $S_{A,f,t} = 1 - e^{-\Omega(n^\delta)}$ *if $t \geq c(n^3 + n^2(\log^5 n + 2^{2m})) = \text{poly}(n)$ for some constants $\delta > 0$ and c, and*
   - $R_{A,f} = O(n \log^3 n(n \log^2 n + 2^{2m})) = \text{poly}(n)$.

Sketch of Proof. (Jansen and Wegener (1999)).
With overwhelming probability, the evolutionary algorithm produces within $O(n^2 \log n)$ steps a population of search points where each search point contains $n - m$ or n ones. This can be proved by a generalization of the solution to the coupons collector's problem. Crossover is not necessary for this but it also does not cause difficulties.

Either we have found the optimum or we have a population of individuals with $n - m$ ones. This property does not change in the future. Let us assume (incorrectly) that the individuals are random and independent strings with $n-m$ ones. If we choose two of them randomly, it is very likely that they have no common position with a zero. Then crossover creates with probability 2^{-2m} the optimal vector 1^n and mutation does not destroy this string with probability $(1 - \frac{1}{n})^n \approx e^{-1}$.

The problem is that crossover and also mutation have the tendency to create positively correlated individuals. If all members of the population are equal (or each pair of individuals has many common zeros), crossover cannot destroy these zeros and this job again has to be done by mutation. Hence, we need crossover for a quick jump over the hole but crossover decreases the diversity in the population. This is the reason why we assume that $p_c(n)$ is very small. (It is conjectured that the theorem also holds for constant values of $p_c(n)$ but we cannot prove this.)

Now a tedious analysis is necessary to ensure that with overwhelming probability we have a population whose diversity is large enough. More precisely, it is proved that for a long period of time the following statement is true for each bit position i. The number of individuals with a zero at position i is bounded above by $\frac{1}{2m}n$. This property implies that, with probability at least $1/2$, two individuals chosen randomly from the current population have no common zero. Since $p_c(n)$ is small enough, we may assume during these calculations that, if crossover does not create an optimal individual, it only has bad effects.   □

References

1. Bäck, T. (1993). Optimal mutation rates in genetic search. 5th Int. Conf. on Genetic Algorithms (ICGA), 2-8.
2. Bäck, T. (1998). An overview of parameter control methods by self-adaptation in evolutionary algorithms. Fundamenta Informaticae 34, 1-15.
3. Droste, S., Jansen, T., and Wegener, I. (1998). A rigorous complexity analysis of the $(1+1)$ evolutionary algorithm for linear functions with boolean inputs. ICEC '98, 499-504.
4. Droste, S., Jansen, T., and Wegener, I. (2000). On the analysis of the $(1+1)$ evolutionary algorithm. Submitted: Theoretical Computer Science.
5. Fogel, D. B. (1995). Evolutionary Computation: Toward a New Philosophy of Machine Intelligence. IEEE Press.
6. Goldberg, D. E. (1989). Genetic Algorithms in Search, Optimization and Machine Learning. Addison Wesley.
7. Hagerup, T. and Rüb, C. (1989). A guided tour of Chernoff bounds. Information Processing Letters 33, 305-308.
8. Holland, J. H. (1975). Adaption in Natural and Artificial Systems. Univ. of Michigan.
9. Horn, J., Goldberg, D. E., and Deb, K. (1994). Long path problems. 3rd Int. Conf. on Parallel Problem Solving from Nature (PPSN), LNCS 866, 149-158.
10. Jansen, T. and Wegener, I. (1999). On the analysis of evolutionary algorithms – a proof that crossover really can help. ESA '99, LNCS 1643, 184-193.
11. Jansen, T. and Wegener, I. (2000). On the choice of the mutation probability for the $(1+1)$ EA. Submitted: PPSN 2000.

12. Jerrum, T. and Sorkin, G. B. (1998). The Metropolis algorithm for graph bisection. Discrete Applied Mathematics 82, 155-175.
13. Motwani, R. and Raghavan, P. (1995). Randomized Algorithms. Cambridge Univ. Press.
14. Mühlenbein, H. (1992). How genetic algorithms really work. I. Mutation and hillclimbing. 2nd Int. Conf. on Parallel Problem Solving from Nature (PPSN), 15-25.
15. Rabani, Y., Rabinovich, Y., and Sinclair, A. (1998). A computational view of population genetics. Random Structures and Algorithms 12, 314-334.
16. Rabinovich, Y., Sinclair, A., and Widgerson, A. (1992). Quadratical dynamical systems. 33rd FOCS, 304-313.
17. Rechenberg, I. (1994). Evolutionsstrategie '94. Frommann-Holzboog, Stuttgart.
18. Rudolph, G. (1997a). How mutations and selection solve long path problems in polynomial expected time. Evolutionary Computation 4, 195-205.
19. Rudolph, G. (1997b). Convergence Properties of Evolutionary Algorithms. Ph.D. Thesis. Dr. Kovač, Hamburg.
20. Schwefel, H.-P. (1995). Evolution and Optimum Seeking. Wiley.

n Points and One Line:
Analysis of Randomized Games
(Abstract of Invited Lecture)

Emo Welzl

Institut für Theoretische Informatik, ETH Zürich, Switzerland. emo@inf.ethz.ch

We are given n points in the plane and a vertical line ℓ (directed upwards) that intersects the convex hull, conv$S$, of S. We want to find 'ways' to the first edge (facet) of conv$S$ met by line ℓ.

Here is a first process for getting to this first edge: Choose arbitrary points a and b from S to the left and right, resp., of ℓ. Now take a random sample R of roughly $\sqrt{n}$ points in S. Determine the first edge e_1 of conv$(R \cup \{a, b\})$ (don't worry how for the moment). If no point of S lies below the line through e_1, then e_1 is already the edge of conv$S$ we were looking for. Otherwise, determine the set V_1 of all points in S below e_1 and look for the first edge e_2 of conv$(V_1 \cup R \cup \{a, b\})$ met by ℓ. Now, either there are no points below e_2 (and we are done), or we determine the set V_2 of points below and continue. How many such phases will we go through until we have found the first edge of conv$S$? What is the expected size of $V_1 \cup V_2 \cup \ldots$? It turns out that two phases suffice, and the expected size of $V_1 \cup V_2$ is $O(\sqrt{n})$.

The second process starts with an edge $e_1 = ab$ connecting arbitrary points a and b as above. Either no point in S lies below the line through e_1 (and we are done), or we choose a random point c below e_1. Either edge ac or edge bc will be intersected by ℓ; let this edge be e_2. Now either there is no point in S below e_2, or we sample a random point below, and so on. How long will this process take on the average? We show a bound of $O((\log n)^2)$, and there are point sets, for which this bound is actually achieved.

It has been known since the early eighties that the problem of computing the first edge of conv$S$ met by ℓ can be solved optimally in linear time by linear programming [2,5]. Our motivation for considering the above processes is actually motivated by the higher-dimensional counterparts. The first process, analyzed in [4], describes what is known as the multiple pricing heuristics in linear programming, and it is a simplified variant of an algorithm previously analyzed by Clarkson [1]. The second process (see [3]) describes the so-called random edge pivot rule for the simplex method.

References

1. K.L. Clarkson, A Las Vegas algorithm for linear and integer programming when the dimension is small. *J. Assoc. Comput. Mach.* **42** (1995) 488–499.
2. M.E. Dyer, Linear algorithms for two and three-variable linear programs. *SIAM J. Comput.* **13** (1984) 31–45.

U. Brandes and D. Wagner (Eds.): WG 2000, LNCS 1928, pp. 11–12, 2000.
© Springer-Verlag Berlin Heidelberg 2000

3. B. Gärtner, J. Solymosi, F. Tschirschnitz, P. Valtr and E. Welzl, Random edge pivots on d-polytopes with $d + 2$ facets, in preparation.
4. B. Gärtner and E. Welzl, Random sampling in geometric optimization: New insights and applications, *Proc. 16th Annual ACM Symposium on Computational Geometry* (2000) 91-99.
5. N. Megiddo, Linear time algorithms for linear time programming in R^3 and related problems, *SIAM J. Comput.* **12** (1983) 759–776.

Approximating Call-Scheduling Makespan in All-Optical Networks

Luca Becchetti[1], Miriam Di Ianni[2], and Alberto Marchetti-Spaccamela[1]

[1] Dipartimento di Informatica e Sistemistica, Università di Roma "La Sapienza".
{becchett,marchetti}@dis.uniroma1.it
[2] Dipartimento di Ingegneria Elettronica e dell'Informazione, Università di Perugia.
diianni@diei.unipg.it

Abstract. We study the problem of routing and scheduling requests of limited durations in an all-optical network. The task is servicing the requests, assigning each of them a route from source to destination, a starting time and a wavelength, with restrictions on the number of available wavelengths. The goal is minimizing the overall time needed to serve all requests. We study the relationship between this problem and minimum path coloring and we show how to exploit known results on path coloring to derive approximation scheduling algorithms for meshes, trees and nearly-Eulerian, uniformly high-diameter graphs. Independently from the relationship with path coloring we propose different approximation algorithms for call scheduling in trees and in trees of rings. As a side result, we present a constant approximation algorithm for star networks. We assume for simplicity that all calls are released at time 0, however all our results hold also for arbitrary release dates at the expense of a factor 2 in the approximation ratio.

1 Introduction

Scheduling a set of communication requests (*calls* in the following) in an all-optical network requires to assign a route, a wavelength chosen from a set of limited size and a starting time to each call, with the constraint that no pair of calls using the same wavelength may simultaneously use the same link.

All-optical networks allow very high transmission rates due to both Wavelength Division Multiplexing (WDM) and the avoidance of opto-electronic conversions. WDM allows the concurrent transmission of multiple data streams on the same optic fiber, subject to the constraint that different data streams use different wavelengths when using the same optical link at the same time and in the same direction [5,11,17]. The avoidance of opto-electronic conversions is guaranteed by optical switches, directly acting on the optical signal. The optical switches we consider in this paper can direct an incoming data stream on any of the available wavelengths to any of their outgoing links on the same wavelength.

In this paper, we consider calls of limited durations (different calls may have different durations) and we address the problem of scheduling calls in order to minimize the overall time required to serve all of them. This problem is known as MINIMUM CALL SCHEDULING [4].

U. Brandes and D. Wagner (Eds.): WG 2000, LNCS 1928, pp. 13–22, 2000.

Related work. Scheduling a set of communication requests in a network in order to achieve minimum completion time is a widely studied problem in several network models [16,15,12,7,8,9]. Concerning all-optical networks, so far the problem has been mostly considered neglecting call durations (i.e. considering them infinite). A popular problem in the literature is minimizing the number of wavelengths used to serve a set of connections having infinite durations. This problem is often referred to as MINIMUM PATH COLORING and has been widely studied. It has been considered in the off-line setting in trees [19,10], in mesh networks [18], in uniformly, high-diameter nearly-Eulerian graphs [14]. In the on-line setting, it has been studied in [2,3]. In [1] the authors propose algorithms to minimize the number of rounds necessary to schedule a set of calls in a mesh. They also show that any permutation can be routed in one round if the number of wavelengths is $O(\log^2 m/\beta^2)$, where β is the edge-expansion of the network and m is the number of vertices. Call scheduling was mainly considered in the non optical case. Erlebach and Jansen study the problem in stars, trees, meshes and rings, both in the on-line and the off-line case [7,8,9]. In particular, they propose constant approximation algorithms for stars and $O(\log m)$-approximation algorithms for trees (where m is the number of nodes). The scheduling of calls of limited durations in all-optical networks has been considered in [4] for chains and rings. The authors propose constant approximation algorithms for both chains and rings.

Results of the paper. In this paper, we consider both the case in which the graph is directed and a call is an ordered pair of vertices requiring a directed path, and the similarly defined undirected case. The former model is more realistic [11], the latter is more frequent in the literature about MINIMUM PATH COLORING. In Section 3 we show how to transform a solution for the one wavelength case into one for the k wavelengths case, and then we study the relationship between MINIMUM CALL-SCHEDULING and MINIMUM PATH COLORING. If m and Δ denote, respectively, the number of vertices in the graph and the ratio between the durations of the longest and the shortest call, then the above results and those of [18,10,19,14] allow to obtain the following for MINIMUM CALL-SCHEDULING: an $\mathbf{O}(\log \Delta \cdot \operatorname{poly}(\log \log m))$ approximation algorithm for meshes in the undirected case, an $\mathbf{O}(\log \Delta \cdot \log m)$-approximation algorithm in undirected, nearly-Eulerian, uniformly high-diameter graphs and $\mathbf{O}(\log \Delta)$ approximation algorithms for trees, both in the directed and in the undirected case. On the other side, the technique we propose is general and can alwayse be used to obtain results for MINIMUM CALL SCHEDULING from bounds for MINIMUM PATH COLORING.

In Sections 4 we provide an alternative approximation algorithm for directed trees that is independent from the results on MINIMUM PATH COLORING. More in detail, we provide an approximation preserving reduction that allows to extend the approximation results of [8,9] for Call Scheduling in non optical stars to MINIMUM CALL SCHEDULING in directed stars; we also use this result to derive a $O(\log m)$ approximation algorithm for directed trees.

In Section 5 trees of rings are considered. In particular, a constant approximation algorithm is also presented for the case of chains of rings. We remark that trees of rings have practical relevance, since they are realistic topologies for LAN interconnection networks [19].

For the sake of simplicity, in this paper we assume that all calls are released at time 0. However all results can be extended to the case of arbitrary release dates at the expense of a factor 2 in the approximation ratio, using a well known result of [20].

2   Definitions

As far as directed graphs are concerned, in the following $G = (V, A)$ denotes a simple, directed graph on the vertex set $V = \{v_0, v_1, \ldots, v_{m-1}\}$ such that arc (v_i, v_j) exists if and only if arc (v_j, v_i) exists. A *network* is a pair $\langle G, k \rangle$ where k is the number of available wavelengths. Since a wavelength may also be viewed as a color, in the sequel we use the term *color* instead of wavelength. A *call* $C = [(s, d), l]$ is an ordered pair of vertices s, d completed by an integer $l > 0$ representing the call duration. More precisely, s is the source vertex originating the data stream to be sent to the destination d.

Similarly, in the undirected case $G = (V, E)$ denotes a simple, undirected graph on the vertex set $V = \{v_0, v_1, \ldots, v_{m-1}\}$ and on the edge set E and a *network* is a pair $\langle G, k \rangle$ where k is the number of available colors. A *call* $C = [\{s, d\}, l]$ is now an unordered pair of vertices s, d completed by an integer $l > 0$ representing the call duration. In this case, s and d represent a pair of vertices between which a communication has to be established.

Given a directed (undirected) network $\langle G, k \rangle$ and a set $\mathcal{C}$ of calls, a *routing* is a function $R : \mathcal{C} \to \mathcal{P}(G)$, where $\mathcal{P}(G)$ is the set of simple paths in G such that, for any $C \in \mathcal{C}$, $R(C)$ is a path connecting the source of C to its destination (its end vertices in the undirected case). Given a routing R, we say that $C_1, C_2 \in \mathcal{C}$ *conflict* in R if $R(C_1) \cap R(C_2) \neq \emptyset$.

Given a directed (undirected) network $\langle G, k \rangle$, a set $\mathcal{C}$ of calls and a routing R for calls in C, a *schedule* is an assignment of starting times and colors to calls such that at every instant no pair of calls with the same color use the same arc (edge). More formally, a schedule for a set $\mathcal{C} = \{C_h : h = 1, \ldots, n\}$ of calls is a pair $\langle S, F \rangle$, where $S : \mathcal{C} \to [0, 1, \ldots \sum_{h=1}^{n} l_h]$ is a function assigning starting times to calls and $F : \mathcal{C} \to \times \{1, \ldots k\}$ is a function assigning colors to calls. In particular, each call C_h has to be scheduled for a *contiguous* interval of l_h time units starting at $S(C_h)$, i.e. no preemption is allowed. S and F must be such that, if $T = \max_{1 \leq h \leq k} \{S(C_h) + l_h\}$ then, for any $0 \leq t \leq T$ and for any $(u, v) \in A$ ($\{u, v\} \in E$), no pair of calls with the same color uses (u, v) ($\{u, v\}$) at t. We call T the *makespan* of schedule S.

In this paper we consider MINIMUM CALL-SCHEDULING (in short, MIN-CS) defined as follows: given an directed (undirected) network $\langle G, k \rangle$ and a set $\mathcal{C} = \{C_h : h = 1, \ldots, n\}$ of calls, find a routing and a schedule for $\mathcal{C}$ having minimum makespan. For the sake of brevity, we sometimes write tree-MIN-CS, ring-MIN-CS and so forth to denote the problem in a particular class of networks. Whether we are considering the directed or the undirected case will be clear from context.

In the special case of one available color, MIN-CS will be shortly denoted as MIN-CS$_1$. In the sequel we assume that $k \le |\mathcal{C}|$, since otherwise the problem is trivial. Henceforth, an instance of the problem will be completely described by the pair $\langle\langle G, k\rangle, \mathcal{C}\rangle$. For ease of notation, an instance of MIN-CS$_1$ will also be denoted by $\langle G, \mathcal{C}\rangle$.

Given an instance of MIN-CS and a routing R, the *load* $L_R(e)$ of edge (arc) e is defined as the sum of the durations of calls that traverse e. Notice that, if $L^* = \min_R(\max_e L(e))$, then L^*/k is a natural lower bound to the makespan of any schedule for a given instance.

In the remainder of this paper Δ will always denote the ratio between the durations of the longest and of the shortest calls in $\mathcal{C}$ and OPT the makespan of an optimum schedule for an instance $\langle\langle G, k\rangle, \mathcal{C}\rangle$ of MIN-CS.

3   From MINIMUM PATH COLORING to MIN-CS

We present two theorems that allow the reduction of MIN-CS to MINIMUM PATH COLORING. Notice that the results hold both in the undirected and the directed case.

MINIMUM PATH COLORING has already been informally described in the introduction. It is now formally defined as follows: given an undirected graph $G = (V, E)$ and a set $\mathcal{F}$ of source-destination pairs (s_h, d_h) in G, $h = 1, \ldots, n$, find a set $\mathcal{P}$ of paths and a coloring for paths in $\mathcal{P}$, such that no pair of paths with the same color share the same edge and the overall number of colors is minimized. MINIMUM PATH COLORING may be similarly defined in the directed case. Notice that, if calls have unit durations, then MIN-CS$_1$ in undirected networks is isomorphic to MINIMUM PATH COLORING. The same observation holds in the case of directed networks.

Theorem 1. *If* MINIMUM PATH COLORING *is α-approximable then* MIN-CS$_1$ *is $2\alpha\lceil\log\Delta\rceil$ approximable.*

Proof. In order to apply the α-approximation algorithm for MINIMUM PATH COLORING to $\langle G, \mathcal{C}\rangle$, we partition the set $\mathcal{C}$ into $\lceil\log\Delta\rceil$ subsets $\mathcal{C}_1, \ldots, \mathcal{C}_{\lceil\log\Delta\rceil}$ as follows. To this aim, we first round up the duration of each call to the next power of 2 and we normalize to 1 the shortest duration. Then, we assign calls of durations 2^{i-1} to $\mathcal{C}_i$, $i = 1, \ldots, \lceil\log\Delta\rceil$. In this way we obtain $\lceil\log\Delta\rceil$ instances $\langle G, \mathcal{F}_i\rangle$, $i = 1, \ldots, \lceil\log\Delta\rceil$, of MINIMUM PATH COLORING, where $\mathcal{F}_i$ is obtained from $\mathcal{C}_i$ in the obvious way. Solving $\langle G, \mathcal{F}_i\rangle$, $i = 1, \ldots, \lceil\log\Delta\rceil$, we obtain a routing for calls in $\mathcal{C}_i$. A schedule for calls in $\mathcal{C}_i$ can be inductively derived. Suppose a schedule has been derived for calls in $\mathcal{C}_1 \cup \ldots \cup \mathcal{C}_{i-1}$ and let T_{i-1} be its completion time; the schedule for calls in $\mathcal{C}_i$ is built as follows: if the path corresponding to a call C in $\mathcal{C}_i$ has received color j in the solution of $\langle G, \mathcal{F}_i\rangle$, then C starts at time $T_{i-1} + (j-1)2^{i-1}$. The resulting schedule has length at most $2\alpha\lceil\log\Delta\rceil$ the length of the optimum schedule, since a factor at most 2 is lost when rounding call durations.   □

Theorem 2. *If there exists a polynomial-time α-approximation algorithm for* MIN-CS$_1$, *there also exists a polynomial-time 2α-approximation algorithm for* MIN-CS.

Proof. Let $\langle\langle G, k\rangle, \mathcal{C}\rangle$ be an instance of MIN-CS and let $\langle G, \mathcal{C}\rangle$ be the corresponding instance of MIN-CS$_1$. Denote by S and T, respectively, an α-approximation schedule for $\langle G, \mathcal{C}\rangle$ and its completion time. Given S, the following algorithm produces a schedule S_k for $\langle\langle G, k\rangle, \mathcal{C}\rangle$:

Algorithm *from_1_to_k_colors*
input: $G, \mathcal{C}, k$, a schedule S for 1 wavelength;
output: a schedule $\langle S_k, F\rangle$ for $G, \mathcal{C}, k$;
begin
    for $j := 1$ **to** k **do**
    begin
        $\mathcal{D}_j := \{C : C \in \mathcal{C} \wedge S(C) \in [(j-1)T/k, jT/k)\}$;
        Assign color j to calls in $\mathcal{D}_j$: $F(C) = j$ **for each** $C \in D_j$;
        for any call $C \in \mathcal{D}_j$ **do**
            $S_k(C) := S(C) - (j-1)T/k$;
    end
    Output $\langle S_k, F\rangle$;
end.

Clearly, since $k \leq |\mathcal{C}|$, algorithm *from_1_to_k_colors* runs in polynomial time.

After the execution of algorithm *from_1_to_k_colors* we are left with a feasible schedule S_k for MIN-CS. In fact, each call in $\mathcal{C}$ has been assigned a color and a starting time. Furthermore, for each j, the relative delays between calls in $\mathcal{D}_j$ have not changed with respect to S, hence S_k is feasible.

Let us now derive the completion time of S_k. For any wavelength j, let T_j be the completion time of calls that are assigned color j. Finally, let OPT_1 and OPT_k respectively denote the optimal values for $\langle G, \mathcal{C}\rangle$ and $\langle\langle G, k\rangle, \mathcal{C}\rangle$. For any j, let $l_{\max}^j = \max\{l : C \in \mathcal{D}_j\}$. We have:

$$T_j \leq \frac{jT}{k} - \frac{(j-1)T}{k} + l_{\max}^j \leq \frac{T}{k} + l_{\max}^j.$$

Two cases may arise:

1. $l_{\max}^j \leq \frac{T}{k}$. In this case $T_j \leq 2\frac{T}{k} \leq 2\alpha\frac{OPT_1}{k} \leq 2\alpha OPT_k$. The last inequality is proved by contradiction: if $\frac{OPT_1}{k} > OPT_k$, we could use an optimal algorithm for MIN-CS to schedule instances of MIN-CS$_1$ and the corresponding completion time should be less than OPT_1).
2. $l_{\max}^j > \frac{T}{k}$. Since $OPT_k > l_{\max}^j$, then $OPT_k > \frac{T}{k}$ and therefore $T_j < 2OPT_k$.

This completes the proof. $\qquad\square$

Corollary 1. *There exist polynomial-time approximation algorithms for* MIN CS *in undirected meshes, directed and undirected trees and undirected nearly-Eulerian, uniformly high-diameter graphs with approximation ratio i)* $\mathbf{O}(\log \Delta \cdot$

$poly(\log \log m))$ *for undirected mesh networks; ii)* $\mathbf{O}(4.4 \log \Delta)$ *for undirected trees and* $(20/3) \log \Delta$ *for directed trees; iii)* $\mathbf{O}(\log \Delta \cdot \log m)$ *for undirected, nearly-Eulerian uniformly high-diameter networks.*

Proof. Consider an instance $\langle \langle G, k \rangle, \mathcal{C} \rangle$ of MIN-CS.

If G is an undirected mesh, we first solve MIN-CS$_1$ by applying Theorem 1 and exploiting the polynomial-time $\mathbf{O}(poly(\log \log m))$-approximation algorithm for PATH COLORING in the mesh proposed in [18]. This yields an $\mathbf{O}(\log \Delta \cdot poly(\log \log m))$-approximated solution for $\langle \langle G, k \rangle, \mathcal{C} \rangle$. By applying Theorem 2 the thesis follows.

The proof for trees is similar to the previous case. For undirected trees we use a result in [7], improving that of Raghavan and Upfal [19], yielding a $1.1OPT+0.8$ approximated algorithm for MINIMUM PATH COLORING in undirected trees. Then, we apply Theorems 1 and 2 to derive an $\mathbf{O}(4.4 \log \Delta)$ approximated schedule for MIN-CS. In the directed case we use a result of Erlebach et al. [10] achieving with the same technique an approximation ratio of $(20/3) \log \Delta$ for MIN-CS.

Nearly-Eulerian, uniformly high-diameter graphs are planar graphs that extend the mesh. MINIMUM PATH COLORING in these graphs was studied in [14], where an $\mathbf{O}(\log m)$-approximation algorithm was presented. Proceeding the same way as in the previous cases we prove the last part of the Corollary.   □

As remarked in the introduction, the results presented in this and the next sections can be extended to the case in which each call has a release date at the expense of a factor 2 in the approximation ratio, using a result of [20].

4   Call Scheduling in Directed Trees

A directed star Σ_m is a directed graph with vertex set $\{u, v_1, \ldots, v_m\}$ and arc set $\{(u, v_i), (v_i, u) : i = 1, \ldots, m\}$. An instance of the problem is now a pair $\langle \Sigma_m, \mathcal{C} \rangle$, where $\mathcal{C}$ is a set of directed calls.

We now show an approximation preserving reduction of call scheduling from directed star networks to undirected stars.

Theorem 3. *Given an instance* $\mathcal{I}_d$ *of* directed star-MIN-CS *there exists an instance* $\mathcal{I}_u$ *of* undirected star-MIN-CS *such that there exists a one-to-one correspondence between feasible solutions of* $\mathcal{I}_d$ *and* $\mathcal{I}_u$ *and the makespan of corresponding solutions is the same.*

Proof. Let $\mathcal{I}_d$ consist of a directed star Σ_m and a set of calls $\mathcal{C}$ in Σ_m. The corresponding instance $\mathcal{I}_u = \langle \Sigma_{2m}^u, \mathcal{C}^u \rangle$ is derived as follows:

1. each vertex v in Σ_m is replaced in Σ_{2m}^u by the pair v_i, v_o of vertices;
2. each call $C = (s, t, d) \in \mathcal{C}$ is replaced in $\mathcal{C}^u$ by a new call $C^u = (s_i, t_o, d)$.

Notice that two calls in $\mathcal{C}$ conflict in Σ_m if and only if the corresponding two calls in $\mathcal{C}^u$ conflict in Σ_m^u. Consider a feasible solution S_d of $\mathcal{I}_d$ with makespan T; the corresponding schedule S_u for $\mathcal{I}_u$ is obtained as follows: if call C is scheduled at time t in S_d, the corresponding call C^u is scheduled at time t in S_u. Hence the makespan of S_u is T as well.

The proof of the converse is analogous. $\qquad\square$

The theorem above yields an algorithm for star-MIN-CS$_1$ in the directed case as stated in the following

Corollary 2. *There is an algorithm for star-min-cs$_1$ in the directed case that produces a schedule whose completion time is at most* 2 *times the optimum.*

Proof. Use the result of [9] for Call Scheduling in non-optical star networks and apply Theorem 3. $\qquad\square$

The generalization to the case of k colors easily follows from theorem 2, as stated in the next corollary.

Corollary 3. *There is an algorithm for star-min-cs in the directed case that produces a schedule whose completion time is at most* 4 *times the optimum.*

In the case of tree networks we use the following well known separator theorem.

Theorem 4. [6,21] *Let T be an m-vertex tree. There exists at least one vertex (called a separator or median) whose removal splits the tree into subtrees, each containing at most $m/2$ vertices.*

Theorem 5. *There exists a polynomial-time algorithm that finds a schedule for the instance $\langle T, \mathcal{C}, k \rangle$ of tree-MIN-CS in the directed case and whose completion time is at most*

$$\min\{4 \log m, \frac{20}{3} \log \Delta\} \cdot OPT.$$

Proof. We first solve tree-MIN-CS$_1$. To this purpose, if the minimum value is $4 \log m$, we use the following algorithm:

1. choose a median v_0 in $\mathcal{T}$. Assign v_0 level 0. Let $T_1, \ldots, T_h$ be the subtrees of T obtained by removing v_0. Schedule calls that traverse v_0 using the MIN-CS algorithm for directed stars.
2. For $i = 1, \ldots, h$, recursively apply the above procedure to each subtree T_i and to the corresponding calls. At the j-th step of the recursion, each new median is assigned level j. The procedure terminates when each vertex in T has been assigned a level.

The above procedure assigns a level l to each vertex in the tree, with $0 \leq l \leq \log m - 1$. This way, each call may also be given a level, uniquely defined as that of the lowest level vertex the call uses. This induces a partition of the set of calls into $\log m$ sets $\mathcal{C}_0, \ldots, \mathcal{C}_{\log m-1}$. Notice that:

i): all calls of the same level using the same median can be scheduled using the MIN-CS algorithm for directed stars;

ii): all calls of the same level, but using different medians do not share any edge along their routes and can therefore be scheduled independently and in parallel.

Calls are scheduled in phases. In particular, during phase j calls in $\mathcal{C}_j$ ($j = 0, \ldots, \log m - 1$) are scheduled. More in detail, for each median v of level j an instance $\langle \Sigma(v), \mathcal{C}(v) \rangle$ of $star$-MIN-CS$_1$ is solved, in which $\Sigma(v)$ is the star in $\mathcal{T}$ with central vertex v and $\mathcal{C}(v)$ is the set containing all calls in $\mathcal{C}_j$ that traverse v. Notice that, if $i < j$, then all calls of level i are terminated before the first call of level j is scheduled. The considerations above and Corollary 3 imply that the partial schedule of each phase has makespan at most $4OPT$. Since there are $\log m$ phases the overall schedule has makespan at most $4 \log m \cdot OPT$.

Instead, if the minimum value is $(20/3) \log \Delta$ we apply Corollary 1.   □

5   Trees and Chains of Rings

A Tree of rings [19] can be obtained from a tree by replacing each vertex of the tree by a ring under the following constraints: i) rings corresponding to non-adjacent vertices of the tree share no vertex; ii) rings corresponding to adjacent vertices of the tree share exactly one vertex.

To solve the directed case we use results of [4]. The undirected case can be solved in the same way using results of [13,4]. The proofs will be given in the full version of the paper, here we only present our results and try to give a flavour of the techniques used. Let $\mathcal{TR}$ denote a tree of rings. We solve the general problem of call scheduling in trees of rings (both in the directed and the undirected case) in two steps:

1. Perform a routing of calls with minimum load.
2. Schedule calls using the routing obtained in step 1.

The main issue in step 1 is routing in the ring. This problem has been solved in both the directed and in the undirected case [13,4]. In both cases polynomial time approximation algorithms have been proposed. Let L^* denote the load of a minimum load routing for $\langle \langle \mathcal{TR}, k \rangle, \mathcal{C} \rangle$. Of course L^* and L^*/k are lower bound to the optimum makespans of MIN-CS$_1$ and MIN-CS respectively.

In step 2 we schedule calls routed in step 1 proceeding similarly as in the case of trees, the main difference being that in each phase we solve instances of MIN-CS in ring networks. In conclusion, we prove the following theorem:

Theorem 6. *There exists a polynomial algorithm to solve* DIR-MIN-CS *in trees of rings with approximation ratio* $O(\log r)$, r *being the number of rings in the networks.*

We observe that while obtaining a constant approximation ratio is still open, not surprisingly the problem is $O(1)$-approximable in the special case of chains of rings, i.e. when the underlying topology is a chain. In this case we use a different algorithm, a major step of which is reducing the instance $\langle \langle \mathcal{TR}, k \rangle, \mathcal{C} \rangle$ to a proper instance of chain-MIN-CS in a non trivial way and then using the results of [4] (directed case) or [13] (undirected case). This allows us to claim:

Theorem 7. *There exist approximation algorithms that solve* MIN-CS *in chains of rings with approximation ratio* $(22 + \epsilon)$, *for any* $\epsilon > 0$.

The constant is high, the main contribution deriving from the scheduling of calls that are internal to the single rings of the network. In [4] it is shown that obtaining a constant lower than 11 for ring-MIN-CS may be a non trivial task.

References

1. Y. Aumann and Y. Rabani. Improved bounds for all-optical routing. In *Proc. of the 6-th ACM Symposium on Discrete Algorithms*, pages 567–576, 1995.
2. Y. Bartal, A. Fiat, and S. Leonardi. Lower bounds for on-line graph problems with application to on-line circuit and optical-routing. In *Proc. of the 28th Annual Symposium on the Theory of Computing*, pages 531–540, 1996.
3. Y. Bartal and S. Leonardi. On-line routing in all-optical networks. In *Proc. of the 24th International Colloquium on Automata, Languages and Programming*, volume 1256 of *Lectures Notes in Computer Science*, pages 516–526, 1997.
4. L. Becchetti, M. Di Ianni, and A. Marchetti-Spaccamela. Approximation Algorithms for Routing and Call Scheduling in all-optical Chains and Rings. In *Proc. of the 19th Conference on Foundations of Software Technology and Theoretical Computer Science*, volume 1738 of *Lectures Notes in Computer Science*, pages 201–212, 1999.
5. C. Brackett. Dense Wavelength Division Multiplexing Networks: Principles and Applications. *IEEE Journal Selected Areas in Comm.*, 8:948–964, 1990.
6. J. Van Leeuwen editor. *Handbook of Theoretical Computer Science. Volume A: Algorithms and Complexity*. Elesevier North-Holland, Amsterdam, 1990.
7. T. Erlebach and K. Jansen. Scheduling Virtual Connections in Fast Networks. In *Proc. of the 4th Parallel Systems and Algorithms Workshop PASA '96*, 1996.
8. T. Erlebach and K. Jansen. Call Scheduling in Trees, Rings and Meshes. In Proc. of the 30th Hawaii International Conference on System Sciences, 1997.
9. T. Erlebach and K. Jansen. Off-line and on-line call-scheduling in stars and trees. In *WG: Graph-Theoretic Concepts in Computer Science, International Workshop WG*, volume 1335 of *Lecture Notes in Computer Science*, 1997.
10. T. Erlebach, K. Jansen, C. Kaklamanis, M. Mihail, and P. Persiano. Optimal Wavelength Routing on Directed Fiber Tree. *Theoretical Computer Science*, 221 (1–2):119–137, 1999.
11. P. E. Green. *Fiber-optic Communication Networks*. Prentice-Hall, 1992.
12. M. Di Ianni. Efficient delay routing. *Theoretical Computer Science*, 196:131–151, 1998.
13. S. Khanna. A Polynomial Time Approximation Scheme for the SONET Ring Loading Problem. *Bell Labs Tech. J.*, Spring, 1997.
14. J. Kleinberg and E. Tardos. Approximations for the disjoint paths problem in high-diameter planar networks. In *Proc. of 27th ACM Symposium on the Theory Of Computing*, pages 26–35, 1995.
15. F.T. Leighton, B.M. Maggs, and A.W. Richa. Fast Algorithms for Finding O(Congestion+Dilation) Packet Routing Schedules. In *Proc. of 28th Annual Hawaii International Conference on System Sciences*, pages 555–563, 1995.
16. T. Leighton, B. Maggs, and S.Rao. Packet routing and jobshop scheduling in O(congestion+dilation) steps. *Combinatorica*, 14, 1994.
17. K. Nosu N. K. Cheung and G. Winzer. Dense Wavelength Division Multiplexing Networks: Principles and Applications. *IEEE Journal Selected Areas in Comm.*, 8, 1990.

18. Y. Rabani. Path coloring on the mesh. In *Proc. of 37th Annual IEEE Symposium Foundations of Computer Science*, pages 400–409, 1996.
19. P. Raghavan and E. Upfal. Efficient Routing in All-Optical Networks. In *Proc. of the 26th Annual Symposium on the Theory of Computing*, pages 134–143, 1994.
20. D. B. Shmoys, J. Wein, and D. P. Williamson. Scheduling parallel machines on-line. *SIAM Journal of Computing*, 24 (6):1313–1331, 1995.
21. B. Zelinka. Medians and Peripherians on Trees. *Arch. Math. (Brno)*, pages 87–95, 1969.

New Spectral Lower Bounds on the Bisection Width of Graphs[*]

Sergei L. Bezrukov[1], Robert Elsässer[2], Burkhard Monien[2], Robert Preis[2], and Jean-Pierre Tillich[3]

[1] University of Wisconsin, Superior, USA. sb@math.uwsuper.edu
[2] University of Paderborn, Germany. {elsa,bm,robsy}@uni-paderborn.de
[3] LRI, Orsay, France. tillich@lri.fr

Abstract. The communication overhead is a major bottleneck for the execution of a process graph on a parallel computer system. In the case of two processors, the minimization of the communication can be modeled by the graph bisection problem. The spectral lower bound of $\frac{\lambda_2 |V|}{4}$ for the bisection width of a graph is well-known. The bisection width is equal to $\frac{\lambda_2 |V|}{4}$ iff all vertices are incident to $\frac{\lambda_2}{2}$ cut edges in every optimal bisection. We discuss the case for which this fact is not satisfied and present a new method to get tighter lower bounds on the bisection width. This method makes use of the level structure defined by the bisection. Under certain conditions we get a lower bound depending on $\lambda_2^{\beta} |V|$ with $\frac{1}{2} \leq \beta < 1$. We also present examples of graphs for which our new bounds are tight up to a constant factor. As a by-product, we derive new lower bounds for the bisection widths of 3- and 4-regular graphs. We use them to establish tighter lower bounds for the bisection width of 3- and 4-regular Ramanujan graphs.

1 Introduction

The relation between the second smallest eigenvalue of the Laplacian and the bisection width of a graph has already been discussed in the literature. A bisection of a graph is an equal distribution of the vertices of a graph into two parts. The cut size is the number of crossing edges between the two parts. The bisection width is the minimal cut size of all bisections. Its calculation for arbitrary graphs is *NP*-complete [8], even when it is restricted to the class of *d*-regular graphs [3]. The bisection of a graph and, more generally, the partitioning of a graph into several balanced parts have a wide range of applications, especially in the field of parallel computation.

Many different methods of calculating partitions of a graph have been proposed in the past by scientists from different fields like mathematics, computer science or engineering. There are also many software libraries available which include the most efficient graph partitioning approaches. Most heuristics have

---

[*] Supported by the German Science Foundation (DFG) Project SFB-376 and by the European Union ESPRIT LTR Project 20244 (ALCOM-IT).

U. Brandes and D. Wagner (Eds.): WG 2000, LNCS 1928, pp. 23–34, 2000.

been used widely on examples in many applications and the partition qualities are satisfying.

This is not the case for lower bounds on the bisection width and there are only few approaches that are known. Leighton [11] proposes a lower bound of the bisection width by calculating a routing scheme between all pairs of vertices such that the congestion is minimized. This lower bound is strict for some graphs such as complete graphs, hypercubes, grids or tori. However, for irregularly structured graph, as they appear in most applications, remains the problem to calculate a routing scheme with a low congestion.

Lower bounds on the bisection width can be derived from algebraic graph theory by relating the bisection problem to an eigenvalue problem. The well known lower bound on the bisection width bw of a graph $G = (V, E)$ with $n = |V|$ and second smallest eigenvalue $\lambda_2(G)$ of the Laplacian is

$$bw \geq \frac{\lambda_2(G) \cdot n}{4} . \tag{1}$$

In theorem 1 we show that the equality holds for (1) iff all vertices are incident to exactly $\frac{\lambda_2}{2}$ cut edges. This is the case for some graphs, e.g. complete graphs, complete bipartite graphs, hypercubes or the Petersen graph. Nevertheless, there is a large gap between this lower bound and the bisection width for most graphs. To give an example, crash or flow simulations can be solved by the Finite Element Method. The mesh used to discretized the object often has a structure similar to a grid. The second smallest eigenvalue of a $\sqrt{n} \times \sqrt{n}$ grid is $\lambda_2 = 2 - 2 \cdot \cos(\pi/\sqrt{n}) \approx \pi^2/n$. Obviously, the bisection width is $\sqrt{n}$. However, the lower bound of equation (1) does only result in a value of $\frac{\pi^2}{4}$. Here, a relation of bw and $\sqrt{\lambda_2}$ would give a tight bound.

Such a quadratical gap also appears in the relation of the isoperimetric number and λ_2 [4,6,14]. The isoperimetric number of a graph $G = (V, E)$ is defined as $i(G) = \min_{S \subset V, 0 < |S| \leq \frac{n}{2}} \frac{|E(S, \overline{S})|}{|S|}$, where $E(S, \overline{S})$ is the set of edges which connect vertices of S with vertices of $\overline{S}$. In [6,14] it is shown that

$$\frac{i^2(G)}{2 \cdot d_{max}(G)} \leq \lambda_2(G) \leq 2 \cdot i(G)$$

with maximal degree $d_{max}(G)$. Additionally, it is $\lambda_2 = O(\frac{1}{n})$ for bounded-degree planar graphs and two-dimensional meshes and $\lambda_2 = O(\frac{1}{n^{2/d}})$ for well-shaped d-dimensional meshes [18].

If the lower bound of equation (1) is not strict, either all vertices are still incident to cut edges but not to exactly $\frac{\lambda_2}{2}$ of them, or there are vertices which are not incident to any cut edge. We will present a new method to determine an upper bound for λ_2. This method makes use of the level structure of a bisection in which every level consists of all vertices which have the same distance to the cut. This will lead to a generalized lower bound of equation (1) depending on the growth of the sizes of these levels on each side of the cut. Let $g : \mathbb{N} \to \mathbb{N}$ be a function. We will introduce the class of Level Structured graphs $LS(g, \sigma)$ which have a bisection with a cut size of σ and a level structure such that there are not more than $\sigma g(i)$ edges between vertices of distance i to the cut and vertices of distance $i+1$ to the cut. We show that if the sum $1+2\sum_{i=2}^{\infty}\frac{1}{g(i-1)}$ is a constant A, then for $\frac{n}{\sigma}\to\infty$ it is

$$\sigma \geq A\frac{\lambda_2(G)n}{4}(1+o(1)) . \tag{2}$$

Thus, we established a relation of σ and $\lambda_2^{\beta}n$ in the range $\frac{1}{2}\leq\beta<1$. For equation (2) we show that there are graphs from $LS(g,\sigma)$ for which the bound is tight The techniques used in this proof might be of special interest for some readers since we develop new methods computing asymptotic exact the second smallest eigenvalue of the laplacian for certain weighted paths.

We get an improved relation of the bisection width and λ_2 if the growth of the level sizes defined by an optimal bisection is bounded by some function $g(i)$. In general, the level structure of an optimal bisection is not known, but there is one notable exception. If G is a graph of maximum degree d, then $G \in LS(g, bw(G))$ with $g(i) = (d-1)^i$. Thus, for $\frac{n}{bw}\to\infty$, equation (2) results in a lower bound of

$$bw \geq \frac{d}{d-2}\frac{\lambda_2(G)n}{4}(1+o(1)).$$

We expect that there exist further classes of graphs for which equation (2) can be used directly to derive a lower bound on the bisection width.

As a by-product, we get better lower bounds on the bisection width of 3- and 4-regular graphs. We use them to establish tighter lower bounds for the bisection width of 3- and 4-regular Ramanujan graphs. Ramanujan graphs of degree d have $\lambda_2 \geq d - 2\sqrt{d-1}$ and, thus, for a fixed d it holds $bw = \Theta(n)$. There are several methods of explicitly constructing Ramanujan graphs [5,12,13,15]. Equation (1) leads to lower bounds of $0.043n$ and $0.134n$ for the bisection widths of 3- and 4-regular Ramanujan graphs. With the use of the level structure of a bisection we can prove that any 3-regular Ramanujan graph has a bisection width of at least $0.082n$ and any 4-regular Ramanujan graph a bisection width of at least $0.176n$. These are the highest lower bounds for explicitly constructible 3- and 4-regular graphs. However, there are 3- and 4-regular graphs with a higher lower bound on the bisection width. Kostochka and Melnikov [10] have shown that the bisection width is at least $\frac{1}{4.95}n \approx 0.101n$ for almost all 3-regular random graphs and Bollobas [2] has shown that the bisection width is at least $\frac{11}{50}n = 0.22n$ for almost all 4-regular random graphs.

Section 2 presents the basic relation of λ_2 and the bisection width. In section 3 we show a new relation between λ_2 and the cut size σ of a bisection. In section 4 we demonstrate that there are graphs for which the new bounds are tight up to a constant.

2   Definitions and Background

A bisection π of an undirected graph $G = (V, E)$ with $|V| = n$ is a partition $V = V_0 \cup V_1$ such that $|V_0| = |V_1| = n/2$. In the following we assume that n is even. For a bisection π, the number of edges between V_0 and V_1 is called the *cut size* of π. The minimum cut size of a bisection is called the *bisection width*

$bw(G)$ of G. For a graph G, the $n \times n$ *Laplacian matrix* $L(G) = \{l_{v,w}\}$ is defined as

$$l_{v,w} = \begin{cases} \deg(v), & \text{if } v = w \\ -1, & \text{if } v \neq w \text{ and } \{v, w\} \in E \\ 0, & \text{otherwise} \end{cases}$$

It is known that all the eigenvalues of $L(G)$ are nonnegative. We denote them by $\lambda_1, \ldots, \lambda_n$ with $0 = \lambda_1 \leq \lambda_2 \leq \ldots \leq \lambda_n$ and pairwise perpendicular eigenvectors $y_1, y_2, \ldots y_n$. The eigenvector to the eigenvalue 0 is $y_1 = \frac{1}{\sqrt{n}}(1, 1, 1, \ldots 1)$ and its multiplicity is equal to the number of connected components in the graph (see e.g. [17]). We consider connected graphs only, so $\lambda_2 > 0$. Let $x = (x_1, x_2, \ldots, x_n)$ be a non-zero vector. It follows from the Courant-Fisher principle that

$$\lambda_2(G) = \min_{x \perp 1} \left\{ \frac{x^t L x}{\|x\|^2} \right\} = \min_{x \perp 1} \left\{ \frac{\sum_{\{u,v\} \in E} (x_u - x_v)^2}{\sum_{v \in V} x_v^{\,2}} \right\}. \tag{3}$$

Furthermore, the minimum in (3) is attained iff x is an eigenvector to λ_2. Using the Lagrange identity, (3) can be rewritten (cf. [7]) as

$$\lambda_2(G) = \min_{x \neq \text{const}} \left\{ \frac{\sum_{\{u,v\} \in E} (x_u - x_v)^2}{\sum_{\{u,v\} \in V^2} (x_u - x_v)^2} n \right\}. \tag{4}$$

Here the minimum runs over all vectors not collinear to $(1, 1, \ldots, 1)$. Now a simple lower bound on $bw(G)$ can be derived by applying (4) to the x-tuple defined by

$$x_v = \begin{cases} a, & \text{if } v \in V_0 \\ b, & \text{if } v \in V_1 \end{cases} \text{ with some } a \neq b. \tag{5}$$

This leads to $\lambda_2 \leq n \frac{bw(a-b)^2}{\frac{n}{2} \cdot \frac{n}{2} \cdot (a-b)^2} = \frac{4 \cdot bw}{n}$, i.e. to the well-known lower bound of $bw \geq \frac{\lambda_2 \cdot n}{4}$, which is strict for some classes of graphs. In the following theorem we completely specify the situation when this bound is attainable.

Theorem 1. *Let $G = (V, E)$ be a graph and λ_2 be the second smallest eigenvalue of $L(G)$. Then the following statements are equivalent:*

a. *$bw(G) = \frac{\lambda_2 n}{4}$;*
b. *there is an eigenvector corresponding to λ_2 which has only -1 and $+1$ entries;*
c. *in any optimal bisection $V = V_0 \cup V_1$ any vertex is incident with $\frac{\lambda_2}{2}$ cut edges.*

Proof. We prove that statement (a) implies (b), (b) implies (c), and (c) implies (a).

Assume $bw(G) = \frac{\lambda_2 n}{4}$ and let $V = V_0 \cup V_1$ be an optimal bisection. Consider the vector $(x_1, \ldots, x_n)$ with $x_u = 1$ for $u \in V_0$, and $x_u = -1$ for $u \in V_1$. Now (4) implies $\lambda_2 \leq \dfrac{\sum_{\{u,v\} \in E} (x_u - x_v)^2}{\sum_{\{u,v\} \in V^2} (x_u - x_v)^2} n = \frac{4 bw(G)}{n} = \lambda_2$. Therefore, x is an eigenvector to λ_2.

Now assume there is an eigenvector x corresponding to λ_2 specified above. Let $u \in V_0$ (a similar argument works well for $u \in V_1$), and let a_u (resp. b_u) denote the number of vertices of V_0 (resp. V_1) incident to u. Then the uth entry of $L(G)x$ equals $\deg(u) - a_u + b_u$. Since $L(G)x = \lambda_2 x$ and the uth entry of x is 1, then $\deg(u) - a_u + b_u = \lambda_2$. This equality along with $a_u + b_u = \deg(u)$ implies $b_u = \frac{\lambda_2}{2}$.

Finally, let $V = V_0 \cup V_1$ be an optimal bisection. Assume any vertex is incident with exactly $\frac{\lambda_2}{2}$ cut edges. Since $|V_0| = |V_1| = n/2$, then the size of the cut equals $\frac{n}{2} \cdot \frac{\lambda_2}{2}$. On the other hand the size of the cut equals $bw(G)$ because the bisection is optimal.   □

Therefore, there are plenty of graphs for which $bw(G) = \frac{\lambda_2 n}{4}$ holds. Examples are the complete graphs, the complete bipartite graphs, hypercubes or the Petersen graph.

3   New Relations between λ_2 and σ Derived by the Level Structure

As it is shown in theorem 1, the lower bound (1) is tight only if any vertex of G is incident to a cut edge. We consider the case where this condition is not satisfied, and show that for such graphs this lower bound can be significantly improved.

Definition 1 (Level Structure). *Let π be a bisection of a graph $G = (V, E)$ with $V = V_0 \cup V_1$ and cut size σ. Denote the subsets V_0^i of V_0 as follows. Let V_0^1 be the set of all vertices in V_0 which are incident to a cut edge. Let V_0^i be the sets of all vertices in V_0 with distance $i - 1$ to a vertex of V_0^i. Denote with V_1^i the according sets on side V_1. Furthermore, denote with E_0^i, $i \geq 1$, the edge sets which connect vertices between sets V_0^i and V_0^{i+1} and with E_1^i the same in part V_1.*

Let $g : \mathbb{N} \to \mathbb{N}$ be a function. We denote with $LS(g, \sigma)$ the class of graphs which have a bisection π with a cut size σ and a level structure such that $|E_0^i| \leq \sigma g(i)$ and $|E_1^i| \leq \sigma g(i)$ for all $i \geq 1$.

In lemma 1 we bound λ_2 from above by some expression depending only on the grow function $g(i)$. Indeed, the proof of the lemma shows that the worst case occurs if $|V_i| = g(i-1)$ holds for any i. This result will be used in theorem 2.

Lemma 1. *Let $G \in LS(g, \sigma)$. Let $l \in \mathbb{N}$ be a number such that $n > 2\sigma \sum_{i=1}^{l-1} g(i-1)$. Then*

$$\lambda_2(G) \leq \min_{1 = a_1 \leq a_2 \leq \cdots \leq a_l} \frac{2\sigma a_1^2 + \sigma \sum_{i=1}^{l-1} g(i)(a_i - a_{i+1})^2}{\sigma \sum_{i=1}^{l-1} a_i^2 g(i-1) + a_l^2 \left(\frac{n}{2} - \sigma \sum_{i=1}^{l-1} g(i-1)\right)}.$$

Proof. We need a non-zero vector x with $x \perp 1$ for equation (3), whereas we do only need a non-constant x for (4). Consider (4) and choose a vector x with entries x_v for $v \in V$ defined by

$$x_v = \begin{cases} a_i, & \text{if } v \in V_0^i \\ -a_i, & \text{if } v \in V_1^i \end{cases} \tag{6}$$

with $1 = a_1 \leq a_2 \leq \cdots \leq a_l$. The upper bound of equation (4) now only depends on the a_i's. Denote by $A(x) := \sum_{\{u,v\} \in E} (x_u - x_v)^2$ the numerator of (4). One has

$$A(x) = 4\sigma a_1^2 + \sum_{i=1}^{l-1} (|E_0^i| + |E_1^i|)(a_i - a_{i+1})^2 \leq 4\sigma a_1^2 + 2\sigma \sum_{i=1}^{l-1} g(i)(a_i - a_{i+1})^2 . \tag{7}$$

Now we estimate the denominator $B(x) := \sum_{\{u,v\} \in V^2} (x_u - x_v)^2$ of (4). It holds $|V_0^j| \leq |E_0^{j-1}| \leq \sigma g(i-1)$. Assume now that $|V_0^j| < \sigma g(i-1)$ for some j, $1 \leq j < l$. Denote by $\tilde{x}$ the corresponding vector for the level structure obtained by moving a vertex from V_0^l to V_0^j. One has

$$B(\tilde{x}) - B(x)$$

$$= -\sum_{i=1}^{l-1} (a_i - a_l)^2 |V_0^i| + \sum_{i=1}^{l} (a_i - a_j)^2 |V_0^i| - (a_l - a_j)^2 + \sum_{i=1}^{l} ((a_i + a_j)^2 - (a_i + a_l)^2)|V_1^i|$$

$$= \sum_{i=1}^{l-1} ((a_i - a_j)^2 - (a_i - a_l)^2 - (a_l - a_j)^2)|V_0^i| - (a_l - a_j)^2$$

$$+ \sum_{i=1}^{l} ((a_i + a_j)^2 - (a_i + a_l)^2 + (a_l - a_j)^2)|V_1^i|$$

$$= \sum_{i=1}^{l-1} 2(a_l - a_j)(a_i - a_l)|V_0^i| - (a_l - a_j)^2 + \sum_{i=1}^{l} 2(a_i + a_j)(a_j - a_l)|V_1^i| \leq 0.$$

To establish the first equality we used $|V_0^l| = \sum_{i=1}^{l} |V_0^i| - \sum_{i=1}^{l-1} |V_0^i|$, while the inequality follows from $a_l \geq a_i$ for $0 \leq i \leq l$. Therefore, $B(x)$ does not increase because of this transformation and the minimum of $B(x)$ (for a fixed V_1 part) is attained if $|V_0^i| = \sigma g(i-1)$ for $1 \leq i < l$.

A similar argument provides $|V_1^i| = \sigma g(i-1)$ for $1 \leq i < l$ is sufficient for $B(x)$ to attain its minimum. Therefore, in this case $|V_0^i| = |V_1^i|$ for $1 \leq i < l$. Thus, $x \perp 1$ and the denominators of (3) and (4) (multiplied with $\frac{1}{n}$) are equal. These arguments imply

$$\sum_{v \in V} x_v^2 \geq 2\sigma \sum_{i=1}^{l-1} a_i^2 g(i-1) + 2a_l^2 \left(\frac{n}{2} - \sigma \sum_{i=1}^{l-1} g(i-1) \right). \tag{8}$$

The lemma follows by substituting (7) and (8) into (3). $\quad\square$

Theorem 2. *Let $G \in LS(g, \sigma)$. Then there exists a function $\gamma : \mathbb{R}^+ \to \mathbb{R}$ with $\gamma(x) \to 0$ for $x \to \infty$ such that*

- *if $A := 1 + 2 \sum_{i=2}^{\infty} \frac{1}{g(i-1)} < \infty$, then $\sigma \geq A \frac{\lambda_2(G)n}{4}(1 + \gamma(\frac{n}{\sigma}))$.*
- *if $g(i) = (i+1)$, then $\sigma \geq LambertW(\frac{4}{\lambda_2(G)}) \frac{\lambda_2(G)n}{4}(1 + \gamma(\frac{n}{\sigma}))$, where $LambertW(x)$ is the inverse function of $x \cdot e^x$.*

- if $g(i) = (i+1)^\alpha$ and $0 \le \alpha < 1$, then $\sigma \ge \delta(\alpha)\lambda_2(G)^{\frac{\alpha+1}{2}}n(1+\gamma(\frac{n}{\sigma}))$ where $\delta(\alpha) = \frac{1+\alpha}{2((1-\alpha)(3-\alpha))^{\frac{\alpha+1}{2}}}$.

Proof. Let $l \in \mathbb{N}$ be defined by $2\sigma\sum_{i=1}^{l-1}g(i-1) < n \le 2\sigma\sum_{i=1}^{l}g(i-1)$. This, in particular, implies $\frac{n}{\sigma} \to \infty$ as $l \to \infty$. We apply lemma 1 with $a_i = 1 + 2\sum_{j=2}^{i}\frac{1}{g(j-1)}$. Since $2a_1^2 + \sum_{i=1}^{l-1}g(i)(a_i - a_{i+1})^2 = 2 + 4\sum_{i=1}^{l-1}\frac{1}{g(i)} = 2a_l$ then

$$\lambda_2 \le \frac{2\sigma a_l}{\sigma\sum_{i=1}^{l-1}a_i^2 g(i-1) + a_l^2\left(\frac{n}{2} - \sigma\sum_{i=1}^{l-1}g(i-1)\right)} \le \frac{2a_l}{\sum_{i=1}^{l-1}a_i^2 g(i-1)} \quad (9)$$

In the following we use some functions γ_i with the property described in the theorem.

- $A := 1 + 2\sum_{i=2}^{\infty}\frac{1}{g(i-1)} < \infty$: Set $r_i := A - a_i$, $1 \le i \le l$, with $r_i > 0$. The first inequality of equation (9) implies

$$\lambda_2 \le \frac{2\sigma(A - r_l)}{\sigma\sum_{i=1}^{l-1}(A^2 - 2Ar_i + r_i^2)g(i-1) + (A^2 - 2Ar_l + r_l^2)\left(\frac{n}{2} - \sigma\sum_{i=1}^{l-1}g(i-1)\right)}$$

$$= \frac{2\sigma(A - r_l)}{\sigma A^2(1+\gamma_1(\frac{n}{\sigma}))\sum_{i=1}^{l-1}g(i-1) + A^2(1+\gamma_2(\frac{n}{\sigma}))\left(\frac{n}{2} - \sigma\sum_{i=1}^{l-1}g(i-1)\right)}$$

$$\le \frac{4\sigma}{A \cdot n(1+\gamma(\frac{n}{\sigma}))}.$$

- $g(i) = (i+1)$: Note that $\ln(n+1) = \int_1^{n+1}\frac{1}{x}dx \le \sum_{i=1}^{n}\frac{1}{i} \le 1 + \int_1^n\frac{1}{x}dx = 1 + \ln(n)$. Therefore, the second inequality in (9) with $g(i) = i+1$ provides

$$\lambda_2 \le \frac{2a_l}{\sum_{i=1}^{l-1}ia_i^2} \le \frac{4\ln(l)+2}{1 + \sum_{i=2}^{l-1}i(2\ln(i+1)-1)^2}$$

$$\le \frac{4\ln(l)+2}{1 + \int_1^{l-1}(4x\ln^2(x+1) - 4x\ln(x+1) + x)dx} \le \frac{2}{l^2\ln(l)(1+\gamma_1(\frac{n}{\sigma}))}.$$

It follows from $n \le 2\sigma\sum_{i=1}^{l}i = 2\sigma\frac{l^2}{2}(1+\gamma_2(\frac{n}{\sigma}))$ that $l \ge \sqrt{\frac{n}{\sigma}(1+\gamma_2(\frac{n}{\sigma}))}$. This leads to $\lambda_2 \le \frac{2\sigma}{n\ln(\sqrt{n/\sigma})}(1+\gamma_2(\frac{n}{\sigma}))$. Solving this equation in σ we get the result of the theorem.

- $g(i) = (i+1)^\alpha$ and $0 \le \alpha < 1$: Since $\sum_{i=1}^{l}i^\beta = \frac{l^{\beta+1}}{\beta+1}(1+o(1))$ for $\beta > -1$, it is

$$\lambda_2 \le \frac{2a_l}{\sum_{i=1}^{l-1}i^\alpha a_i^2} \le \frac{4\frac{l^{1-\alpha}}{1-\alpha}}{\sum_{i=1}^{l-1}i^\alpha 4(\frac{i^{1-\alpha}}{1-\alpha})^2(1+\gamma_1(\frac{n}{\sigma}))} = \frac{(1-\alpha)l^{1-\alpha}}{\sum_{i=1}^{l-1}i^{2-\alpha}(1+\gamma_1(\frac{n}{\sigma}))} \le \frac{(1-\alpha)(3-\alpha)}{l^2(1+\gamma_2(\frac{n}{\sigma}))}.$$

From $n \le 2\sigma\sum_{i=1}^{l}i^\alpha = 2\sigma\frac{l^{\alpha+1}}{\alpha+1}(1+\gamma_3(\frac{n}{\sigma}))$ it follows that $l \ge (\frac{n(\alpha+1)}{2\sigma(1+\gamma_3(\frac{n}{\sigma}))})^{\frac{1}{\alpha+1}}$. This leads to

$$\lambda_2 \le \frac{(1-\alpha)(3-\alpha)}{(\frac{n(\alpha+1)}{2\sigma})^{\frac{2}{\alpha+1}}(1+\gamma_4(\frac{n}{\sigma}))} \quad , \text{ i. e. } \quad \sigma \ge \frac{1+\alpha}{2((1-\alpha)(3-\alpha))^{\frac{\alpha+1}{2}}}\lambda_2^{\frac{\alpha+1}{2}}n(1+\gamma(\frac{n}{\sigma})). \quad \square$$

If G is a graph of maximum degree d, then $G \in LS(g, bw(G))$ with $g(i) = (d-1)^i$, because $\max\{|V_0^1|, |V_1^1|\} \le bw(d-1)^i$. Thus, theorem 2 implies

$$bw(G) \ge A \frac{\lambda_2 n}{4}(1 - o(1)) = \frac{d}{d-2} \frac{\lambda_2 n}{4}(1 - o(1)), \quad \text{as } \frac{n}{bw} \to \infty. \tag{10}$$

In the following, we derive bounds for all 3- and 4-regular graphs. One example are the d-regular Ramanujan graphs. For these graphs it holds $\lambda_2 \ge d - 2\sqrt{d-1}$. Notice that the latter value is asymptotically the largest possible for d-regular graphs, since $\lambda_2 \le d - 2\sqrt{d-1} + \frac{4\sqrt{d-1}}{\log_{d-1}(n) - O(1)}$ [16] (see also [1, 12]). There are known construction of infinite families of d-regular Ramanujan graphs for any d of the form $d = p^k + 1$, where p is any prime number, and k is an arbitrary positive integer (see [5,12,13,15]).

Theorem 3. *The bisection width of any*

1. *4-regular graph with $\lambda_2 \le 2$ is at least* $\min\{\frac{n}{2}, \frac{5-\lambda_2}{7-(\lambda_2-1)^2} \cdot \frac{\lambda_2 n}{2}\}$.
2. *3-regular graph with $\lambda_2 \le 2$ is at least* $\min\{\frac{n}{2}, \frac{4-\lambda_2}{4-\lambda_2^2+2\lambda_2} \cdot \frac{\lambda_2 n}{2}\}$.
3. *3-regular graph with $\lambda_2 \le \frac{5-\sqrt{17}}{2}$ is at least* $\min\{\frac{n}{6}, \frac{10+\lambda_2^2-7\lambda_2}{8+3\lambda_2^3-17\lambda_2^2+10\lambda_2} \cdot \frac{\lambda_2 n}{2}\}$.

Proof. As mentioned above, for any d-regular graph G it holds $G \in LS(g, bw(G))$ with $g(i) = (d-1)^i$. Solving the equation of lemma 1 for $bw(G)$ we get

$$bw(G) \ge \max_{1=a_1 \le \dots \le a_l} \frac{a_l^2 \lambda_2 \frac{n}{2}}{2a_1^2 + \sum_{i=1}^{l-1}[(d-1)^i(a_i - a_{i+1})^2 + \lambda_2(d-1)^{i-1}(a_l^2 - a_i^2)]} \tag{11}$$

with optimal values $a_1 = 1$, $a_2 = \frac{1+d-\lambda_2}{d-1}$ and $a_{j+1} = a_j \frac{d-\lambda_2}{d-1} - a_{j-1}\frac{1}{d-1}$. We only have to make sure that $1 = a_1 \le a_2 \le \dots \le a_l$ provided that $n \ge 2bw \sum_{i=1}^{l-1} g(i-1)$.

4-regular, $\lambda_2 \le 2$ and $bw \le \frac{n}{2}$: We apply equation (11) with $d = 4$ and $l = 2$. We have $1 = a_1 \le a_2 = \frac{5-\lambda_2}{3}$ and $n \ge 2bw = 2bw \sum_{i=1}^{l-1} g(i-1)$. We get $bw \ge \frac{(5-\lambda_2)\lambda_2}{7-(\lambda_2-1)^2} \cdot \frac{n}{2}$ in this case.

3-regular, $\lambda_2 \le 2$ and $bw \le \frac{n}{2}$: We apply equation (11) with $d = 3$ and $l = 2$. We have $1 = a_1 \le a_2 = \frac{4-\lambda_2}{2}$ and $n \ge 2bw = 2bw \sum_{i=1}^{l-1} g(i-1)$. We get $bw \ge \frac{(4-\lambda_2)\lambda_2}{4-\lambda_2^2+2\lambda_2} \cdot \frac{n}{2}$ in this case.

3-regular, $\lambda_2 \le \frac{5-\sqrt{17}}{2}$ and $bw \le \frac{n}{6}$: We apply equation (11) with $d = 3$ and $l = 3$. We have $1 = a_1 \le a_2 = \frac{4-\lambda_2}{2} \le a_3 = \frac{\lambda_2^2-7\lambda_2+10}{4}$ and $n \ge 6bw = 2bw \sum_{i=1}^{l-1} g(i-1)$. We get $bw \ge \frac{(\lambda_2^2-7\lambda_2+10)\lambda_2}{3\lambda_2^3-17\lambda_2^2+10\lambda_2+8} \cdot \frac{n}{2}$ in this case. $\square$

From $\lambda_2 \le d - 2\sqrt{d-1} + O(\log_{d-1}(n))^{-1}$ [16,1,12] follows that the conditions $\lambda_2 \le 2$ and $\lambda_2 \le \frac{5-\sqrt{17}}{2} \approx 0.44$ of cases 1 and 3 of theorem 3 hold true for sufficiently large graphs. The results of theorem 3 can not generally be improved by using more levels.

The bounds of theorem 3 are at least as good as the bound of equation (1). For the first two cases, the new bounds are identical to the traditional bound for $\lambda_2 \to 2$. For $\lambda_2 \to 0$ the first case is by a factor of $\frac{10}{7}$ and the second case is by a factor of 2 higher than the traditional bound. For $\lambda_2 \to 0$ the third case is by a factor of $\frac{5}{2}$ higher than equation (1) and for $\lambda_2 \to \frac{5-\sqrt{17}}{2}$ it is higher by a factor of $\frac{2\sqrt{17}+6}{3\sqrt{17}-3} \approx 1.52$.

Theorem 3 can be used to derive stronger lower bounds on the bisection width of Ramanujan graphs.

Corollary 1. *The bisection width of any sufficiently large 3-regular Ramanujan graph (such that $\lambda_2 \leq \frac{5-\sqrt{17}}{2}$) is at least $0.082n$. The bisection width of any sufficiently large 4-regular Ramanujan graph (such that $\lambda_2 \leq 2$) is at least $0.176n$.*

This improves upon the previously known lower bounds of $0.042n$ for 3-regular and of $0.133n$ for 4-regular Ramanujan graphs derived by equation (1).

4   Worst Case Graphs

In this section we present examples for which the results of the previous section are asymptotically tight. Firstly, we introduce a new class of edge-weighted graphs. They belong to the class $LS(g, \sigma)$ with $g(i) = (i + 1)^\alpha$. We show that the results of theorem 2 are tight for this graph class if $\alpha \geq 1$ and tight up to a constant factor if $0 \leq \alpha < 1$. Secondly, we analyze the double-root trees as a more realistic example.

4.1   Worst Case Graphs with Edge Weights

The Laplacian of a graph can be generalized for edge weighted graphs where the off-diagonal entries contain the negative values of the corresponding edge weight.

Definition 2. *Denote with B_l^α, $\alpha \geq 0$ and $l \geq 1$, the edge-weighted graph obtained as follows. B_l^α consists of two symmetric sides. Each side consists of l levels of vertices with i^α vertices in level i, $1 \leq i \leq l$. Edges connect every pair of vertices in consecutive levels. Each edge between vertices in levels i and $i + 1$ is weighted by $\frac{1}{i^\alpha}$. Moreover, the single vertices on level 1 of each side are connected by an edge of weight 1.*

The graphs B_l^α belong to the class $LS(g, \sigma)$ with $g(i) = (i + 1)^\alpha$. We need the following lemma to prove the main theorems.

Lemma 2. *The eigenvalues of a matrix of the structure $\begin{bmatrix} A & C \\ C & A \end{bmatrix}$ are the union of the eigenvalues of the matrices $A + C$ and $A - C$*

Proof. Let x be an eigenvector of the matrix $A+C$. Then (x, x) is an eigenvector to the matrix $\begin{bmatrix} A & C \\ C & A \end{bmatrix}$ for the same eigenvalue. If y is an eigenvector to the matrix $A-C$ then $(y, -y)$ is an eigenvector of the matrix $\begin{bmatrix} A & C \\ C & A \end{bmatrix}$ to the same eigenvalue.

$\square$

The following lemma can be shown using the separation theorem of [19].

Lemma 3. *Let $\lambda_2(B_l^\alpha) < 2$, then the entries in the eigenvector of B_l^α to the eigenvalue λ_2 are equal for all vertices on the same level. Moreover $x_i^0 = -x_i^1$, with x_i^s being the entry of the corresponding eigenvector on level i on side $s \in \{0, 1\}$.*

Now we are going to calculate lower bounds for $\lambda_2(B_l^\alpha)$ and it follows from theorem 2, that there exists an l_0, such that for any $l \geq l_0$ it holds $\lambda_2(B_l^\alpha) < 2$. In the following we consider $l \geq l_0$. We denote with x_i the entry of the eigenvector of level i to the eigenvalue λ_2 of B_l^α. In section 3 we used a_i as an approximation for x_i. Now we set $x_i = a_i + r_i$, where $r_i = -\lambda_2 p_i$ and $a_i = 1 + 2\sum_{j=1}^{i-1} \frac{1}{g(j)}$. Here r_i represents an error and in the following our method consists in showing that the r_i's are small compared with a_i.

First, we consider the case $\alpha > 1$. For this case we can show in general for a Graph G that the upper bound for λ_2 is asymptotically tight if the results of Lemma 3 hold for G and $A = 1 + 2\sum_{j=1}^{\infty} \frac{1}{g(j)} < \infty$. Let q_i be defined as $q_{i+1} = q_i + \alpha(i)(q_i - q_{i-1} + A)$, $q_1 = 0$, $q_2 = \frac{1}{g(1)}$. From the definition of q_i, we obtain $q_i = \sum_{j=1}^{i-1} \frac{1}{g(j)} + \sum_{(p,q),1\leq p<q\leq i-1} \frac{g(p)}{g(q)} A$ and $q_{i+1} - q_i = \frac{1}{g(i)} + \sum_{p=1}^{i-1} \frac{g(p)}{g(i)} A$.

Lemma 4. *For a graph G with $A = 1 + 2\sum_{j=1}^{\infty} \frac{1}{g(j)} < \infty$ it holds $q_{i+1} - q_i \geq p_{i+1} - p_i \geq 0$ for any $1 \leq i \leq l - 1$. Moreover $\lambda_2 q_l \to \infty$*

Proof. We first show the second statement of our lemma.

$$q_i = \sum_{j=1}^{i-1} \frac{1}{g(j)} + \sum_{1\leq p<q\leq i-1} \frac{g(p)}{g(q)} A \leq A + \sum_{q=1}^{i-1}\sum_{p=1}^{q-1} \frac{g(p)}{g(q)} A \leq A + \sum_{p=1}^{i-1}\left(\sum_{q=p+1}^{i-1} \frac{1}{g(q)}\right)g(p)$$

Conform theorem 2 it holds $\lambda_2 \leq O(l^{-2})$ and therefore the second statement of the lemma holds. For the first statement let $\delta_i = q_i - p_i$. We use induction on i. For $i = 1$ the lemma holds. We assume that the lemma holds for any $j \leq i$. Then

$$\delta_{i+1} = q_{i+1} - p_{i+1} = q_i + \alpha(i)(q_i - q_{i-1} + A) - p_i - \alpha(i)((1 - \lambda_2)p_i - p_{i-1} + a_i)$$
$$= \delta_i + \alpha(i)(\delta_i - \delta_{i-1}) + \alpha(i)(A - a_i) + \alpha(i)(p_{i-1} + \lambda_2 p_i) \geq \delta_i$$

Now we have to show, that $p_{i+1} - p_i \geq 0$. Our assumption was, that the lemma holds for any j, $0 \leq j \leq i$. This implies that $0 \leq p_i \leq q_i$. Then $p_{i+1} - p_i \geq \alpha(i)(a_i - \lambda_2 q_i) \geq 0$, where $\lambda_2 \leq O(l^{-2})$ because of theorem 2 and therefore the last inequality holds. □

We state a lower bound for $\lambda_2(G)$ in the next theorem with $A = 1 + 2\sum_{j=1}^{\infty} \frac{1}{g(j)} < \infty$.

Theorem 4. *If For a graph G with $A = 1 + 2\sum_{j=1}^{\infty} \frac{1}{g(j)} < \infty$ if the results of lemma 3 hold, then we obtain*

$$\lambda_2(G) \geq \frac{4}{A} \cdot \frac{1}{n(1 + o(1))}.$$

Proof. Let us consider $\lambda_2 = \frac{f_1(x)}{f_2(x)}$, where x is the eigenvector to the eigenvalue λ_2. Now using lemma 3 it holds $f_1(x) = 4x_1^2 + 2\sum_{i=1}^{l-1}(x_{i+1} - x_i)^2 g(i)$ and $f_2(x) = 2\sum_{i=1}^{l} x_i^2 g(i-1)$. We get

$$f_1(x) \geq 4 + 2\sum_{i=1}^{l-1}\frac{4}{g(i)} - 8\lambda_2\sum_{i=1}^{l-1}(q_{i+1} - q_i) \geq 4a_l - o(l)$$

and

$$f_2(x) \leq 2\sum_{i=1}^{l-1}(a_i^2 + \lambda_2^2 q_i^2)g(i) \leq 2\sum_{i=1}^{l-1} a_i^2 g(i) + o(l^2) \leq 2\sum_{i=1}^{l-1} a_l^2 g(i) + o(l^2)$$

Calculating the bounds for $f_1(x)$ and $f_2(x)$ we obtain the result of the theorem.
$\square$

This theorem provides also results for the graph class B_l^α which is a subclass of the graphs considered above.

In the next theorem we consider the case $0 \leq \alpha < 1$.

Theorem 5. *If $0 \leq \alpha < 1$ then there exists a constant $c(\alpha)$ such that $\lambda_2(B_l^\alpha) \geq c(\alpha)\frac{1}{l^2(1+o(1))}$ with $c(\alpha) = \frac{2(1-\alpha^2)(3-\alpha)(7-\alpha)}{4(7-\alpha)+(1-\alpha)^2(3-\alpha)}$.*

The technique which is used is the same as for theorem 4.

We summarize the results of this section and compare them to the results of theorem 2. For the case $\alpha > 1$, the upper and lower bounds on λ_2 of theorems 2 and 4 are asymptotically tight. In the case $0 < \alpha < 1$, the bounds of theorems 2 and 5 are tight up to a constant factor. To see this, it is known that $n = 2\sum_{i=1}^{l} g(i) = \frac{2}{\alpha+1}l^{\alpha+1}(1 + o(1))$ and thus there exits a $d(\alpha)$ such that $\lambda_2 \geq d(\alpha)\frac{1}{n^{\frac{2}{\alpha+1}}(1+o(1))}$. A special case is $\alpha = 0$. The graph B_l^0 is a path of length $2l$ and it holds $c(0) = \frac{42}{31}$ from theorem 5. From theorem 2 we get $\lambda_2 \leq 3\frac{1}{l^2(1+o(1))}$. However, for the path we know that $\lambda_2 \approx \frac{\pi^2}{4l^2}$. Thus, there is a gap in a constant factor between the lower and the upper bound for λ_2.

4.2 The Double-Root Trees as Worst Case Graphs

We define the double-root tree $T_{d,l}$ as follows. It has a central edge c, each vertex of the tree has either degree 1 or d, and the distance between c and any vertex of degree 1 is $l - 1$. Let $n = 2\frac{(d-1)^l-1}{d-2}$ denote the number of vertices of $T_{d,l}$ and we denote with $L(T_{d,l})$ the Laplacian of $T_{d,l}$.

In this paper we make use of the proof of a lemma in [9], where a lower and upper bound is given for the second smallest eigenvalue of the double-root tree. In [9] the authors show that the entries of the eigenvector to the second smallest eigenvalue of $L(T_{d,l})$ depends only on the levels Indeed for any $1 \leq i \leq l$ it holds $x_i^0 = -x_i^1$. Moreover $T_{d,l}$ belongs to the class $L(g,1)$ with $g(i) = (d-1)^i$ and $A = 1 + 2\sum_{j=1}^{i-1}\frac{1}{g(j)} < \infty$ and therefore we can use theorem 4 and obtain the following theorem

Theorem 6. *The smallest non-zero eigenvalue λ_2 of the Laplacian $L(T_{d,l})$ satisfies*

$$\lambda_2 = 4\frac{d-2}{d \cdot n(1 + o(1))}, \quad as \ n \to \infty.$$

Theorem 6 shows that the bound of equation (10) is asymptotically tight.

References

1. N. Alon. Eigenvalues and expanders. *Combinatorica*, 6(2):83–96, 1986.
2. B. Bollobas. The isoperimetric number of random regular graphs. *Europ. J. Combinatorics*, 9:241–244, 1988.
3. T.N. Bui, S. Chaudhuri, F.T. Leighton, and M. Sisper. Graph bisection algorithms with good average case behaviour. *Combinatorica*, 7(2):171–191, 1987.
4. J. Cheeger. A lower bound for the smallest eigenvalue of the laplacian. *Problems in analysis*, pages 195–199, 1970.
5. P. Chiu. Cubic ramanujan graphs. *Combinatorica*, 12(3):275–285, 1992.
6. J. Dodziuk and W.S. Kendall. Combinatorial laplacians and isoperimetric inequality. *Pitman Res. Notes Math. Ser.*, pages 68–74, 1986.
7. M. Fiedler. A property of eigenvectors of nonnegative symmetric matrices and its application to graph theory. *Czechoslovak Mathematical J., Praha*, 25(100):619–633, 1975.
8. M.R. Garey, D.S. Johnson, and L. Stockmeyer. Some simplified NP-complete graph problems. *Theoretical Computer Science*, 1:237–267, 1976.
9. S. Guattery and G. Miller. On the performance of spectral graph partitioning methods. In *Proc. Sixth Annual ACM-SIAM Symposium on Discrete Algorithms*, pages 233–242, 1995.
10. A.V. Kostochka and L.S. Melnikov. On a lower bound for the isoperimetric number of cubic graphs. In *Probabilistic Methods in Discrete Mathematics*, pages 251–265. TVP/VSP, 1993.
11. F.T. Leighton. *Introduction to Parallel Algorithms and Architectures: Arrays, Trees, Hypercubes.* Morgan Kaufmann Publishers, 1992.
12. A. Lubotzky, R. Phillips, and P. Sarnak. Ramanujan graphs. *Combinatorica*, 8(3):261–277, 1988.
13. G. A. Margulis. Explicit group-theoretical constructions of combinatorial schemes and their application to the design of expanders and concentrators. *Probl. Inf. Transm.*, 24(1):39–46, 1988.
14. B. Mohar. Isoperimetric numbers of graphs. *J. Combin. Theory*, 47(3):274–291, 1989.
15. M. Morgenstern. Existence and explicit constructions of $q + 1$ regular ramanujan graphs for every prime power q. *J. Comb. Theory, Ser. B*, 62(1):44–62, 1994.
16. A. Nilli. On the second eigenvalue of a graph. *Discrete Mathematics*, 91:207–210, 1991.
17. A. Pothen, H.D. Simon, and K.P. Liu. Partitioning sparse matrices with eigenvectors of graphs. *SIAM J. on Matrix Analysis and Applications*, 11(3):430–452, 1990.
18. D. A. Spielman and S.-H. Teng. Spectral partitioning works: Planar graphs and finite element meshes. In *Proc. 37th Conf. on Foundations of Computer Science*, pages 96–105, 1996.
19. J. H. Wilkinson. *The Algebraic Eigenvalue Problem.* Oxford University Press, 1965.

Traversing Directed Eulerian Mazes

(Extended Abstract)

Sandeep Bhatt[1], Shimon Even[2,*], David Greenberg[3], and Rafi Tayar[2]

[1] Akamai Technologies, 201 Broadway, Cambridge, MA 02139, USA.
bhatt@akamai.com
[2] Computer Science Deptartment, Technion - Israel Institute of Technology,
Haifa, Israel. {even,cstayar}@cs.technion.ac.il
[3] Center for Computing Science, Maryland, USA. dsg@super.org

Abstract. Two algorithms for threading directed Eulerian mazes are
described. Each of these algorithms is performed by a traveling robot
whose control is a finite-state automaton. Each of the algorithms puts
one pebble in one of the exits of every vertex. These pebbles indicate an
Eulerian cycle of the maze. The simple algorithm performs $O(|V| \cdot |E|)$
edge traversals, while the advanced one traverses every edge three times.
Both algorithms use memory of size $O(\log d_{out}(v))$ in every vertex v.

1   Introduction

We consider the problem of traversing a directed finite Eulerian maze, $G(V, E)$,
by a robot R whose movements are controlled by a finite automaton. Initially
R contains no information about the maze, other than the fact that it is at a
vertex of the maze, called the *root*. R's duty is to traverse every edge of the
maze, and in the process to leave signs which indicate an Eulerian cycle; i.e. a
cycle in which every edge appears once, and only once.

It is assumed that the vertices carry no names. R initially marks the root
as r, but no other vertex will be given a name. Yet, in each vertex, the exits
to the outgoing edges are cyclically numbered. If the outdegree of vertex v is
$d_{out}(v)$, then the exits to the edges are marked $1, 2, \ldots, d_{out}(v)$. The entrances
(of incoming edges) are not marked, and R, when entering a vertex has no
awareness of which entrance it has come through. We call this situation *non-
input-awarness* in contrast to *input-awareness* in which R knows from which
entrance it has come through.

In 1967, Michael Rabin [7] considered the problem of threading undirected
connected graphs by means of a finite automaton which has some fixed number of
pebbles it can leave at vertices and which it can move from one vertex to another.
He proved that no such automaton can thread all finite undirected graphs. Blum
and Sakoda [2] extended the result. It follows that no finite automaton threads
all finite, degree bounded, connected graphs.

---

* Supported by the Fund for the Promotion of Research at the Technion.

In view of this impossibility result, it is no wonder that the classical (19-th century) algorithms for threading undirected mazes, such as that of Trémaux's and Tarry's (see, for example Even's book [4], Chapter 3) store data in the vertices. More about threading undirected mazes can be found in the book of Hemmerling [6]. However, these algorithms are not easy to implement on directed graphs due to the fact that backtracking edges is not allowed.

Notice that every connected undirected graph can be transformed to a directed Eulerian graph by replacing every edge u—v with two directed edges $u \longrightarrow v$ and $v \longrightarrow u$. Thus, the impossibility result for undirected graphs implies the impossibility for directed Eulerian graphs.

Therefore, in order to thread all finite directed Eulerian graphs, it is natural to allow our robot, R, to store some information in the vertices.

Even, Litman and Winkler [5] presented an algorithm of time complexity $O(|V|^2)$ to thread directed networks. However, their algorithm assumes a different computational model — each vertex is a finite automaton and the directed edges are communication links. Afek and Gafni [1] present a fairly complicated solution which uses essentially Depth-First Search (like Trémaux's algorithm), and includes a method to effectively backtrack on directed edges. Their method uses $O(|V| \cdot |E| + |V|^2 \cdot \log |V|)$ edge traversals. We shall later say more about the comparison of our results with those of Afek and Gafni.

Deng and Papadimitriou [3] use a more powerful model. They assume that vertices have distinct identities and that the robot is a full power computer (such as a Turing machine). They describe a recursive algorithm, for the case of directed Eulerian graphs, which traverses each edge exactly twice, and they build on this algorithm to attack the problem for general strongly connected directed graphs.

We describe two algorithms. The first, simple algorithm, uses one pebble in each vertex v. The pebble's task is to mark one of the exits. Alternatively, this is like having a memory of $O(\log d_{out}(v))$ at v. The algorithm uses also a one bit memory at each vertex to remember if all its outgoing edges have been traversed. This variable is initially assumed to be set to NEW, and later R will switch it to OLD. The pebble is moved by R from the exit it is about to use, to the next one in a round-robin fashion. This is the only mechanism used for deciding on the order in which edges are traversed. R carries a flag which, in conjunction with the NEW/OLD signals at the vertices helps R decide when to halt. The Simple algorithm may use $O(|V| \cdot |E|)$ edge traversals before it halts.

Afek and Gafni argue that their scanning method is essentially optimal, since there are (nonEulerian) directed graphs which require $\Omega(|V| \cdot |E|)$ edge traversals. However, as we shall see in Sect. 2.3, there are (Eulerian) graphs for which any Depth-First Search approach, including that of Afek and Gafni, takes $\Omega(|V| \cdot |E|)$ edge traversals, while ours takes $O(|V|^2)$.

Next, we describe a generic algorithm which has two parts, Preparatory and Stroll. Preparatory does not adhere to the traveling robot model, and is non-deterministic. During Preparatory a pebble is placed in each vertex at one of its exists. Stroll is performed by a traveling robot R. It uses the pebble, placed initially by Preparatory, exactly as in the Simple algorithm. When R has traversed all outgoing edges of the root, it halts. It is proved that R's route is an Eulerian cycle, and that when it is done the pebbles are back where Preparatory put them.

Finally, we describe an advanced algorithm. It is deterministic and is performed by a traveling robot. Its end result is in accord with Preparatory. It uses two more pebbles per vertex, as well as two variables that use 3 bits of memory in each vertex. Each edge is traversed exactly 3 times.

The Recursive algorithm of Deng and Papadimitriou can also be performed by a traveling finite robot. In this case if we allow input awareness then the algorithm will have the advantage that it scans every edge only twice . However, the memory in each vertex v will be $\Theta(d_{out}(v) \cdot \log d_{out}(v))$. If the traveling robot has no input awareness, then each edge is scanned exactly 3 times and the memory in each vertex v is $O(d_{out}(v) \cdot \log d_{out}(v))$. This was shown by Tayar [8].

It is worth mentioning that all three algorithms (Simple, Advanced and Recursive) may yield different Eulerian cycles, as they do if applied to the maze shown in Fig. 3.

2   The Simple Traversal Algorithm

Let $G(V, E)$ be a finite Eulerian directed graph. Namely, G's underlying undirected graph is connected, and for every vertex $v \in V$, $d_{in}(v) = d_{out}(v)$; i.e. its indegree is equal to its outdegree. Euler's Theorem implies that G has a directed cycle in which every edge of G appears exactly once.

Since $d_{in}(v) = d_{out}(v)$ we shall simply use $d(v)$ to denote $d_{out}(v)$. In each vertex $v \in V$, the $d(v)$ exits to its outgoing incident edges are numbered $1, 2, \dots ,$ $d(v)$.

2.1   The Algorithm

The traversal of the graph is to be performed by a finite-state robot R. The algorithm which governs the behavior of R consists of two mechanisms: The *scanning mechanism*, which determines the order in which the edges are traversed, and the *halting mechanism*.

The Scanning Mechanism   In each vertex v there is one pebble, initially placed at exit No. 1. The robot, when in vertex v, leaves through the exit marked with the pebble, but before going out R moves the pebble to the next exit, in a round-robin fashion.

Clearly, the scanning continues indefinitely, unless the halting mechanism stops it.

The Halting Mechanism For the purpose of deciding when R stops scanning, the vertices are marked as follows:

- The vertex in which R is initially placed is called the *root* and is marked r. This label is never changed.
- Every other vertex $v \neq r$ is initially marked NEW. (No need to have a vertex name.) This label is eventually changed to OLD and will never change again.

In addition, the robot carries a flag F. The flag may be UP or DOWN, and its initial value is DOWN. Upon reaching a vertex v the marking of v and the flag are changed as shown in Table 1.

Table 1. The halting mechanism.

```
if v ≠ r then do
        if v is marked NEW then do
                F ← DOWN
                if the pebble is at exit No. d(v) then
                set mark to be OLD
        if v = r and the pebble is at exit No. 1 then do
        (This event is called the end of the phase)
                if F = UP then HALT
                F ← UP
```

2.2   Validity

Theorem 1. *The robot R will halt (at the root r) after having traversed all edges of G. This will happened within $O(|V| \cdot |E|)$ edge traversals. In its last $|E|$ edge traversals R makes an Eulerian tour.*

Note that upon termination the pebbles are not necessarily back in their initial condition. In fact, their position is exactly what has been "learned". If now the traversal is restarted, with $F = $ UP, R will perform an Eulerian tour and stop, and the pebbles will again be in the "learned" position.

Proof. Note the definition of a phase, as in Table 1. First, let us examine the route of R during the first phase. Clearly, every edge incident with r has been traversed exactly once. Furthermore, every edge on the route of the first phase is traversed once only. This is easily proved by induction on the order in which the edges are traversed: When a vertex $v \neq r$ is entered by an edge (traversed for the first time) the number of used incoming edges is one greater than the number of used outgoing edges. Thus, the exit in which the pebble is placed leads to an edge which has not been traversed yet.

Note that if during the first phase every incident edge of $v \neq r$ is traversed then v's label is OLD. Other vertices are still marked NEW. However, for every vertex the number of untraversed incoming edges is equal to the number of untraversed outgoing edges.

Now consider the second phase. As long as the route goes through vertices labeled OLD, the route of the second phase is identical to that of the first phase. This follows from the fact that at the end of the first phase, the pebble is back at exit No. 1. If not all edges of G have been traversed in the first phase, there must be vertices on the first route for which some of their outgoing edges have not been traversed and such vertices are still labeled NEW. This follows from the fact that G is strongly connected. If this is the case, let v be the first vertex labeled NEW which is encountered in the second phase. A new tour starts at v, using only edges untraversed in the first phase, and ending in v. Now v is marked OLD and its pebble is at exit No. 1. The route of the first phase is resumed. Since every vertex which is still marked NEW and which is on the route of the first phase is similarly treated, at the end of the second phase all vertices on the first route are now labeled OLD. Every traversed edge has been traversed exactly once in the second phase, which is the second time for those of the first route.

In the following phases, the algorithm will maintain (as in the first two phases) the following property: at the beginning of every $i'th$ phase, in every OLD vertex , the pebble's location is exactly where it was at beginning of the $(i-1)'st$ phase. This ensures that R's behavior in the $i'th$ phase will be similar to that in the second phase. In other words, this property ensures that R, in the $i'th$ phase, will retrace the route of the $(i-1)'st$ phase and will expand it upon reaching every NEW vertex ,u ,that belongs to the route of the $(i-1)'st$ phase. After the expansion in u, R will exit through exit No. 1, which is, again, where the pebble was in the beginning of the $(i-1)'st$ phase.

If no NEW vertex is encountered during a phase, R will return to r with its flag UP, and the process is halted. During each phase, other than the last and possibly the first, at least one NEW vertex becomes OLD. Thus, there are at most $|V|+1$ phases. Hence, the number of edge traversals is $O(|V| \cdot |E|)$.   □

2.3 Lower Bounds on the Simple Algorithm

Let us consider the scanning mechanism by itself and discuss its complexity of edge traversals, up to the point that every edge is scanned.

First consider a *simple chain*, as shown in Fig. 1. Let us denote by $f_s(n)$ the

Fig. 1. A simple chain

number of edge traversals, starting in vertex 1 and ending with the last edge to be traversed, namely, the edge from vertex n to vertex $n-1$. Clearly, $f_s(2) = 2$ and $f_s(n+1) = f_s(n) + 2n - 1$. Thus,

$$f_s(n) = n^2 - 2n + 2. \tag{1}$$

It follows that in this case, the bound of $O(|V| \cdot |E|)$ is tight. However, the simple chain is sparse. To see that this may happen for dense graphs too, consider the following directed graph $G_p(V, E)$, where p is an odd prime.

$$V = \{0, 1, \cdots, 2p - 1\},$$

The set of directed edges consists of two types:

- **body**
  For every two vertices, $i, j < p$, let $\delta(i, j) \overset{\triangle}{=} j - i \ mod(p)$. If $\delta(i, j) < \frac{p}{2}$ then there is an edge $i \longrightarrow j$, and the exit at i is labeled $\delta(i, j)$.
- **tail**
  - There is an edge $0 \longrightarrow p$ labeled $\frac{p+1}{2}$.
  - For every $p \le i < 2p - 1$ there is an edge $i \longrightarrow i + 1$, labeled 2.
  - For every $p < i \le 2p - 1$ there is an edge $i \longrightarrow i - 1$, labeled 1.
  - There is an edge $p \longrightarrow 0$ labeled 1.

The case of G_5 is shown in Fig. 2.

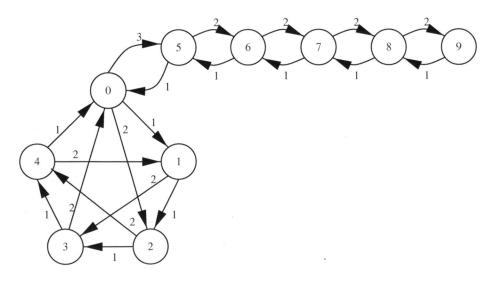

Fig. 2. G_5

Note that the scanning mechanism, starting in vertex 0, scans all $\frac{p(p-1)}{2}$ edges of the body; i.e. the edges between vertices whose name is less than p, before it takes the edge to p. In fact, every time it returns to vertex 0, via the edge from

p, it will repeat the scan of all edges of the body. This will occur p times. Thus the whole tour until the last edge (from $2p-1$ to $2p-2$) is traversed is of length

$$p \cdot \frac{p(p-1)}{2} + p^2 - 2p + 2.$$

Again, this is of the order $|V| \cdot |E|$, and G_p is dense.

Finally, we examine an important property of the Simple algorithm: if in a constant number of phases all vertices are visited, the edge traversals complexity is $O(|E|)$. There are many Eulerian graphs in which R will pass through all vertices in a constant number of phases. In such cases, this property makes the Simple algorithm at least as good as any DFS based algorithm. In particular, there are graphs in which the Simple algorithm is superior.

An example in which the Simple algorithm has a better edge traversals complexity is its execution on the a directed graph which is composed of a directed cycle with a self-loop in each vertex (assume that for every vertex u, exit No.2 leads to u's self-loop). The Simple algorithm's edge traversal complexity is $O(|V|)$. In the first phase R visits all vertices and in the second phase an Eulerian tour is performed. However, if we execute any DFS based algorithm on the same graph, R will have to make $\Omega(|V|^2)$ edge traversals.

Another example of its superiority is in the case of a complete directed graph of n vertices; namely for every two vertices a and b there is an edge $a \longrightarrow b$. In the first phase all vertices are visited, since the root has an edge to every other vertex. Therefore, all edges not traversed in the first phase are traversed in the second, and the third phase is the last phase. Thus, the total number of edge traversals is $O(|V|^2)$. However, if one applies a DFS based algorithm, it takes $\Omega(|V|^3)$ steps. (The DFS tree is a directed path of length $|V|-1$, and each time a back-edge is traversed, the tree must be traversed again, at least up to the parent of the start vertex of the back-edge.).

3 A Generic Traversal Algorithm

In effort to construct a more efficient algorithm for finding an Eulerian directed cycle in $G(V, E)$, in terms of the number of edge traversals, we describe first a *generic algorithm*. This is a nondeterministic algorithm which deviates from our computational model; i.e. it is not limited to actions performed by a traveling finite automaton. Later on we show how to remove the nondeterminism and run the algorithm by means of a finite automaton which threads the directed graph.

3.1 Definitions

- An edge is *new* if it has not been traversed yet. It is *old* otherwise.
- A vertex is *new* if all its outgoing edges are new. It is *old* otherwise.
- An **exploration** from vertex v is a directed path which starts in v and continues via a new directed edge, as long as there is one. It is assumed that in each vertex exits are chosen consecutively according to their exit numbers, beginning with the least exit number which leads to a new edge.

3.2   The Preparatory Algorithm

Consider Preparatory, as stated in Table 2. Note that the first exploration ends at the vertex it has started. This observation of Euler, follows from the fact that when another vertex u is entered, the number of old incoming edges is greater (by 1) than the number of old outgoing edges, and since $d_{in}(u) = d_{out}(u)$, there is at least one new outgoing edge. When the first exploration ends, for every vertex the number of new incoming and outgoing edges is the same, and this invariant holds every time an exploration ends.

Table 2. Preparatory

choose a vertex r
run an **exploration** from r
while there are old vertices which have new exits *do*
      choose such a vertex u
      run an **exploration** from u
for every vertex v *do*
      if only one exploration passed through v
            then put v's pebble at the first exit of this exploration
      else put the pebble at the first exit of the
            second exploration which passed through v

For every vertex, Preparatory places a pebble at one of its exits. The position of these pebbles is the information *learned* by running Preparatory.

3.3   The Stroll Algorithm

After Preparatory has terminated, the tracing of an Euler cycle is to be performed by a finite automaton R. R does not need to recognize vertices, except the need to recognize the start vertex r. R uses the pebbles, initially placed by Preparatory. By using this information R can now trace an Euler cycle; see Table 3.

Table 3. Stroll

put R in r
let e be the outgoing edge indicated by exit number 1
(This is the exit where the pebble is located.)
R moves the pebble to the next exit and traverses e to its endpoint
while the vertex, R is at, is not r or the pebble is not at exit number 1 *do*
      let e be the outgoing edge indicated by the pebble
      R moves the pebble to the next exit and traverses e to its endpoint

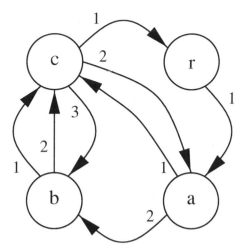

Fig. 3. Example for running Generic

Let us demonstrate the generic algorithm on the example shown in Fig. 3. For each vertex, the exits are numbered, and for convenience, all vertices, not just r are given names. Let us denote the i-th exit of vertex v by vi. The first exploration yields a cycle $C_1 : r1, a1, c1$. Assume the second exploration starts at a; it yields a cycle $C_2 : a2, b1, c2$. Assume the third cycle is $C_3 : b2, c3$. Now the pebbles are placed at exits $r1, a2, b2$ and $c2$. This completes the operation of Preparatory.

Now, during Stroll, the following Eulerian cycle is traversed: $r1, a2, b2, c2, a1$, $c3, b1$ and $c1$. Note that the edges of a cycle may not be strolled consecutively, and they may even appear in an order different from the order in which they have been traversed in the exploration.

Lemma 1. *In the directed path indicated by Stroll, no edge is repeated.*

Proof. Omitted from the extended abstract.   □

Lemma 2. *Let u be a vertex and C_i, $i \geq 1$, is the cycle of the first exploration to pass through u. If R has traversed all edges of C_i which exit from u, then R has traversed all outgoing edges of u.*

Proof. Omitted from the extended abstract.   □

Lemma 3. *R traverses all edges of G.*

Proof. Omitted from the extended abstract.   □

Theorem 2. *The robot R, moving according to algorithm Stroll, traverses an Eulerian cycle.*

Proof. By Lemma 1 no edge is traversed more than once, and by Lemma 3, every edge is traversed. □

Note that after Stroll halts, the pebbles are back where Preparatory has put them.

4   An Advanced Algorithm

In this section we describe an algorithm in which the operation of Preparatory is performed by a traveling robot R, whose control is a finite automaton. The amount of memory required in each vertex v is $O(\log d(v))$, and each edge is traversed 3 times.

Each vertex has 3 pebbles, named *explore-pebble*, *retrace-pebble* and *tour-pebble*. Initially, all three pebbles are at exit No. 1.

There are two variables stored in each vertex v.

visited indicates the number of cycles (explorations) which have passed through v; its value is 0, 1, or 2. Initially *visited = 0*. If its value is 2 then there are at least two cycles which have passed through v.

flag indicates if v has previously been passed in the current cycle. Initially, $flag = 0$, and when R reaches v for the first time during the current exploration, it changes the variable *flag* to 1. Once all edges of an exploration are found, R retraces the cycle and assigns all flags on the cycle to be 0 again. Thus, the flag of a vertex v is changed to 1, and back to 0, as many times as there are cycles which pass through it.

The vertex in which R is initially placed is the root r.

The main procedure, which governs the actions of R, is **create-eulerian-cycle**; see Table 6. It employs two subroutines, **explore** (Table 4) and **retrace** (Table 5). Also, we use two macros:

- **exit(X-pebble)**: R exits the vertex it is at via the exit in which the X-pebble is located now, but first it moves the X-pebble to the next exit, in a round-robin fashion.
- **number(X-pebble)**: the macro returns the number of the exit in which the X-pebble is located.

Subroutine **explore** is activated in order to perform an exploration.

If $flag = 0$ then the current vertex is visited during this exploration for the first time. If in addition, *visited = 1* then the running exploration is not the first to pass through the current vertex, and the tour-pebble is put where explore-pebble is at, which is the first exit of the second exploration to pass through the vertex. If *visited < 2* then it is incremented. Now *flag* is set to 1. R leaves the current vertex through the exit indicated by the explore-pebble, not before moving it to the next exit, in a round robin fashion.

If $flag = 1$ then the number of the exit where the explore-pebble is at is checked. If it is exit No. 1, the subroutine is halted; R is back at the vertex where the exploration has started, and all outgoing edges are old. If it is not exit No. 1, then there are new outgoing edges and the exploration continues.

Table 4. Subroutine **explore**

```
while TRUE
     if flag = 0 then do
          if visited = 1 then
               put tour-pebble at number(explore-pebble)
          if visited < 2 then
               visited ← visited + 1
          flag ← 1
          exit(explore-pebble)
     else   if number(explore-pebble) = 1 then HALT
          exit(explore-pebble)
```

Table 5. Subroutine **retrace**

```
while TRUE
     if flag = 0 then do
          if number(retrace-pebble) = 1 then HALT
          else exit(retrace-pebble)
     else do    flag ← 0
          exit(retrace-pebble)
```

Subroutine **retrace** is activated after the exploration of a cycle is done. Its purpose is to reset all flags of the vertices of the newly found cycle to 0. This is done as follows.

If $flag = 0$ at the current vertex, it is checked if the retracing is done. This is indicated by checking the position of the retrace-pebble. If it is at exit No. 1, R is back at the vertex in which the exploration has started, and thus the retrace subroutine is halted. Otherwise, the retracing continues.

If $flag = 1$, the flag is reset to 0, and the retracing continues.

Table 6. Main procedure **create-eulerian-cycle**

```
call explore
call retrace
exit(tour-pebble)
while R is not at r or number(tour-pebble) ≠ 1 do
     if number(explore-pebble) ≠ 1 then do
          call explore
          call retrace
     exit(tour-pebble)
```

Note that **create-eulerian-cycle** is deterministic. R starts an exploration from r, first performing **explore** and then **retrace**. Now it is back at r, while all three pebbles are at exit No. 1. R behaves now as Stroll, advancing the tour-pebble only, but this behavior is stopped when R hits a vertex v which has new outgoing edges; this is detected by the fact that the exit where the explore-pebble is at is not exit No. 1. When this happens, R suspends the Stroll behavior. It performs an exploration from v, again by first running **explore** and then **retrace**. While doing that, the tour-pebble of every vertex encountered on this cycle is moved from exit No. 1 to the first exit of the new exploration if the present exploration is the second to pass through that vertex. When **retrace** terminates, R resumes the Stroll behavior.

It is easy to see that **create-eulerian-cycle** causes R to place the tour-pebbles in accord with Preparatory, and when R uses the exits where the tour-pebbles are at, it traverses every edge once. When **create-eulerian-cycle** terminates, the tour-pebbles are back where Preparatory could have placed them. Thus, at that time, one can use Stroll, using the tour-pebbles, to traverse an Eulerian cycle.

References

1. Y. Afek and E. Gafni, *Distributed Algorithms for Unidirectional Networks*, SIAM J. Comput., Vol. 23, No. 6, 1994, pp. 1152-1178.
2. M. Blum and W.J. Sakoda, *On the Capability of Finite Automata in 2 and 3 Dimensional Space*. In Proceeding of the Eighteenth Annual Symposium on Foundations of Computer Science, 1977. pp. 147-161.
3. X. Deng and C.H. Papadimitriou, *Exploring an Unknown Graph*. In Proceeding of the Thirty First Annual Symposium on Foundation of Computer Science, 1990, pp. 355-361.
4. S. Even, **Graph Algorithms**, Computer Science press, 1979.
5. S. Even, A. Litman and P. Winkler, *Computing with Snakes in Directed Networks of Automata*. J. of Algorithms, Vol. 24, 1997, pp. 158-170.
6. A. Hemmerling, **Labyrinth Problems; Labyrinth-Searching Abilities of Automata**, Teubner-Texte zur Mathematik, Band 114, 1989.
7. M.O. Rabin, *Maze Threading Automata*. An unpublished lecture presented at MIT and UC Berkeley, 1967.
8. R. Tayar, *Scanning Directed Eulerian Mazes by a Finite-State Robot*, Master thesis, Computer Science Department, Technion, Haifa, Israel. In preparation .

On the Space and Access Complexity of Computation DAGs*

Gianfranco Bilardi[1], Andrea Pietracaprina[1], and Paolo D'Alberto[2]

[1] Dipartimento di Elettronica e Informatica, Università di Padova, Italy.
{bilardi,andrea}@art.dei.unipd.it
[2] Information and Computer Science, University of California at Irvine. USA.
paolo@ics.uci.edu

Abstract. We study the space and the access complexity of computations represented by Computational Directed Acyclic Graphs (CDAGs) in hierarchical memory systems. First, we present a unifying framework for proving lower bounds on the space complexity, which captures most of the bounds known in the literature for relevant CDAGs, previously proved through *ad-hoc* arguments. Then, we expose a close relationship between the notions of space and access complexity, where the latter represents the minimum number of accesses performed by any computation of a CDAG at a given level of the memory hierarchy. Specifically, we present two general techniques to derive bounds on the access complexity of a CDAG based on the space complexity of certain subgraphs. One technique, simpler to apply, provides only lower bounds, while the other provides (almost) matching lower and upper bounds and improves upon previous well-known result by Hong and Kung.

1 Introduction

The substantial fraction of the cost of a computing system accounted for by memory has motivated the study of space-efficient computations since the early times of automatic computing. However, a more fundamental reason to strive for space efficiency arises directly from the pursuit of performance: due to basic physical principles, access time is bound to increase with the size of memory. Ultimately, smaller memory translates into faster computation, making space efficiency a crucial objective, even when the cost of memory is negligible.

When a large overall memory is required, it becomes convenient to organize it hierarchically, into a sequence of levels whose size and access time increase progressively. Currently, the levels of the memory hierarchy include the register file, two or three caches, main memory, and disks. Compared to registers, main memory is a few hundred times slower and disks are a few million times slower. Hence, effective use of the faster levels of the memory hierarchy is becoming a

---

* This work was supported, in part, by CNR and MURST of Italy. Part of the work of G. Bilardi was done while this author was a visitor at the IBM T. J. Watson Research Laboratory.

U. Brandes and D. Wagner (Eds.): WG 2000, LNCS 1928, pp. 47–58, 2000.

key concern in the design of algorithms. As in all endevours where performance is systematically pursued, it is important to be able to evaluate the distance from optimality of a proposed solution. The present paper focuses on the development of general techniques that can be employed to prove bounds on certain relevant measures related to the usage of the memory hierarchy by a computation.

We consider the problem of finding an optimal *implementation* of a computation which has been specified in terms of a *Computational Directed Acyclic Graph* (CDAG), whose vertices represent operations (of both input and processing type) and whose arcs represent data dependencies. The degrees of freedom of the implementor are essentially the *schedule* of execution of operations, possibly including repetitions, and the *memory management*, that is, the assignment of a memory location to each value arising in the computation during the time between production and last use of such value. An extensive literature on the space complexity of CDAGs, typically formulated in terms of so-called *pebbling games*, has developed in the past (see e.g., [9,5,7,10,8,11,12]). A pioneering paper by Hong and Kung [6] begun exploring complexity in two-level memory hierarchies and proposed a technique to bound from below the number of accesses performed at the slowest level of the hierarchy for a given CDAG. To date, this paper remains the main point of departure of most lower bound analysis for hierarchical memory performance (see, e.g., [1,2]). Savage [13] developed an interesting reformulation of the Hong and Kung's technique as well as an extension to multilevel hierarchies. (A comprehensive survey of all of the above results can be found in [14, Ch.10,11].) In a recent work, Bilardi and Preparata [3] proposed a new lower bound technique based on the notion of *dicothomy-width*, which is restricted to implementations where every operation is executed exactly once. Such technique has been successfully extended to distributed memory hierarchies.

In this paper, we study both the space and the access complexity of CDAGs, and, in particular, expose a tight relationship between the two. Models and relevant quantities are defined in Section 2. In Section 3, we introduce a family of graph-theoretic measures that can be employed to prove lower bounds to the space complexity of CDAGs. Each measure is associated with a suitable class of permutations of the vertices, and it is defined in terms of a suitable notion of boundary size for the prefixes of such permutations. More specifically, the measure is the minimum, over all the permutations in the class, of the maximum boundary size of a prefix. The key link between the space complexity of a CDAG and each measure is established by showing that each execution of the CDAG must embed a permutation of the operations belonging to the associated class and that, at each stage of the execution, the values of nodes in the current prefix must be stored in memory. We show how most of the results in the literature on space complexity, which are derived by *ad-hoc* arguments for the CDAG at hand, can be systematically derived in the outlined framework.

In Section 4, we turn our attention to the *access complexity* $Q_G(s)$ of a CDAG G defined as minimum number of accesses to locations with address $x \geq s$, over all executions of G, on a memory with locations addressed by the natural numbers starting from 0. First, we discuss a simple but effective technique that affords the derivation of lower bounds on $Q_G(s)$ from the space complexity of disjoint subdags of G. Second, we show, in a sense that can be made precise, that $Q_G(s)$ is closely related, by both lower and upper bounds, to the minimum number of subdags of space complexity at most $2s$ that "cover" the given CDAG. On the side of lower bounds, our results encompass those of both [6] and [14]; on the side of upper bounds, we are not aware of any previous comparable results. In spite of their greater generality, our arguments are somewhat simpler than the previous ones. Among other things, they show that the concept of *minimum set* featuring in the analysis of Hong and Kung is not essential. In Section 5, we briefly indicate some directions for further research.

2   Preliminaries

Let $G = (V, E)$ be a CDAG where the nodes represent values produced by unit-time operations, and the arcs represent data dependencies. For every arc $(u, v) \in E$ we say that u is a *predecessor* of v and v is a *successor* of u. The set of predecessors of a node v are the operands of the operation that produces v. The *in-degree* (resp., *out-degree*) of a node is the number of its predecessors (resp., successors). Nodes of in-degree (resp., out-degree) 0 are regarded as *inputs* (resp., *outputs*) of the CDAG. A *computation* of G specifies a particular scheduling of the operations associated with its nodes, which satisfies data dependences, and a particular memory management.

We study CDAG computations on the RAM model with a memory of unbounded size whose cells are addressed by the natural numbers starting from 0. A *standard computation* of a CDAG G starts with the values of all input nodes in memory and must determine the values of the output nodes by performing a sequence of *node evaluations*, where a node evaluation performs the operation associated with a node v, provided that the operands of v are in memory, and stores the result in memory.

In general, a node can be evaluated more than once. The space needed by a computation is defined as the maximum number of values concurrently residing in memory at any one time during the computation. We define the *space complexity* of G, denoted by $S(G)$, as the minimum space over all standard computations of G.

In the paper we will also refer to a slightly different class of CDAG computations, called *free-input computations*, that adopt the following rule: the computation of a CDAG G starts with an initially empty memory and produces the value of each input node, every time it is needed, by invoking a special `load` instruction. It is easy to argue that the space complexity of the free-input computations of G, which we denote by $S_{\text{free}}(G)$, is not larger than $S(G)$, hence any lower bound to $S_{\text{free}}(G)$ applies to $S(G)$ as well. We observe that $S_{\text{free}}(G)$ is also the measure of space captured by the *Red Pebble Game* model [7,14].

Suppose that the memory is partitioned into two levels, namely L1 consisting of the s cells of address $x \in [0, s)$, and L2 consisting of the cells of address $x \geq s$, featuring respectively low and high access costs. For a CDAG G we define the *access complexity* $Q_G(s)$ as the minimum number of accesses to L2 performed by any computation of G. When studying the access complexity we will refer only to standard computations.

3   Space Complexity

If we rule out multiple executions of the same operation, the evaluation schedules of a CDAG are in one-to-one correspondence with the topological orderings of its nodes. Furthermore, for a given topological ordering, say $\phi = \phi_1\phi_2\cdots\phi_n$, one can easily argue that, immediately after evaluation of node ϕ_i, every $v \in \{\phi_1,\dots,\phi_i\}$ with at least one successor in $\{\phi_{i+1},\dots,\phi_n\}$ must be in memory. Repetition of operations greatly complicates the analysis, because the simple and convenient correspondence between topological orderings and valid schedules no longer holds, and it is difficult to identify what must be in memory at any given point of a schedule.

The contribution of this section is a framework that associates arbitrary schedules (where repetitions are allowed) with suitable permutations of the vertices, while maintaining a grip on space requirements. Informally the approach works as follows. Consider an arbitrary n-node CDAG G. We introduce the notion of marking rule f as a criterion to associate with each node v of G a family $f(v)$ of subsets of its successors. Given a computation of G, the marking rule can be employed to "mark" exactly one evaluation for each node of G. The marked evaluations, in order of execution, define a permutation $\phi = \phi_1\phi_2\cdots\phi_n$ of the nodes of G with the following property. For $1 \leq i \leq n$, there is a time during the computation such that every ϕ_j with $j \leq i$ and with at least one of the subsets belonging to $f(\phi_j)$ included in $\{\phi_{i+1},\dots,\phi_n\}$ must be in memory. (Such ϕ_j's are said to form the boundary of the prefix.) We should observe that, when applied to a given CDAG G, a marking rule provides a lower bound to $S_{free}(G)$. In general, different rules generate different classes of permutations as well as different types of boundaries, often with opposite impact on the lower bound measure. The approach outlined above is formalized next and then applied to some well-known CDAGs.

3.1   A General Lower Bound

Let $G = (V, E)$ be an n-node CDAG. Let also $O \subseteq V$ denote the set of output nodes, and $N(v)$ the set of successors of v, for every $v \in V$. A *marking rule* for G is a function f which maps every $v \in V$ to a family of subsets of $N(v)$ such that if $v \in O$ then $f(v)$ contains only the empty set, while if $v \in V\backslash O$ then $f(v)$ cannot contain the empty set and must contain at least one (nonempty) subset of $N(v)$. In the paper we will often refer to a particular marking rule, the *singleton marking rule* $f_G^{(sing)}$, where for every $v \in V\backslash O$ $f_G^{(sing)}(v) = \{\{u\} \, : \, u \in N(v)\}$.

A permutation $\phi = \phi_1\phi_2\cdots\phi_n$ of the nodes of G is an *f-marking* of G iff for every $1 \leq i \leq n$ there exists $q \in f(\phi_i)$ such that $q \subseteq \{\phi_j \, : \, i < j \leq n\}$.

Moreover, we define the *i-boundary* of any *f-marking* ϕ as the set

$$B_\phi^f(i) = \{v \in \{\phi_1, \ldots, \phi_i\} \backslash O \ : \ \exists q \in f(v) \text{ s.t. } q \subseteq \{\phi_{i+1}, \ldots, \phi_n\}\}.$$

In other words, $B_\phi^f(i)$ represents the set of nodes $v \in V \backslash O$ such that $v\phi_{i+1} \cdots \phi_n$ is the suffix of a legal f-marking of G.

The following theorem shows a relation between the space complexity of the free-input computations of G and the size of the boundaries of its f-markings. Let F_G denote the set of marking rules for G and, for $f \in F_G$, let $\Phi(f)$ denote the set of f-markings of G.

Theorem 1. *The space complexity of the free-input computations of G satisfies:*

$$S_{\text{free}}(G) \geq \max_{f \in F_G} \ \min_{\phi \in \Phi(f)} \ \max_{1 \leq i \leq n} |B_\phi^f(i)|.$$

Proof. Consider an arbitrary function $f \in F_G$ and a T-step free-input computation C of G. Let v_t be the node evaluated at step t of C, for $1 \leq t \leq T$. We can construct an f-marking $\phi = \phi_1 \phi_2 \cdots \phi_n$ of G by sweeping C backwards through the following loop.

$j = n;$
for $t = T$ down-to 1 do
    if $(v_t \notin \{\phi_{j+1}, \ldots, \phi_n\})$ and $(\exists q \in f(v_t) : q \subseteq \{\phi_{j+1}, \ldots, \phi_n\})$
    then $\{$set $\phi_j = v_t; \ j = j - 1;\}$

It is easy to verify that at the end of the loop the resulting sequence ϕ forms indeed an f-marking of G. For $1 \leq j \leq n$ let $\tau(j)$ denote the index such that the node ϕ_j is set at iteration $\tau(j)$ (hence $\phi_j = v_{\tau(j)}$). For an arbitrary index i we now show that the values of the nodes in $B_\phi^f(i)$ must be in memory at the end of step $\tau(i)$ of C. Consider, in fact, a node $\phi_j \in B_\phi^f(i)$. The definition of $B_\phi^f(i)$ implies that $j \leq i$ and that $\{\phi_{i+1}, \ldots, \phi_n\}$ includes a subset $q \in f(\phi_j)$. Let $k \geq i + 1$ be the smallest index such that $\phi_k \in q$. It is easy to see that $\tau(j) \leq \tau(i) < \tau(k)$ and that for every $\tau(j) < t < \tau(k)$ $v_t \neq \phi_j$. Hence the value of node ϕ_j, which is used to evaluate ϕ_k, must be in memory at step $\tau(i)$. Since i was chosen arbitrarily, we may conclude that the space required by C is not less than $\max_{1 \leq i \leq n} |B_\phi^f(i)|$. The theorem follows by minimizing over all possible $\phi \in \Phi(f)$ and by maximizing over all possible $f \in F_G$. $\qquad\square$

3.2 Application to Known CDAGs

In this subsection we illustrate how the marking approach constitutes a uniform and effective framework for proving lower bounds on the space complexity of CDAGs. Specifically, we will show how most of the lower bounds known for relevant CDAGs can be obtained by applying Theorem 1.

Complete binary tree. Let $G = (V, E)$ be a *complete binary tree* with $n = 2m - 1$ nodes, m leaves and arcs directed from each internal node to its parent. Thus, there are m inputs (the leaves) and one output (the root). The following argument is a reformulation of the one given in [9], within the marking framework. Let $\phi = \phi_1\phi_2 \cdots \phi_n$ be an arbitrary $f_G^{(sing)}$-marking of G and let $\phi_1\phi_2 \cdots \phi_j$ be the smallest prefix that includes all leaves. Hence ϕ_j must be a leaf. It is easy to see that all nodes in the unique path π from ϕ_j's father to the root must be included in the suffix $\phi_{j+1} \cdots \phi_n$. There are $\log m + 1$ paths from distinct leaves to the root which intersect only at π. Since all leaves are in the prefix and all nodes of π are in the suffix, every such path includes a distinct node in $B_\phi^{f_G^{(sing)}}(j)$. By Theorem 1, we get

$$S_{\text{free}}(G) \geq \log m + 1,$$

which is tight, as shown in [9].

Pyramid. Let $G = (V, E)$ be a *pyramid CDAG* with $n = m(m + 1)/2$ nodes obtained by taking the half of an $m \times m$ mesh above and including the nodes on the main diagonal, and by directing all edges towards the upper-right corner of the mesh. The graph has m inputs (the nodes on the mesh diagonal) and one output (the upper-right corner of the mesh). Let $\phi = \phi_1\phi_2 \cdots \phi_n$ be an arbitrary $f_G^{(sing)}$-marking of G. Let also $\phi_1\phi_2 \cdots \phi_j$ be the smallest prefix that includes all inputs. Hence ϕ_j must be an input. By reasoning as for the binary tree (thus reformulating the argument in [5] within the marking framework) we can show that $|B_\phi^{f_G^{(sing)}}(j)| \geq m$. By Theorem 1, this implies

$$S_{\text{free}}(G) \geq m,$$

which is tight [14].

Composed CDAGs. Let $G = (V, E)$ be a CDAG and $G' = (V', E')$ a subdag of G, where $|V| = n$ and $|V'| = m \leq n$. Clearly, any free-input computation of G includes (at least) one free-input computation of G', hence $S_{\text{free}}(G) \geq S_{\text{free}}(G')$.

This fact is reflected in the lower bound stated in Theorem 1, as explained below. It is easy to see that for every marking rule $f_{G'}$ for G' there exists a marking rule f_G for G which is an *extension* of $f_{G'}$ in the sense that $f_G(v) = f_{G'}(v)$ for every node $v \in V'$ which is not an output for G'. Moreover, the $f_{G'}$-markings of G' are in one-to-one correspondence with the projections of the f_G-markings of G on the nodes of V'. Therefore, we have that

$$\max_{f_G \in F_G} \min_{\phi \in \Phi(f_G)} \max_{1 \leq i \leq n} |B_\phi^{f_G}(i)| \geq \max_{f_{G'} \in F_{G'}} \min_{\phi \in \Phi(f_{G'})} \max_{1 \leq i \leq m} |B_\phi^{f_{G'}}(i)|. \qquad (1)$$

Let G be the *FFT CDAG* with m inputs, m outputs, $m(\log m + 1)$ nodes and arcs directed from the inputs towards the outputs (a formal definition can be found in [14]). Such a CDAG contains several copies of the m-leaf complete

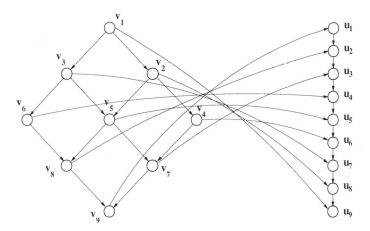

Fig. 1. The diamond/linear array CDAG for $m = 9$.

binary tree G' as subgraph. Relation 1 implies that the marking framework provides a lower bound $S_{\text{free}}(G) \geq \log m + 1$ which, as argued in [14], is tight. As another example consider a *diamond CDAG* G with one input, one output, and $n = m^2$ nodes, obtained by gluing together two $(m(m + 1)/2)$-node pyramids at the inputs (corresponding inputs are coalesced into one node) and by reversing the direction of the arcs in the top pyramid. Again, Relation 1 implies that the marking framework provides a lower bound $S_{\text{free}}(G) \geq m$, which is easily seen to be tight.

Finally, we want to show that marking rules can be defined in such a way to bring forward the presence of space demanding subgraphs within a CDAG, and that, in fact, in some cases this is necessary in order to obtain good lower bounds. Let G be the $n = 2m$-node CDAG formed by an m-node diamond CDAG D and an m-node linear array L, connected as follows (see Figure 1). Let $v_1, v_2, \ldots, v_m$ be a topological ordering of the diamond nodes and let $u_1, u_2, \ldots, u_m$ be the (unique) topological ordering of the array nodes. In addition to the arcs of the diamond and those of the linear array, E contains the arcs (v_i, u_{m-i+1}), for $1 \leq i \leq m$. Therefore, G has one input, namely the diamond input v_1, and one output, namely the linear array output u_m. It is easy to see that the space complexity of the free-input computations of G is dominated by that of the diamond component. By selecting $f \in F_G$ as a suitable extension of $f_D^{(sing)}$, one can prove $S_{\text{free}}(G) \geq \sqrt{m}$, which is tight. However, the following claim shows that the singleton marking rule $f_G^{(sing)}$ relative to the entire CDAG G yields only a trivial bound on $S_{\text{free}}(G)$.

Claim. $\min_{\phi \in \Phi(f_G^{(sing)})} \max_{1 \leq i \leq n} |B_\phi^{f_G^{(sing)}}(i)| \leq 2.$

Proof. Let $\phi = \phi_1 \phi_2 \cdots \phi_{2m}$ be such that for $1 \leq i \leq m$

$$\phi_{2i-1} = v_{m-i+1},$$

$$\phi_{2i} = u_i.$$

It is easy to see that ϕ is an $f_G^{(sing)}$-marking. Moreover, $B_\phi^{f_G^{(sing)}}(1) = \{v_m\}$ and, for $1 \leq i < m$,

$$B_\phi^{f_G^{(sing)}}(2i) = \{u_i\},$$
$$B_\phi^{f_G^{(sing)}}(2i+1) = \{v_{m-i}, u_i\}.$$

Thus, the lower bound provided by $f_G^{(sing)}$ is at most 2.   □

Unlike the marking rule f mentioned above, which coincides with $f_D^{(sing)}$ on the diamond, $f_G^{(sing)}$ is unable to bring forward the complexity of the diamond due to the "disturbance" introduced by the arcs between D and L.

4   Access Complexity

The notion of access complexity (closely related to the I/O complexity studied in [6]) was introduced in [3] to capture the cost of CDAG computations in memory hierarchies by counting the number of accesses directed above a certain level. In this section we develop bounds on the access complexity of a CDAG in terms of the space complexity of certain subgraphs. A first method is based on a partition of the given CDAG G chosen a priori and relates the access complexity of G to the space complexity of the subgraphs of such partition. The method is conceptually simple to apply, since only the chosen subgraphs need to be analyzed, and, as an example shows, it may be employed successfully to get tight bounds. Building upon [6,13], the remaining two subsections develop a more sophisticated analysis which, while possibly more involved when studying individual CDAGs, does provide almost matching lower and upper bounds on the access complexity of a CDAG G in terms of the minimum cardinality of certain coverings of G into (non necessarily disjoint) subgraphs with limited space complexity.

4.1   The Prepartition Approach

We begin with a simple observation. A standard computation C of a CDAG G can be transformed into a free-input computation C' which starts from an initially empty memory and uses the same schedule of operations as C except that every node evaluation that requires an input as operand is preceded by a load of that input. It is easy to see that at least $S_{\text{free}}(G)$ distinct memory cells are accessed during C' and, therefore, during C. Since $\max\{0, S_{\text{free}}(G) - s\}$ of these cells must belong to L2 we have:

$$Q_G(s) \geq \max\{0, S_{\text{free}}(G) - s\}.$$

The following theorem generalizes this argument.

Theorem 2. *Let G be h node-disjoint subgraphs $G_1, G_2, \ldots, G_h$ with space complexities (for free input computations) $S_{\text{free}}(G_i)$, $1 \le i \le h$. We have:*

$$Q_G(s) \ge \sum_{i=1}^{h} \max\{0, S_{\text{free}}(G_i) - s\}.$$

Proof. Consider a standard computation C of G and let C_i be the subcomputation of C relative to subgraph G_i, $1 \le i \le h$. By reasoning as before we can easily conclude that at least $S_{\text{free}}(G_i)$ distinct memory cells are accessed to read/write values of nodes of G_i. This accounts for at least $\max\{0, S_{\text{free}}(G_i) - s\}$ accesses to L2 relative to nodes of G_i, and the theorem follows.   □

As an example, consider a diamond CDAG G with $n = m^2$ nodes. Note that G can be partitioned into $\Theta\left(n/(4s^2)\right)$ node-disjoint diamond subgraphs with $4s^2$ nodes and space complexity (relative to free-input computations) $2s$ each, thus yielding $Q_G(s) = \Omega\left(n/s\right)$. Moreover, one can easily orchestrate a computation of G which indeed performs $O\left(n/s\right)$ access at addresses $x \ge s$.

4.2   The Hong and Kung's Lower Bound Method

Below, we review the methodology proposed by [6] for proving lower bounds on the access complexity of CDAGs, based on the notion of s-partition. Let $G = (V, E)$ be an n-node CDAG. A partition of V into subsets $\{V_1, V_2, \ldots, V_h\}$ is called a *topological partition* if, for each $(v, v') \in E$, $v \in V_i$ and $v' \in V_j$ implies $i \le j$.

Definition 1. *An s-partition of G is a topological partition $\{V_1, V_2, \ldots, V_h\}$ of subsets of V such that:*

1. *For every $1 \le i \le h$ there is a set $D_i \subseteq V$ of size at most s intersected by every path from an input to a node in V_i. D_i is said to be a* dominator set *for V_i.*
2. *For every $1 \le i \le h$ the set $M_i \subset V_i$ of nodes of V_i with no successors in V_i has size at most s. M_i is called the* minimum set *of V_i.*

The following theorem is proved in [6, Theorem 3.1].

Theorem 3. *Let $h_G(2s)$ be the minimum cardinality of a $2s$-partition for G. Then, the access complexity of G satisfies*

$$Q_G(s) \ge s(h_G(2s) - 1).$$

We observe that the above result, although successfully employed in [6] to prove strong lower bounds for several CDAGs, is rather weak if applied to CDAGs with few inputs and outputs. Specifically, if a CDAG G has at most $2s$ inputs and at most $2s$ outputs, as is the case, for example, of the diamond CDAG, the entire node set V has a dominator set (the set of inputs) and a minimum set (the set of outputs) of size at most $2s$, hence it forms an $2s$-partition of cardinality

$h_G(2s) = 1$. In this case the $s(h_G(2s) - 1)$ lower bound on the access complexity becomes meaningless.[1]

Savage [14] presents a version of the Hong and Kung result, indicating that his version is weaker but simpler to state and prove. Strictly speaking, Savage's version is not comparable with Hong and Kung's (i.e., each version might lead to stronger bound than the other one on suitable CDAGs). In the following subsection we further develop the ideas of both [6] and [14], achieving a result strictly stronger than both, and show that in an interesting sense the resulting bound is tight.

4.3   The Covering Method: Lower and Upper Bounds

Let $G = (V, E)$ be an n-node CDAG. We observe that, when partitioning V into subsets, the crucial measure for a subset V_i is the amount of space needed to compute the nodes in V_i starting from a suitably chosen subset D_i of their ancestors (which is easily seen to be a dominator set of V_i). Generally, in the process, a larger set U_i of nodes is evaluated.

Definition 2. *A topological partition* $\{V_1, V_2, \ldots, V_k\}$ *of V is s-coverable if for* $1 \leq i \leq k$, *there exists a set U_i, with $V_i \subseteq U_i \subseteq \cup_{j=1}^i V_j$, such that $S(U_i) \leq s$, considering as inputs of U_i all nodes of in-degree 0 or with a predecessor outside U_i. (Note that the inputs of U_1 are a subset of the inputs of G.)*

Theorem 4. *Let $k_G(2s)$ be the minimum cardinality of a $2s$-coverable partition for G. Then the access complexity of G satisfies*

$$Q_G(s) \geq s(k_G(2s) - 1)$$

Proof. Consider a T-step computation of G. For every $v \in V$ we mark the step of its first evaluation (inputs are all marked at the beginning of the computation). Clearly, each node of V is associated with *exactly* one marked step. Partition the interval $[1, T]$ relative to the computation of G into K subintervals such that in the first $K - 1$ subintervals exactly s accesses to L2 are performed, and in the K-th subinterval at most s accesses to L2 are performed. For $1 \leq i \leq K$ we let V_i consist of the nodes associated with the steps marked in the i-th subinterval I_i. It is easy to see that $\{V_1, V_2, \ldots, V_K\}$ is a topological partition. To show that it is a $2s$-coverable, we let U_i be the union of two (possibly nondisjoint) sets of nodes: the set of nodes whose values, at the beginning of I_i, reside in L1 or in the (at most) s cells of L2 accessed during I_i; and the set of nodes evaluated during I_i. It is easy to see that the inputs of U_i are included in the former set and that $V_i \subseteq U_i \subseteq \cup_{j=1}^i V_j$. Moreover, it is easy to argue that there exists a standard computation of U_i that uses space at most $2s$. The theorem follows since the evaluation of G performs at least $s(K - 1)$ accesses to L2 and K is bounded from below by $k_G(2s)$.   □

---

[1] In fact, an $\Omega(n/s)$ lower bound on the access complexity of a variant of the n-node diamond is proved in [6], assuming that all $2\sqrt{n} - 1$ nodes on the upper sides of the diamond are inputs.

The following theorem provides an upper bound on the access complexity of general CDAGs.

Theorem 5. *Let G have a 2s-coverable partition of cardinality k. Then,*

$$Q_G(2s) \leq 4sk.$$

Proof. Let $\{V_1, V_2, \ldots, V_k\}$ be a $2s$-coverable partition of G. Let L1 be the set of memory locations of address smaller than $2s$ and L2 the set of the remaining locations. We can compute G in k *stages*, where in Stage i the following activities are performed, $1 \leq i \leq k$:

1. Download the inputs of U_i from L2 to L1 (except for Stage 1);
2. Compute U_i in L1, uploading to L2 the value of every node v belonging to some U_j, with $j > i$, that was not uploaded in previous stages.

The correctness of the computation follows immediately from the definition of $2s$-coverable partition. Moreover, it is clear that each input of U_i contributes at most two accesses to L2 (one associated with its upload and one at the beginning of Stage i), while nodes that are not inputs of some U_i never pass through L2. Let $D_i \subseteq U_i$ be the set of inputs of U_i and note that by definition of $2s$-partition $|D_i| \leq 2s$. We have that the access complexity of the computation is at most

$$2\left(\sum_{i=1}^{k} |D_i|\right) \leq 4sk.$$

$\square$

As a straighforward corollary we have that $Q_G(2s) \leq 4sk_G(2s)$. Moreover, from Theorem 4 we have that $Q_G(s) \geq s(k_G(2s) - 1)$. As observed in several studies (e.g., [1]), relations on Q_G of this type translate into tight upper and lower bounds on running times for memory hierarchies where access time varies smoothly with the address.

5 Conclusions

In this paper, we have studied some issues related to space and access complexity. We have developed a lower bound technique for space of wide applicability. One interesting open issue remains whether our technique could reproduce the $\Omega(n/\log n)$ lower bound obtained in [11] for a particular CDAG recursively constructed out of superconcentrators. Attempts to resolve this issue might shed further light on the relation of space complexity to other graph-theoretic properties of CDAGs. We have also explored various relations between access complexity and space complexity. Here the attention was focused on minimizing $Q_G(s)$ for a fixed value of s. When studying computations in hierarchical memory models, such as the HMM of [1] or the HRAM of [3], the objective is the minimization of running time, which can be viewed as a functional of the function $Q_G(\cdot)$. A number of interesting questions arise in this connection, which can benefit of our results on the access complexity, but also from a deeper investigation of the possible shapes for the function Q (e.g., see [4] for issues related to the portability of performance across different hierarchical machines).

References

1. A. Aggarwal, B. Alpern, A.K. Chandra, and M. Snir. A model for hierarchical memory. In *Proc. of the 19th ACM Symp. on Theory of Computing*, pages 305–314, 1987.
2. A. Aggarwal and J.S. Vitter. The input/output complexity of sorting and related problems. *Communications of the ACM*, 31(9):1116–1127, 1988.
3. G. Bilardi and F.P. Preparata. Processor-time tradeoffs under bounded-speed message propagation: Part II, lower bounds. *Theory of Computing Systems*, Vol. 32, 531-559, 1999.
4. G. Bilardi and E. Peserico. Efficient portability across memory hierarchies. *Manuscript*, 2000.
5. S.A. Cook. An observation on time-storage trade off. *Journal of Computer and System Sciences*, 9:308–316, 1974.
6. J.W. Hong and H.T. Kung. I/O complexity: The red-blue pebble game. In *Proc. of the 13th ACM Symp. on Theory of Computing*, pages 326–333, 1981.
7. J. Hopcroft, W. Paul, and L. Valiant. On time versus space. *Journal of the ACM*, 24(2):332–337, April 1977.
8. T. Lengauer and R.E. Tarjan. Asymptotically tight bounds on time-space tradeoffs in a pebble game. *Journal of the ACM*, 29(4):1087–1130, October 1982.
9. M.S. Paterson and C.E. Hewitt. Comparative schematology. In *Proc. of Project MAC Conf. Concurrent Systems and Parallel Computation*, pages 119–127, Woods Hole, MA, 1970.
10. N. Pippenger. Pebbling. Technical report, IBM T.J. Watson Research Center, Yorktown Heights, NY 10598, 1980. See also *5th IBM Symposium of Mathematical Foundations of Computer Science*.
11. W.J. Paul, R.E. Tarjan, and J.R. Celoni. Space bounds for a game on graphs. *Mathematical Systems Theory*, 10:239–251, 1977.
12. A.L. Rosenberg and I.H. Sudborough. Bandwidth and pebbling. *Computing*, 31:115–139, 1983.
13. J.E. Savage. Space-time tradeoffs in memory hierarchies. Technical Report CS-93-08, Department of Computer Science, Brown University, 115 Waterman St. Providence, RI 02912-1910, 1993.
14. J.E. Savage. *Models of Computation – Exploring the Power of Computing*. Addison Wesley, Reading, MA, USA, 1998.

Approximating the Treewidth of AT-Free Graphs

Vincent Bouchitté and Ioan Todinca

LIP-École Normale Supérieure de Lyon, 46 Allée d'Italie, 69364 Lyon Cedex 07, France. {Vincent.Bouchitte,Ioan.Todinca}@ens-lyon.fr

Abstract. Using the specific structure of the minimal separators of AT-free graphs, we give a polynomial time algorithm that computes a triangulation whose width is no more than twice the treewidth of the input graph.

1 Introduction

The *treewidth* of graphs, introduced by Robertson and Seymour [11], has been intensively studied in the last years, mainly because many NP-hard problems become solvable in polynomial and even in linear time when restricted to graphs with small treewidth. These algorithms use a tree-decomposition of small width of the graph. A tree-decomposition or a *triangulation* of a graph is a chordal supergraph, i.e. all the cycles of the supergraph of length strictly more than three have a chord. Computing the treewidth of a graph corresponds to finding a triangulation with the smallest cliquesize. In particular, we can restrict ourselves to triangulations minimal by inclusion, that we call *minimal triangulations*.

Computing the treewidth of arbitrary graphs is NP-hard. Nevertheless, the treewidth can be computed in polynomial time for several well-known classes of graphs, for example the chordal bipartite graphs, the circle and circular-arc graphs, and permutation graphs. All these algorithms use the *minimal separators* of the graph and the fact that these classes of graphs have "few" minimal separators, in the sense that the number of the separators is polynomially bounded in the size of the graph. By studying the potential maximal cliques of a graph, which are the maximal cliques appearing in at least one minimal triangulation of the graph, we proved in [3,4,5] that the minimal separators are sufficient to compute the treewidth of a graph.

It has been proved by Bodlaender et al [2] that the treewidth can be approximated within a $\mathcal{O}(\log n)$ multiplicative factor. In addition, they showed that the treewidth is not approximable within an additive constant unless $P = NP$.

The existence of a polynomial time approximation algorithm which is no more than a constant times the optimal value is a question that still remains open. In this paper, we consider this problem for the class of AT-free graphs. Computing the treewidth is still NP-hard for these graphs [1]. Using the particular structure of their minimal separators, we give a 2-approximation algorithm for the treewidth of AT-free graphs.

U. Brandes and D. Wagner (Eds.): WG 2000, LNCS 1928, pp. 59–70, 2000.

2   Preliminaries

Throughout this paper we consider connected, simple, finite, undirected graphs. Let $G = (V, E)$ be a graph. For a vertex set $V' \subseteq V$ of G, we denote by $N_G(V')$ the neighborhood of V' in $G \backslash V'$.

A graph H is *chordal* (or *triangulated*) if every cycle of length at least four has a chord. A *triangulation* of a graph $G = (V, E)$ is a chordal graph $H = (V, E')$ such that $E \subseteq E'$. H is a *minimal triangulation* if for any intermediate set E'' with $E \subseteq E'' \subset E'$, the graph (V, E'') is not triangulated. A *clique* of G is a complete subgraph of G.

Definition 1. *Let G be a graph. The* treewidth *of G, denoted by* tw(G)*, is the minimum, over all triangulations H of G, of $\omega(H) - 1$, where $\omega(H)$ is the the maximum cliquesize of H.*

In other words, computing the treewidth of G means finding a triangulation with smallest cliquesize. In particular, the treewidth is always achieved by some minimal triangulation of the graph.

2.1   Minimal Separators and Chordal Graphs

The *minimal separators* play a crucial role in the characterization of the minimal triangulations of a graph.

A subset $S \subseteq V$ is an *a, b-separator* for two nonadjacent vertices $a, b \in V$ if the removal of S from the graph separates a and b in different connected components. S is a *minimal a, b-separator* if no proper subset of S separates a and b. We say that S is a *minimal separator* of G if there are two vertices a and b such that S is a minimal a, b separator. Notice that a minimal separator can be strictly included into another. We denote by Δ_G the set of all minimal separators of G.

Let G be a graph and S a vertex set of G. We note $\mathcal{C}_G(S)$ the set of connected components of $G \backslash S$. A component $C \in \mathcal{C}_G(S)$ of $G \backslash S$ is a *full component associated to S* if every vertex of S is adjacent to some vertex in C. For the following lemma, we refer to [7].

Lemma 1. *A set S of vertices of G is a minimal a, b-separator if and only if a and b are in different full components associated to S.*

Let S be a minimal separator of G. If $C \in \mathcal{C}_G(S)$, we say that $(S, C) = S \cup C$ is a *one-block* associated to S. A one-block (S, C) is called *full* if C is a full component associated to S. If (S, C) is a full one-block, then $S = N_G(C)$. If (S, C) is not full, then $S^* = N_G(C)$ is a minimal separator of G, strictly contained in S.

Let S be a minimal separator of G. We say that S *crosses* a set of vertices A if S separates two vertices $x, y \in A$ (i.e. S is an x, y-separator). We say that

S *separates* two sets of vertices A and B if S separates each vertex of $A \backslash S$ from each vertex of $B \backslash S$.

Let S and T be two minimal separators. If S crosses T, we write $S \# T$. Otherwise, S and T are called *parallel*, denoted by $S \| T$. It is easy to prove that these relations are symmetric. Remark that S and T cross if and only if then T intersects each full component associated to S. Conversely, S and T are parallel if and only if T is contained in some one-block (S, C_T) associated to S. In particular, if $T \subseteq S$, then S and T are parallel.

Let $S \in \Delta_G$ be a minimal separator. We denote by G_S the graph obtained from G by *completing* S, i.e. by adding an edge between every pair of non-adjacent vertices of S. If $\Gamma \subseteq \Delta_G$ is a set of separators of G, G_Γ is the graph obtained by completing all the separators of Γ. The results of [8], concluded in [10], establish a strong relation between the minimal triangulations of a graph and its minimal separators.

Theorem 1. *Let $\Gamma \in \Delta_G$ be a maximal set of pairwise parallel separators of G. Then $H = G_\Gamma$ is a minimal triangulation of G and $\Delta_H = \Gamma$.*

Let H be a minimal triangulation of a graph G. Then Δ_H is a maximal set of pairwise parallel separators of G and $H = G_{\Delta_H}$.

In other terms, every minimal triangulation of a graph G is obtained by considering a maximal set Γ of pairwise parallel separators of G and completing the separators of Γ. The minimal separators of the triangulation are exactly the elements of Γ.

It is important to know that the elements of Γ, who become the separators of H, have strictly the same behavior in H as in G. Indeed, the connected components of $H \backslash S$ are exactly the same as in $G \backslash S$, for every $S \in \Gamma$. Moreover, the full components associated to S are the same in the two graphs.

2.2   Blocks

The following definitions are strongly related with the *blocking sets* and the *blocks* introduced in [6].

Definition 2. *Let G be a graph and $\mathcal{S} \subseteq \Delta_G$ a set of pairwise parallel separators such that for any $S \in \mathcal{S}$, there is a one-block $(S, C(S))$ containing all the elements of $\mathcal{S}$. Suppose that $\mathcal{S}$, ordered by inclusion, has no greatest element. We define the* piece between *the elements of $\mathcal{S}$ by*

$$P(\mathcal{S}) = \bigcap_{S \in \mathcal{S}} (S, C(S))$$

Notice that for any $S \in \mathcal{S}$ the one-block of S containing all the separators of $\mathcal{S}$ is unique: if $T \in \mathcal{S}$ is not included in S, there is a unique connected component of $G \backslash S$ containing $T \backslash S$.

Definition 3. *Let B be a set of vertices of a graph G. We denote by $C_1, \ldots, C_p$ the connected components of $G \backslash B$ and by S_i the neighborhood of C_i. We will say that B is a* block *of G if the sets S_i are minimal separators of G and one of the following conditions holds:*

- *$B = G$.*
- *There is an $i \in [1, p]$ such that B is a one-block (S_i, C).*
- *$B = P(S_1, \ldots, S_p)$.*

If B is a block, we say that the minimal separators $S_1, \ldots, S_p$ border *B. The block B is* full *if $G = B$ or, for each S_i, B is contained in a full one-block associated to S_i.*

Remark 1. Observe that if $\{T_1, T_2, \ldots T_q\}$ is a set of minimal separators such that the piece between them exists, then $B = P(T_1, \ldots, T_p)$ is a block. Moreover, any of the minimal separators bordering B is contained in one of $T_1, \ldots, T_q$. Also if B is a one-block (S, C), then any of the minimal separators bordering B is contained in S.

A block is also characterized by the following property:

Proposition 1. *Let B be a set of vertices of G. Consider the connected components $C_1, \ldots, C_p$ of $G \backslash B$ and their neighborhoods $S_1, \ldots, S_p$. If for all i, $1 \le i \le p$ we have that S_i is a minimal separator, B is contained in some one-block (S_i, D_i) and $B \ne S_i$, then B is a block of G.*

Proof. Suppose that the set $\{S_1, S_2, \ldots, S_p\}$ has no greatest element. We show that $P(S_1, \ldots, S_p)$ exists and is equal to B. For each S_i, all the minimal separators $S_1, S_2, \ldots, S_p$ are contained in a one-block (S_i, D_i), so $P(S_1, \ldots, S_p) = \bigcap_i (S_i, D_i)$. In particular, $B \subseteq P(S_1, \ldots, S_p)$. Conversely, let us prove that $P(S_i, \ldots, S_p)$ is contained in B. Let x be a vertex of $G \backslash B$, so x is contained in some C_i. Hence, $x \notin (S_i, D_i)$, so $x \notin P(S_1, \ldots, S_p)$.

Suppose now that $\{S_1, S_2, \ldots, S_p\}$ has a greatest element, say S_1. We show that $B = (S_1, D_1)$. Clearly, $B \subseteq (S_1, D_1)$. Suppose there is a vertex x in $(S_1, D_1) \backslash B$ and let C_i the component of $G \backslash B$ containing x. We have that $S_i \subseteq S_1$, so C_i is also a connected component of $G \backslash S_1$. Since $x \in D_1$, it follows that $D_1 = C_i$, so D_1 is a connected componend of $G \backslash B$. Therefore, $D_1 \cap B = \emptyset$, contradicting $B \ne S_1$.   □

Proposition 2. *Let G be a graph and S, T two minimal separators of G such that $S \# T$. Then $S \cup T$ is a full block of G.*

Proof. Let $B = S \cup T$. The assertion is true if $B = G$. Suppose that $B \ne G$. Consider the connected components $C_1, \ldots, C_p$ of $G \backslash B$ and their neighborhoods $S_1, \ldots, S_p$.

We prove first that each S_i is a minimal separator of G and B is contained in a full one-block (S_i, D_i). The vertex set C_i is contained in a connected component D' associated to S and in a connected component E' associated to T. Let D, E be two full components associated to S, respectively T, such that $D \ne D'$ and $E \ne E'$. Let us show that $D \cup E$ is contained in a same full component associated to S_i. Since $C_i \subseteq D'$ and D' is a connected component of $G\backslash S$, $N(C_i)$ is contained in $D' \cup N(D') \subseteq (S, D')$. It follows that $S_i = N(C_i)$ does not intersect D. Thus, D is contained in a connected component C_D of $G\backslash S_i$. For similar reasons, E is contained in a connected component C_E of $G\backslash S_i$. It remains to prove that $C_D = C_E$. Since $S\#T$ and D is a full component associated to S, there is an $x \in T \cap D$. Then $x \in T$ has at least one neighbor y in the full component E associated to T. We have two adjacent vertices $x \in C_D$ and $y \in C_E$, so $C_D = C_E$. We denote $D_i = C_D$. Since $S_i \subseteq S \cup T$ and D, E are full components associated to S, respectively T, we deduce that each vertex of S_i has a neighbor in $D_i \supseteq D \cup E$. So D_i is a full component associated to S_i, different from C_i. By lemma 1, S_i is a minimal separator of G. We show now that $B \subseteq (S_i, D_i)$. Indeed, $D \subseteq D_i$, so $S = N(D)$ is contained in $N(D_i) \cup D_i = (S_i, D_i)$. In the same way, $T \subseteq (S_i, D_i)$. We conclude that B is contained in the full one-block (S_i, D_i). Also notice that $D_i \cap B \neq \emptyset$, because D_i contains D and $D \cap T \neq \emptyset$. Thus, $B \neq S_i$.

By proposition 1, B is a full block of G. $\qquad\square$

Proposition 3. *Let B be a block of G and $S \subseteq B$ be a minimal separator of G. For any connected component C associated to S which intersects B we have that $(S, C) \cap B$ is a block. In other words, if $S \subseteq B$ crosses B, then S splits the block B into smaller blocks.*

We omit here the proof of proposition 3. If B is a block $S \cup T$ with $S\#T$, then any full component associated to S intersects T. We obtain:

Corollary 1. *Let $S\#T$ be two minimal separators of G. Let C be a full component associated to S. Then $S \cup (T \cap C)$ is a block of G.*

3   AT-Free Graphs and Minimal Separators

We say that three vertices (x, y, z) of a graph form an *asteroidal triple* of a graph G if between every two of them there exists a path avoiding the neighborhood of the third. A graph is *AT-free* if it has no asteroidal triple.

For characterizing AT-free graphs we rather use the notion of *asteroidal triple of separators*.

Definition 4. *Let S, T and U be three minimal separators of G, none of them being contained into another. We say that (S, T, U) form an asteroidal triple of separators if $P(S, T, U)$ exists.*

Proposition 4 ([6]). *A graph G is AT-free if and only if it has no asteroidal triple of separators.*

In particular, an AT-free graph cannot have three-blocks, that is blocks of type $P(S, T, U)$ with none of S, T, U contained into another. So any block B of an AT-free graph is a one-block (S, C), a two-block $P(S, T)$ or the graph G itself.

Let us reconsider corollary 1 for the case of AT-free graphs.

Proposition 5. *Consider an AT-free graph G, a minimal separator S of G and a full component C associated to S. Let T be a minimal separator crossing S. Suppose that $C \backslash T \neq \emptyset$ and let $U = N(C \backslash T)$. Then $S \cup (T \cap C) = P(S, U)$.*

Proof. By corollary 1, $B = S \cup (T \cap C)$ is a block of G.

Let $C_1, C_2, \ldots, C_p$ be the connected components of $G \backslash B$. The C_i's are either connected components of $G \backslash S$ different from C or connected components of $G[C \backslash T]$. Let $S_1, \ldots, S_p$ be the neighborhoods of $C_1, \ldots, C_p$. Notice that, if C_i is a connected component of $G \backslash S$ different from C, then $S_i \subseteq S$. In particular, if C_i is a full component associated to S with $C_i \neq C$, we have $S_i = S$.

Let us show that S is maximal by inclusion in $\{S_1, \ldots, S_p\}$. Suppose that S is strictly contained in some S_i, so $S \subset N(C_i)$. Then C_i is a connected component of $G[C \backslash T]$. Let D be the connected component of $G \backslash T$ containing C_i. We have $N(C_i) \subseteq D \cup N(D) \subseteq (T, D)$, so $S \subseteq (T, D)$. This contradicts the fact that S and T cross.

Let C_i be any connected component of $G[C \backslash T]$ (such a C_i exists because $C \backslash T$ is non-empty). Let us show that $S_i \not\subseteq S$. If $S_i \subseteq S$, then C_i is also a connected component of $G \backslash S$, so $C_i = C$. This implies $B \cap C = \emptyset$, contradicting the fact that T intersects the full component C associated to S. It follows that $S_i \not\subseteq S$.

Thus, the set of minimal separators bording B has at least two elements maximal by inclusion, S and S'. Since an AT-free graph has only one-blocks and two-blocks, we deduce $B = P(S, S')$. It remains to show that $S' = U$. For any connected component C_i of $G[C \backslash T]$, we have $S_i \not\subseteq S$, so $S_i \subseteq S'$ by remark 1. It follows that S' contains $U = N(C \backslash T)$. It is easy to see that any minimal separator bordering B is contained in S or in $N(C \backslash T)$, so $S' \subseteq U$. We conclude that U is a minimal separator of G and $S \cup (T \cap C) = P(S, U)$. □

Notice that the minimal separator U of proposition 5 is also one of the minimal separators bordering the block $S \cup T$. In particular, U is parallel to both S and T. Since U intersects at most one component associated to T, there is at least one full component E associated to T such that $U \cap E = \emptyset$.

Corollary 2. *Let S and T be two crossing separators of an AT-free graph G. Let C be a full component associated to S such that $C \backslash T \neq \emptyset$. Consider the block $S \cup (T \cap C) = P(S, U)$. Let E be a full component associated to T such that $E \cap U = \emptyset$. Then $T \cap C$ is contained in the neighborhood of $S \cap E$.*

Proof. We denote $E_C = E \cap C$. Let us prove that $E_C = \emptyset$. Suppose that E_C is non-empty. The neighborhood of E_C is contained in $S \cup T$. We show first that $N(E_C)$ is not contained in T. If $N(E_C) \subseteq T$, then E_C is a connected component of $G \backslash T$, so we must have $E_C = E$. But $E_C \subseteq C$, so the neighborhood of E_C is included in $C \cup N(C) = (S, C)$. Let now D be a full component associated to S, different from C and let $y \in T \cap D$. We have $y \notin N(E_C)$, contradicting the fact that $E_C = E$ is a full component associated to T. Thus, $N(E_C)$ is not contained in T, so there is a vertex y in $N(E_C) \backslash T$. Clearly, $y \in E$. By proposition 5, $U = N(C \backslash T)$, so $y \in U$. This contradicts the fact that E does not intersect U.

We have proved that $E \cap C = \emptyset$. Let now x be any vertex of $T_C = T \cap C$, we show that x has a neighbor in $S_E = S \cap E$. E is a full component associated to T, so x has a neighbor x' in E. Since $x \in C$, we have that x' is in C or in $N(C) = S$. But $E \cap C = \emptyset$, so the only choice remaining is $x' \in S \cap E$.   □

Proposition 6. *Let S be a minimal separator of the AT-free graph G and let C be a full component associated to S. Consider any minimal separator T, crossing S. Let $x, y \in C \backslash T$ such that x and y have neighbors in $S \backslash T$. Then x and y are in a same connected component of $G \backslash T$.*

Proof. By proposition 5, we know that $U = N(C \backslash T)$ is a minimal separator and the block $B = S \cup (T \cap C)$ can be written $P(S, U)$. Recall that T and U are parallel. Suppose that T separates x and y and let $x', y' \in S \backslash T$ be two neighbors of x, respectively y. Clearly, T separates x' and y'. But $x', y' \in N(C \backslash T) = U$, so T crosses U – contradiction.   □

4   The Algorithm

We begin by an informal description of our algorithm. We are given an AT-free graph G and an integer k and we want to decide if tw$(G) > k$ or to find a triangulation of width at most $2k$. The time complexity of the algorithm we obtain is polynomial and does not depend on the value of k, so it is easy to derive a polynomial algorithm for a 2-approximation of the treewidth of AT-free graphs.

The algorithm relies on this property of AT-free graphs which will be proved in the next section. Given an AT-free graph G of treewidth at most k and a one-block (S, C) with $|S| \leq k$ and $|C| > k + 1$, there is a minimal separator $U \subseteq (S, C)$ such that $|P(S, U)| \leq 2k + 1$ and $|U| \leq k$. Moreover, the separator U can be computed in polynomial time.

Consider any minimal separator S of G and any connected component C of $G \backslash S$. We denote by $R(S, C)$ the realization of (S, C), which is the graph $G_S[S \cup C]$ (see [9]). We have to show that if $|S| \leq k$ then either we can compute a triangulation $H(S, C)$ of $R(S, C)$ of width at most $2k$ or we can decide that tw$(G) > k$. Indeed, a triangulation of G is obtained by taking $\bigcup_C H(S, C)$ [9].

The algorithm works in this way. If $|C| \leq k + 1$, we simply complete (S, C) into a clique. If $|C| > k + 1$, we are looking for a minimal separator $U \subseteq (S, C)$ such that $|U| \leq k$ and $|P(S, U)| \leq 2k + 1$. If such a separator does not exist we assert that tw$(G) > k$. If U exists, observe that U splits the block (S, C) into smaller blocks, namely the block $P(S, U)$ and blocks of the form (U, C') with $C' \subset C$. We then apply this process recursively on each sub-block (U, C') to find a triangulation $H(U, C')$ of $R(U, C')$ of width at most $2k$ or output that tw$(G) > k$. The block $P(S, U)$ will be simply completed into a clique $K(S, U)$. Then $H(S, C)$ is $\bigcup_{C'} H(U, C') \cup K(S, U)$.

Finding the first separator S with $|S| \leq k$ is easy, it is sufficient to look for a separator of minimum size. It is well-known that this computation can be done in polynomial time.

More delicate is the search for the minimal separator U in the general framework and we use a quite indirect method. We will consider the co-bipartite graph $G' = G_{\{S,C\}}[S \cup C]$ obtained from $G[S \cup C]$ by completing S and C. We show that if $\mathrm{tw}(G) < k$, there is a minimal separator X of G' of size at most k. Conversely, we prove that if G' has a minimal separator X with $|X| \le k$, the vertex set $S \cup X$ is actually a block $P(S, U)$ of G, with $U = N_G(C \backslash X)$. The minimal separator U will satisfy $|P(S, U)| \le 2k + 1$ and $|U| \le k$.

We give now the pseudo-code of the algorithm. The main program, **approximate_triangulation**, uses the procedure **triangulate_realization** which, given G, k and a one-block (S, C) computes a triangulation of $R(S, C)$ of width at most $2k$ or outputs that $\mathrm{tw}(G) > k$.

> **approximate_triangulation**
> Input: an AT-free graph G and an integer k
> Output: $\mathrm{tw}(G) > k$ or a triangulation H of width at most $2k$
> begin
>   compute a minimal separator S of G of minimum size
>   if $|S| > k$ then
>   return "$\mathrm{tw}(G) > k$"
>   for all one-blocks (S, C) associated to S do
> $H(S, C) \leftarrow$ **triangulate_realization**(G, S, C, k)
>   if the call outputs "$\mathrm{tw}(G) > k$" then
>   return "$\mathrm{tw}(G) > k$"
>   return $\bigcup_C H(S, C)$
> end

> **triangulate_realization**
> Input: G, S, C and k
> Output: $\mathrm{tw}(G) > k$ or a triangulation $H(S, C)$ of $R(S, C)$ of width $\le 2k$
> begin
>   if $|S \cup C| \le 2k + 1$ then
>   return the clique on vertex set $S \cup C$
>   /* we have to find the separator U*/
> $G' \leftarrow G_{\{S,C\}}[S \cup C]$
>   compute a minimum size separator X of G'
>   if $|X| > k$ then
>   /* we still have the possibility that $\mathrm{tw}(R(S, C)) \le k$ */
>   return **approximate_triangulation**$(R(S, C), k)$
> $U \leftarrow N_G(C \backslash X)$
>   for all the one-blocks (U, C') contained in (S, C) do
> $H'(U, C') \leftarrow$ **triangulate_realization**(G, U, C', k)
>   if the call outputs "$\mathrm{tw}(G) > k$" then
>   return "$\mathrm{tw}(G) > k$"
>   return $\bigcup_{C'} H'(U, C') \cup G_{P(S,U)}[S \cup U]$
> end

Our goal is to implement the procedure **triangulate_realization** in polynomial time. It remains to discuss a point we have hidden until now. It is possible that the algorithm fails to find the separator U even if $\text{tw}(G) \leq k$. This case occurs only when the minimal separator S belongs to all the minimal triangulations of width k. So to handle this problem, we have just to decide if $\text{tw}(R(S,C)) > k$ or to find a triangulation of $R(S,C)$ of width at most $2k$, that is we have to apply the main algorithm on the realization of the block.

5 Main Theorem

In this section we consider an AT-free graph G of treewidth at most k. We take any minimal triangulation H of G, of width at most k (so $\omega(H) \leq k+1$). Consider now any minimal separator S of G and a full component C associated to S. Suppose that S is not a minimal separator of H and that $|C| > k+1$.

Under these conditions, our main result is the following:

Theorem 2. *There is a minimal separator $U \subset (S,C)$ of G such that $|U| \leq k$ and $|P(S,U)| \leq |S| + k$. Moreover, the minimal separator U can be computed in polynomial time.*

The proof of this theorem will be made in several steps. We will first concentrate on the existence of the minimal separator U. Our proofs will characterize U and will give some further information about the block $P(S,U)$. Then we obtain an algorithm computing U.

Since S is not a minimal separator of H and Δ_H is a maximal set of pairwise parallel separators of G by theorem 1, we have at least one minimal separator $T \in \Delta_H$ that crosses S. Take T among all the minimal separators crossing S such that $T \cap C$ is maximal by inclusion. Observe that $C \backslash T \neq \emptyset$, because $|C| > k+1$ and $|T| \leq k$. By proposition 5, B is a block of G and there is some $U \subseteq B$ such that $B = P(S,U)$. We denote $X = U \cup (T \cap C)$. We distinguish the "easy" case when $|X| \leq k$, and the "difficult" case, when $|X| > k$.

5.1 Easy Case: $|X| \leq k$

Proposition 7. *Consider the co-bipartite graph $G' = G_{\{S,C\}}[S \cup C]$ obtained from $G[S \cup C]$ by completing the sets S and C into cliques. Then X is a minimal separator of G'.*

Proof. We show that $C \backslash X$ and $S \backslash X$ are full components associated to X in G'.

Observe that $C \backslash X \neq \emptyset$. Indeed, $|X| \leq k$, and $|C| > k+1$, so $C \backslash X \neq \emptyset$. We prove that $N_{G'}(C \backslash X) = X$. By proposition 5, the neighborhood of $C \backslash T$ in G is U. Clearly, the neighborhood of $C \backslash T$ in G' is $(T \cap C) \cup U = X$, so $C \backslash T = C \backslash X$ is a full component associated to X in G'.

Since S is a clique of G', $S \backslash X$ is contained in a connected component D of $G' \backslash X$. This component is different from $C \backslash X$, and since $V(G') = (S \backslash X) \cup X \cup (C \backslash X)$ the only choice is $D = S \backslash X$. Since T and U are parallel in G, there is a full component E associated to T in G such that $E \cap U = \emptyset$. Notice that $S \backslash X$ is not empty, in particular it contains $S_E = S \cap E$. Let us show that X is in the neighborhood of $S \backslash X$ in G'. According to corollary 2, $T \cap C$ is contained in the neighborhood of S_E in the graph G, so $X \backslash S = T \cap C$ is contained in $N_{G'}(S \backslash X)$. Also $X \cap S$ is contained in $N_{G'}(S \backslash X)$, because S is a clique in G'. Hence, $N_{G'}(S \backslash X) = X$. We conclude that $S \backslash X$ is a full component associated to X in G'.

We have proved that $G' \backslash X$ has two full components associated to X. By lemma 1, X is a minimal separator of G'. □

So there is a minimal separator X of G' of size at most k. Conversely, we show that if G' has a minimal separator Y of size at most k, we will be able to compute a minimal separator U like in theorem 2.

Proposition 8. *Let Y be a minimal separator of the graph $G' = G_{\{S,C\}}[S,C]$. Then $S \cup Y$ is a two-block $P(S,U)$ of G. Moreover, $U = N_G(C \backslash Y)$ and U is contained in Y.*

Proof. We denote $B = S \cup Y$. We show that B is not contained in S or in Y. Since Y is a minimal separator of the cobipartite graph G', the connected components of $G' \backslash Y$ are exactly $S \backslash Y$ and $C \backslash Y$. In particular, both of them are non empty, so $S \nsubseteq Y$. Also, $Y \nsubseteq S$. Indeed, if $Y \subset S$, let x be a vertex of $S \backslash Y$ and let $y \in C$ be a neighbor of x in the graph G. Clearly, x and y are also adjacent in G', contradicting the fact that $S \backslash Y$ and $C \backslash Y$ are in different connected components of $G' \backslash Y$. Thus, $B \nsubseteq S$ and $B \nsubseteq Y$.

Let $C_1, \ldots, C_p$ be the connected components of $G \backslash B$ and let $S_1, \ldots, S_p$ be their neighborhoods in G. The C_i's are either connected components of $G \backslash S$ different from C or connected components of $G[C \backslash Y]$. In particular, if C_i is a full component associated to S, different from C, then $S_i = S$.

We will see that each S_i is a minimal separator of G and that B is contained in some full block (S_i, D_i). Consider first the case when C_i is a connected component of $G \backslash S$, different from C. Then $S_i \subseteq S$. Clearly, $C \cap S_i = \emptyset$, so C is contained in some connected component D_i of $G \backslash S_i$, with $D_i \neq C_i$. Since each vertex of S has a neighbor in C, it follows that every vertex of S_i has a neighbor in D_i. Thus, D_i is a full component associated to S_i. Both C_i and D_i are full components associated to S_i, so S_i is a minimal separator by lemma 1. Since $C \subseteq D_i$, we have that $S = N_G(C)$ is contained in $D_i \cup N_G(D_i) = (S_i, D_i)$. We conclude that $B \subseteq (S, C) \subseteq (S_i, D_i)$. Consider now the case when C_i is a connected component of $G[C \backslash Y]$. Notice that S_i is contained in Y. Let D be a full component associated to S, different from C. Since $S_i \subseteq (S, C)$, S_i does not intersect D, so D is contained in a connected component D_i associated to S_i. Let us prove first that $S \subseteq S_i \cup D_i$. Indeed, $S = N_G(D)$, so S is contained in $D_i \cup N_G(D_i) \subseteq S_i \cup D_i$. We show now that $Y \backslash S$ is also contained in $S_i \cup D_i$. In the graph G', the vertex set $S \backslash Y$ is a full component associated to Y, so any vertex $x \in Y$ has a neighbor $y \in S \backslash Y$. In particular, if $x \in Y \backslash S$, then y is also a neighbor of x in the graph G (the edges between $S \backslash Y$ and $Y \backslash S$ are the same in G and in G'). Thus, $Y \backslash S$ is contained in $N_G(S \backslash Y)$. Since $S_i \subseteq Y$ we have $S \backslash Y \subseteq S \backslash S_i \subseteq D_i$, so $Y \backslash S \subseteq D_i \cup N_G(D_i) \subseteq S_i \cup D_i$. Therefore, $B \subseteq S_i \cup D_i$. Observe now that D_i is a full component associated to S_i. Indeed, each vertex of $S_i \cap S$ has a neighbor in D, so in D_i, and each vertex of $S_i \backslash S \subseteq Y \backslash S$ has a neighbor in $S \backslash Y$, so in D_i. Thus, D_i is a full component associated to S_i. Clearly $D_i \neq S_i$, so by lemma 1, S_i is a minimal separator.

We have proved that each S_i is a minimal separator of G and that B is contained in some full one-block (S_i, D_i). Moreover B is not contained in any S_i: recall that each S_i is contained in S or in Y, but $B \not\subseteq Y$ and $B \not\subseteq S$, so $B \not\subseteq S_i$. By proposition 1, B is a block of G.

S is one of the minimal separators bordering B. We show now that S is maximal by inclusion among $\{S_1, \ldots, S_p\}$. Any minimal separator S_i is contained in S or in Y. Since $S \not\subseteq Y$, S is inclusion maximal among the minimal separators bordering B. We want to prove that S is not the only inclusion-maximal element of $\{S_1, \ldots, S_p\}$. Let C_i be any connected component of $G[C \backslash Y]$. We show that $S_i \not\subseteq S$. Indeed, if $S_i \subseteq S$, then C_i is also a connected component of $G \backslash S$, so the only choice is $C_i = C$. This implies $Y \cap C = \emptyset$, contradicting $Y \not\subseteq S$. The set of minimal separators bordering B has at least two inclusion maximal elements, S and S'. The AT-free graph G has only one-blocks and two-blocks, so B is a two-block $P(S, S')$. It remains to prove that $S' = N_G(C \backslash Y)$. Since $S' \not\subseteq S$, S' is the neighborhood of a connected component C_j of $G[C \backslash Y]$. Thus, $S' \subseteq N_G(C \backslash Y)$. For any connected component $C_i \subseteq C \backslash Y$, we have $S_i \not\subseteq S$, so $S_i \subseteq S'$ by remark 1. Consequently, S' contains $N_G(C \backslash Y)$. We conclude that $S' = N_G(C \backslash Y) = U$. Thus, $B = P(S, U)$. Moreover, $U = N_G(C \backslash Y) \subseteq N'_G(C \backslash Y)$, so $U \subseteq Y$.   □

Corollary 3. *Let G be an AT-free graph and let (S, C) be a full block of G. We denote by G' the cobipartite graph $G_{\{S,C\}}[S \cup C]$. If G' has a minimal separator of size at most k, then we can compute in polynomial time a minimal separator U of G such that $P(S, U) \leq |S| + k$ and $|U| \leq k$.*

Proof. We compute a minimum size separator Y of G'. This can be done in polynomial time. Clearly, Y is a minimal separator of G' and $|Y| \leq k$. According to proposition 8, the vertex set $S \cup Y$ is a two-block $P(S, U)$ of G. Therefore, $|P(S, U)| \leq |S| + |Y| \leq |S| + k$. By proposition 8, $U \subseteq Y$ so $|U| \leq k$. Moreover, $U = N_G(C \backslash Y)$, so U can be computed in polynomial time.   □

5.2   Difficult Case: $|X| > k$

Let us consider now the case when $X = (T \cap C) \cup U$ has strictly more that k vertices. Our goal is to show that also in this case there is a minimal separator Y of $G' = G_{\{S,C\}}[S \cup C]$ such that $|Y| \leq k$. Thus, the corollary 3 will allow us to compute the minimal separator U like in theorem 2.

Due to space restriction, we only give here an outline of the proof. A complete description is given in the full version of this paper.

The construction of the vertex set Y is rather technical and will be made in several steps. We will show that there is a minimal separator T' of H such that T' crosses X. We prove then that $P(T, T')$ exists. We will consider the block $B = P(T, T')$ and we put $Y = (B \cap C) \cup N_G(C \backslash B)$.

The first crucial observation is that Y is strictly contained in the block B. Moreover, Y is a minimal separator of the co-bipartite graph $G' = G_{\{S,C\}}[S \cup C]$.

The second step is to prove that $|Y| \leq k$. This is true if our block B is a clique of H, because the cliques of H have at most $k+1$ vertices. If B is not a clique, we can split it into smaller blocks, and we show that Y is strictly contained in one of these smaller blocks. We reiterate this process until we obtain a block B' that can not be splitted. This implies that B' is a clique, and since $Y \subset B$ we conclude that $|Y| \leq k$.

Using corollary 3, we deduce that in this case we can also compute the minimal separator U such that $|U| \leq k$ and $|P(S,U)| \leq |S| + k$, which proves the theorem 2.

References

1. S. Arnborg, D.G. Corneil, and A. Proskurowski. Complexity of finding embeddings in a k-tree. *SIAM J. on Algebraic and Discrete Methods*, 8:277–284, 1987.
2. H. Bodlaender, J.R. Gilbert, H. Hafsteinsson, and T. Kloks. Approximating treewidth, pathwidth, and minimum elimination tree height. *J. of Algorithms*, 18:238–255, 1995.
3. V. Bouchitté and I. Todinca. Minimal triangulations for graphs with "few" minimal separators. In *Proceedings 6th Annual European Symposium on Algorithms (ESA'98)*, volume 1461 of *Lecture Notes in Computer Science*, pages 344–355. Springer-Verlag, 1998.
4. V. Bouchitté and I. Todinca. Treewidth and minimum fill-in of weakly triangulated graphs. In *Proceedings 16th Symposium on Theoretical Aspects of Computer Science (STACS'99)*, volume 1563 of *Lecture Notes in Computer Science*, pages 197–206. Springer-Verlag, 1999.
5. V. Bouchitté and I. Todinca. Listing all potential maximal cliques of a graph. In *Proceedings 17th Annual Symposium on Theoretical Aspects of Computer Science (STACS 2000)*, volume 1770 of *Lecture Notes in Computer Science*, pages 503–515. Springer-Verlag, 2000.
6. H. Broersma, T. Kloks, D. Kratsch, and H. Müller. A generalization of AT-free graphs and a generic algorithm for solving triangulation problems. In *Workshop on Graphs WG'98*, volume 1517 of *Lecture Notes in Computer Science*, pages 88–99. Springer-Verlag, 1998.
7. M. C. Golumbic. *Algorithmic Graph Theory and Perfect Graphs*. Academic Press, New York, 1980.
8. T. Kloks, D. Kratsch, and H. Müller. Approximating the bandwidth for asteroidal triple-free graphs. *Journal of Algorithms*, 32:41–57, 1999.
9. T. Kloks, D. Kratsch, and J. Spinrad. On treewidth and minimum fill-in of asteroidal triple-free graphs. *Theoretical Computer Science*, 175:309–335, 1997.
10. A. Parra and P. Scheffler. Characterizations and algorithmic applications of chordal graph embeddings. *Discrete Appl. Math.*, 79(1-3):171–188, 1997.
11. N. Robertson and P. Seymour. Graphs minors. II. Algorithmic aspects of treewidth. *J. of Algorithms*, 7:309–322, 1986.

Split-Perfect Graphs: Characterizations and Algorithmic Use

Andreas Brandstädt and Van Bang Le

Fachbereich Informatik, Universität Rostock, A.-Einstein-Str. 21, 18051 Rostock, Germany. {ab,le}@informatik.uni-rostock.de

Abstract. Two graphs G and H with the same vertex set V are P_4-*isomorphic* if every four vertices $\{a, b, c, d\} \subseteq V$ induce a chordless path (denoted by P_4) in G if and only if they induce a P_4 in H. We call a graph *split-perfect* if it is P_4-isomorphic to a split graph (i.e. a graph being partitionable into a clique and a stable set). This paper characterizes the new class of split-perfect graphs using the concepts of homogeneous sets and p-connected graphs, and leads to a linear time recognition algorithm for split-perfect graphs, as well as linear time algorithms for classical optimization problems on split-perfect graphs based on the primeval decomposition of graphs. These results considerably extend previous ones on smaller classes such as P_4-sparse graphs, P_4-lite graphs, P_4-laden graphs, and (7,3)-graphs. Moreover, split-perfect graphs form a new subclass of brittle graphs containing the superbrittle graphs for which a new characterization is obtained leading to linear time recognition.

1 Introduction

Graph decomposition is a powerful tool in designing efficient algorithms for basic algorithmic graph problems such as maximum independent set, minimum coloring and many others. Recently, the modular, the primeval and the homogeneous decomposition of graphs attracted much attention. The last two types of decomposition were introduced by Jamison and Olariu [34] and are based on their Structure Theorem and the concept of P_4-connectedness. A P_4 is an induced path on four vertices. A graph $G = (V, E)$ is P_4-connected (p-connected for short) if for every partition V_1, V_2 of V with nonempty V_1, V_2, there is a P_4 of G with vertices in V_1 and in V_2.

We follow this line of research by introducing and characterizing a new class of graphs - the *split-perfect graphs* - for which the p-connected components have a simple structure generalizing split graphs. As usual, a graph is called a *split graph* if its vertex set can be partitioned into a clique and a stable set.

The p-connected components represent the nontrivial leaves in the primeval decomposition tree, and thus some basic algorithmic problems can be solved in linear time along the primeval decomposition tree.

The primeval tree is a generalization of the *cotree* representing the structure of the well-known *cographs* i.e. the graphs containing no induced P_4. A cograph or its complement is disconnected, and the cotree expresses this in terms of

U. Brandes and D. Wagner (Eds.): WG 2000, LNCS 1928, pp. 71–82, 2000.
© Springer-Verlag Berlin Heidelberg 2000

corresponding co-join and join operations. The cotree representation of a cograph is essential in solving various NP-hard problems efficiently for these graphs; see [14,15] for more information on P_4-free graphs.

Natural extensions of P_4-free graphs are graphs with few P_4's, such as

1. P_4-*reducible graphs* [29,32] (no vertex belongs to more than one P_4),
2. P_4-*sparse graphs* [25,31,33,36] (no set of five vertices induces more than one P_4),
3. P_4-*lite graphs* [30] (every set of at most six vertices induces at most two P_4's or a "spider"), and
4. P_4-*laden graphs* [22] (every set of at most six vertices induces at most two P_4's or a split graph).

Note that in this order, every graph class mentioned in this paragraph is a subclass of the next one.

Recently, Babel and Olariu [3] considered graphs in which no set of at most q vertices induces more than t P_4's, called (q,t)-*graphs*. The most interesting case is $t = q - 4$: $(4,0)$-graphs are exactly the P_4-free graphs, $(5,1)$-graphs are exactly the P_4-sparse graphs, and it turns out that P_4-lite graphs form a subclass of $(7,3)$-graphs. For all these graphs, nice structural results lead to efficient solutions for classical NP-hard problems. Our new class of split-perfect graphs extends all of them.

Another motivation for studying graph classes with special P_4 structure stems from the *greedy coloring heuristic*: Define a linear order $<$ on the vertex set, and then always color the vertices along this order with the smallest available color. Chvátal [13] called $<$ a *perfect order* of G if, for each induced subgraph H of G, the greedy heuristic colors H optimally. Graphs having a perfect order are called *perfectly orderable* (see [26] for a comprehensive survey); they are NP-hard to recognize [39]. Because of the importance of perfectly orderable graphs, however, it is natural to study subclasses of such graphs which can be recognized efficiently. Such a class was suggested by Chvátal in [12]; he called a graph G *brittle* if each induced subgraph H of G contains a vertex that is not an endpoint of any P_4 in H or not a midpoint of any P_4 in H. Brittle graphs are discussed in [27,42, 43]. Babel and Olariu [3] proved that $(7,3)$-graphs are brittle, and Giakoumakis [22] proved that P_4-laden graphs are brittle. A natural subclass of brittle graphs, called *superbrittle*, consists of those graphs G in which *every* vertex is not an endpoint of any P_4 in G or not a midpoint of any P_4 in G. Split graphs are superbrittle since in a split graph with clique C and stable set S, every midpoint of a P_4 is in C and every endpoint of a P_4 is in S. Superbrittle graphs are characterized in terms of forbidden induced subgraphs in [40]. We will show that our new class of split-perfect graphs is a subclass of brittle graphs, containing all superbrittle graphs. Moreover, we construct a perfect order of a split-perfect graph efficiently, and we obtain a new characterization of superbrittle graphs leading to a linear time recognition.

Yet another motivation for studying split-perfect graphs stems from the theory of perfect graphs. A graph G is called *perfect* if, for each induced subgraph

H of G, the chromatic number of H equals the maximum number of pairwise adjacent vertices in H. For example, all the above-mentioned graphs are perfect. For more information on perfect graphs, see [6,11,23]. Recognizing perfect graphs in polynomial time is a major open problem in algorithmic graph theory. Two graphs G and H with the same vertex set V are P_4-*isomorphic* if, for all subsets $S \subseteq V$, S induces a P_4 in G if and only if S induces a P_4 in H. Chvátal [18] conjectured and Reed [41] proved that two P_4-isomorphic graphs are both perfect or both are imperfect. Thus, to recognize perfect graphs it is enough to recognize the P_4-structure of perfect graphs: Given a 4-uniform hypergraph $\mathcal{H} = (V, \mathcal{E})$. Is there a perfect graph $G = (V, E)$ such that $S \in \mathcal{E}$ if and only if S induces a P_4 in G? This was done for the case when the perfect graph G is a tree [20,9,10], a block graph [7], the line graph of a bipartite graph [44] or a bipartite graph [2].

Another question arising from Reed's theorem is the following: Which (perfect) graphs are P_4-isomorphic to a member of a given class of perfect graphs? Let $\mathcal{C}$ be a class of perfect graphs. Graphs P_4-isomorphic to a member in $\mathcal{C}$ are called $\mathcal{C}$-*perfect graphs*. By Reed's theorem, $\mathcal{C}$-perfect graphs are perfect. Moreover, they form a class of graphs which is closed under complementation and contains $\mathcal{C}$ as a subclass. Thus, it is interesting to ask the following question: Assume that there is a polynomial time algorithm for testing membership in $\mathcal{C}$. Can $\mathcal{C}$-perfect graphs be recognized in polynomial time, too? First results in this direction are good characterizations of tree-perfect graphs, forest-perfect graphs [8], and bipartite-perfect graphs [35]. This paper will give a good characterization of split-perfect graphs.

Definition 1. *A graph is called* split-perfect *if it is P_4-isomorphic to a split graph.*

Trivial examples of split-perfect graphs are split graphs and P_4-free graphs. Nontrivial examples are induced paths $P_n = v_1 v_2 \cdots v_n$ for any integer n. To see this we need some definitions, following [8]. Let $(v_1, \ldots, v_n)$ be a vertex order of a graph G. Then $N_{>i}(v_i)$ denotes the set of all neighbors v_k of v_i with $k > i$.

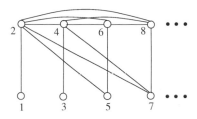

Fig. 1: Elementary graphs illustrated

A vertex order $(v_1, \ldots, v_n)$ of G is said to be *elementary* if for all i:

$$N_{>i}(v_i) = \begin{cases} \{v_{i+2}, v_{i+3}, \ldots, v_n\} & \text{for even } i \\ \{v_{i+1}\} & \text{for odd } i \end{cases}$$

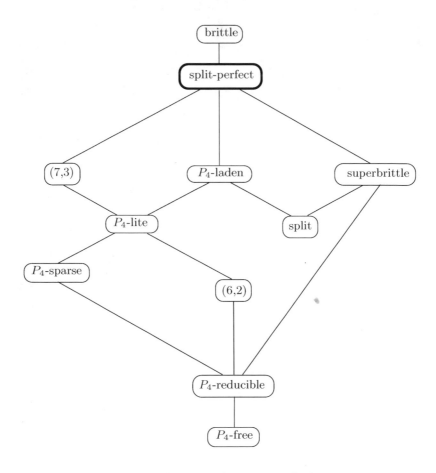

Fig. 2: Relationship between graph classes

Graphs having elementary orders are split graphs in which the "odd vertices" v_{2k+1} form a stable set and the "even vertices" v_{2k} form a clique. A graph is said to be *elementary* if it has an elementary order. If the elementary graph has at least 4 vertices then its partition into a clique and a stable set is unique and can be determined using its degree sequence. Thus, as split graphs in general [24], elementary graphs can be recognized in linear time.

Obviously, $P_n = v_1 v_2 \cdots v_n$ is P_4-isomorphic to the elementary graph consisting of the elementary order $(v_1, \ldots, v_n)$. It can be seen that, for $n \geq 7$, this elementary graph is the only split graph (up to "complementation" and "bipartite complementation") that is P_4-isomorphic to P_n. In Section 4 we will extend this example to the so-called *double-split graphs* which play a key role for characterizing split-perfect graphs.

In section 2 we will show that the class of split-perfect graphs contains all P_4-laden graphs and all (7,3)-graphs (hence all P_4-reducible, P_4-sparse and P_4-lite graphs). The relationship between the above-mentioned graph classes is shown in Figure 2.

In section 3, we describe forbidden induced subgraphs of split-perfect graphs which are needed to characterize split-perfect graphs.

In section 4, we introduce double-split graphs and show that they are split-perfect. As already mentioned, double-split graphs are of crucial importance for a good characterization of split-perfect graphs.

In section 5, we characterize split-perfect graphs in terms of forbidden subgraphs and in terms of their p-connected components: It turns out that for split-perfect graphs without homogeneous sets, the p-connected components are double-split graphs or their complements.

In the last section, we will point out how classical optimization problems such as maximum clique, minimum coloring, maximum independent set, and minimum clique cover can be solved efficiently, in a divide and conquer manner, on split-perfect graphs using the primeval decomposition tree. These results are based on our good characterization of p-connected split-perfect graphs.

Due to the space limitation for this extended abstract, we omit all proofs; some of them are quite long and technically involved. The interested reader will find them in the full version of our paper.

2   Preliminaries

Our notions are quite standard. The neighborhood of the vertex v in a graph G is denoted by $N_G(v)$; if the context is clear, we simply write $N(v)$. The path (respectively, cycle) on m vertices $v_1, v_2, \ldots, v_m$ with edges v_iv_{i+1} (respectively, v_iv_{i+1} and v_1v_m) $(1 \le i < m)$ is denoted by $P_m = v_1v_2 \cdots v_m$ (respectively, $C_m = v_1v_2 \cdots v_mv_1$). The vertices v_1 and v_m are the *endpoints* of the path P_m, and for a P_4 $v_1v_2v_3v_4$, v_2 and v_3 are the *midpoints* of the P_4. Graphs containing no induced subgraphs isomorphic to a given graph H are called H-*free graphs*. It is well-known that split graphs are exactly the $(C_4, \overline{C_4}, C_5)$-free graphs [21].

For convenience, for $S \subseteq V(G)$, let S denote the subgraph $G[S]$ induced by S.

A set S of at least two vertices of a graph G is called *homogeneous*, if $S \ne V(G)$ and every vertex outside S is adjacent to all vertices in S or to no vertex in S. A homogeneous set M is *maximal* if no other homogeneous set properly contains M. It is well-known that in a connected graph G with connected complement $\overline{G}$, every two different maximal homogeneous sets are disjoint (see, e.g. [37]). In this case, the graph G^* obtained from G by contracting every maximal homogeneous set to a single vertex is called the *characteristic graph* of G. Clearly, G^* is connected and has no homogeneous set. We shall use the following useful fact for later discussions (see Figure 3 for the graphs G_i).

Lemma 1 ([28]). *Every graph containing an induced C_4 and no homogeneous set contains an induced $\overline{P_5}$ or G_3 or G_4.*

Throughout this paper, we use the following basic property: every homogeneous set contains at most one vertex of every P_4.

Following Jamison and Olariu [34], a graph is called P_4-*connected* (or *p-connected*) if, for every partition of its vertex set into two nonempty disjoint parts, there exists a P_4 containing vertices from both parts. It is easy to see that every graph has a unique partition into maximal induced p-connected subgraphs (called *p-connected components*) and vertices belonging to no P_4. For the subsequent Structure Theorem of Jamison and Olariu we need the following notion: A p-component H of G is called *separable* if it has a partition into nonempty sets H_1, H_2 such that every P_4 with vertices from both H_i's has its midpoints in H_1 and its endpoints in H_2. Note that a p-connected graph is separable if and only if its characteristic graph is a split graph [34].

Theorem 1 (Structure Theorem, [34]). *For an arbitrary graph G, precisely one of the following conditions is satisfied:*

(i)   *G is disconnected*
(ii) *$\overline{G}$ is disconnected*
(iii) *G is p-connected*
(iv) *There is a unique proper separable p-component H of G with a partition (H_1, H_2) such that every vertex outside H is adjacent to all vertices in H_1 and misses all vertices in H_2.*

Based on this theorem, Jamison and Olariu define the primeval decomposition which can be described by the primeval decomposition tree and leads to efficient algorithms for a variety of problems if the p-connected components are sufficiently simple. We will show that this is the case for split-perfect graphs.

Note that dividing a graph into p-connected components can be done in linear time (see [5]). This fact and the subsequent Proposition 1 below allows us to restrict our attention to p-connected split-perfect graphs only.

Proposition 1. *A graph is split-perfect if and only if each of its p-connected components is split-perfect.*

Observation 1. *Let G be split-perfect and let $H = (C_H, S_H, E_H)$ be a split graph P_4-isomorphic to G. Assume that each of the sets $\{a, b, c, u\}$ and $\{a, b, c, v\}$ induces a P_4 in G. Then exactly one of the following conditions holds:*

(i)   *a, b, c induce a path P_3 in H, and u and v are both adjacent in H to an endpoint of the path $H[a, b, c]$. In particular, u and v both belong to the stable-part S_H of H.*
(ii) *The statement (i) holds in $\overline{H}$ instead of H. In particular, u and v both belong to the clique-part C_H of H.*

Proposition 2. *Let G be a p-connected split-perfect graph. Then every homogeneous set of G induces a P_4-free graph.*

Proposition 3. *Let G be a p-connected graph. G is split-perfect if and only if*

(i) *every homogeneous set of G induces a P_4-free graph, and*
(ii) *G^* is split-perfect.*

Propositions 1 and 3 allow us to consider only p-connected split-perfect graphs without homogeneous sets.

Corollary 1. (i) *P_4-laden graphs are split-perfect.*
(ii) *$(7,3)$-graphs are split-perfect.*

3   Forbidden Induced Subgraphs for Split-Perfect Graphs

As a consequence of Observation 1, we give a list of forbidden induced subgraphs of split-perfect graphs: These are the induced cycles C_k of length $k \geq 5$, the graphs G_i ($1 \leq i \leq 8$) shown in Figure 3, and their complements. It turns out (Theorem 2) that these forbidden induced graphs also characterize split-perfect graphs that have no homogeneous set.

We need some notions. Let G and G' be two graphs with the same vertex set. An induced P_4 in G is *bad* if its vertices do not induce a P_4 in G' (thus, P_4-isomorphic graphs do not have bad P_4's.). Another useful notion is suggested by Observation 1: Let G be a split-perfect graph and H a corresponding split graph having the same P_4-structure. We call the clique and the stable set of H the two *classes* of H. Two vertices x, y in G are called *equivalent* ($x \sim y$) if they are in the same class of H.

Now, Observation 1 means that in a split-perfect graph G, vertices x and y with the property that there are vertices $a, b, c \in V(G) - \{u, v\}$ such that $\{a, b, c, x\}$ and $\{a, b, c, y\}$ both induce a P_4 are in the same class i.e. $x \sim y$.

Therefore, a set of pairwise equivalent vertices induces a P_4-free graph in G.

Lemma 2. *No split-perfect graph has a C_k, $\overline{C_k}$ ($k \geq 5$), G_i, $\overline{G_i}$ ($1 \leq i \leq 8$) as an induced subgraph.*

4   Double-Split Graphs

We define now the class of double-split graphs generalizing the split graphs and playing a key role in the subsequent characterization of split-perfect graphs. As an important step towards this characterization, we will show that double-split graphs are split-perfect.

Definition 2. *A graph is called double-split if it can be obtained from two disjoint (possibly empty) split graphs $G_L = (Q_L, S_L, E_L)$, $G_R = (Q_R, S_R, E_R)$ and an induced path $P = P[x_L, x_R]$, possibly empty, by adding all edges between x_L and vertices in Q_L and all edges between x_R and vertices in Q_R.*

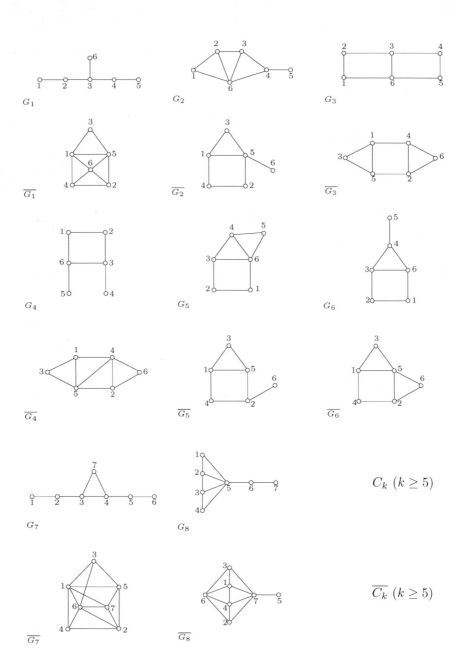

Fig. 3: Forbidden induced subgraphs

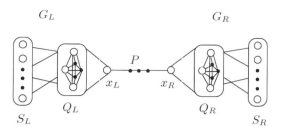

Fig. 4: Double-split graphs illustrated

Remark: Every split graph is double-split as the case of an empty path P and an empty split graph G_R shows.

Lemma 3. *Double-split graphs are split-perfect.*

Double-split graphs and their complements can be recognized in linear time due to their simple structure using techniques from [19].

5   The Structure of Split-Perfect Graphs

We now are able to describe split-perfect graphs without homogeneous sets. They can be characterized as follows.

Theorem 2. *Let G be a graph having no homogeneous set. Then the following statements are equivalent:*

(i)   *G is split-perfect;*
(ii) *G has no induced subgraphs C_k, $\overline{C_k}$ ($k \geq 5$), G_i, $\overline{G_i}$ ($1 \leq i \leq 8$);*
(iii) *G or $\overline{G}$ is a double-split graph.*

Theorem 2 and Propositions 1 and 3 immediately yield:

Theorem 3. *A graph G is split-perfect if and only if each of its p-connected component H has the following properties: Every homogeneous set in H induces a P_4-free graph, and H^* is a double-split graph or the complement of a double-split graph.*

Corollary 2. *Split-perfect graphs can be recognized in linear time.*

This follows from Theorem 3, and the following tasks can be carried out in linear time: finding the p-connected components of a graph [5], finding all maximal homogeneous sets of a (p-connected) graph [37,38], recognizing P_4-free graphs [16], and recognizing double-split graphs and their complements.

Theorem 4. *A graph G is superbrittle if and only if for each of its p connected components H of G,*

(i) *the homogeneous sets of H are cographs and*
(ii) *the characteristic graph H^* is a split graph.*

Corollary 3. *Superbrittle graphs are split-perfect and can be recognized in linear time.*

Corollary 4. *Split-perfect graphs are brittle. Moreover, a perfect order of a split-perfect graph can be constructed efficiently.*

6   Optimization in Split-Perfect Graphs

As already mentioned, Theorem 1 implies a decomposition scheme, called *primeval decomposition*, for arbitrary graphs. The corresponding tree representation, called *primeval tree*, has the p-connected components and vertices not belonging to any P_4 of the considered graph as its leaves. The important features of the primeval tree of a given graph G are:

1. If an optimization problem such as maximum clique, minimum coloring, maximum independent set, and minimum clique cover can be solved efficiently on the p-connected components of G, then one can also efficiently solve the problem on the whole graph G; see for example [1].
2. The primeval tree can be constructed in linear time; see [5].

Based on these facts, linear time or at least polynomial time algorithms have been found for classical NP-hard problems on many graph classes such as $(q, q - 4)$-graphs and various subclasses.

We now point out, how to compute the maximum clique size $\omega(G)$ for p-connected split-perfect graphs G. First, let H be a homogeneous set in G. Then

$$\omega(G) = \max(\omega(H) + \omega(N(H)), \omega(G/H)),$$

where G/H is the graph obtained from G by contracting H to a single vertex. Thus, if $\omega(G^*)$ and $\omega(H)$ for all homogeneous sets H in G can be computed in linear time, then also $\omega(G)$.

Now, if G is a p-connected split-perfect graph, then by Theorem 2, G^* is a double-split graph or the complement of a double-split graph. In any case, as G^* is a chordal graph or the complement of a chordal graph, $\omega(G^*)$ can be computed in linear time. Further, by Proposition 2, every homogeneous set H of G induces a P_4-free graph, hence $\omega(H)$ can be computed in linear time. This and the facts that the primeval tree of G as well as all maximal homogeneous sets of G can be found in linear time show that $\omega(G)$ can be computed in linear time. The problem of minimum coloring can be solved similarly; we omit the details. Note that for perfect graphs in general and in particular for split-perfect graphs, the two parameters coincide.

Since split-perfect graphs are closed under complementation, the independence number and the clique cover number of a split-perfect graph can be computed via the clique number, resp., the chromatic number of the complement of the graph considered. Thus, we can state the following result.

Theorem 5. *The clique number, the chromatic number, the independence number, and the clique cover number of a split-perfect graph can be computed in linear time.*

The same holds for the weighted version of clique number and chromatic number.

References

1. L. BABEL, On the P_4-structure of graphs, *Habilitationsschrift*, TU München, 1997
2. L. BABEL, A. BRANDSTÄDT, V.B. LE, Recognizing the P_4-structure of bipartite graphs, *Discrete Appl. Math.* 93 (1999) 157–168
3. L. BABEL, S. OLARIU, On the structure of graphs with few P_4, *Discrete Appl. Math.* 84 (1998) 1–13
4. L. BABEL, S. OLARIU, On the p-connectedness of graphs: A survey, *Discrete Appl. Math.* 95 (1999) 11–33
5. S. BAUMANN, A linear algorithm for the homogeneous decomposition of graphs, *Report No.* M–9615, Zentrum Mathematik, TU München, 1996
6. C. BERGE, V. CHVÁTAL (eds.), Topics on perfect graphs, *Annals of Discrete Math.* 21 (1984)
7. A. BRANDSTÄDT, V.B. LE, Recognizing the P_4-structure of block graphs, *Discrete Appl. Math.* 99 (2000) 349–366
8. A. BRANDSTÄDT, V.B. LE, Tree- and forest-perfect graphs, *Discrete Appl. Math.* 95 (1999) 141–162
9. A. BRANDSTÄDT, V.B. LE, S. OLARIU, Linear-time recognition of the P_4-structure of trees, *Rutcor Research Report RRR* 19 – 96, *Rutgers University*, 1996
10. A. BRANDSTÄDT, V.B. LE, S. OLARIU, Efficiently recognizing the P_4-structure of trees and of bipartite graphs without short cycles, *Graphs and Combinatorics*, to appear
11. A. BRANDSTÄDT, V.B. LE, J. SPINRAD, Graph Classes: A Survey, *SIAM Monographs on Discrete Math. Appl.*, *Vol.* 3, SIAM, Philadelphia (1999)
12. V. CHVÁTAL, Perfect graphs seminar, McGill University, Montreal, 1983
13. V. CHVÁTAL, Perfectly ordered graphs, *Annals of Discrete Math.* 21 (1984) 63–65
14. D.G. CORNEIL, H. LERCHS, L. STEWART-BURLINGHAM, Complement reducible graphs, *Discrete Appl. Math.* 3 (1981) 163–174
15. D.G. CORNEIL, Y. PERL, L.K. STEWART, Cographs: recognition, applications, and algorithms, *Congressus Numer.* 43 (1984) 249–258
16. D.G. CORNEIL, Y. PERL, L.K. STEWART, A linear recognition algorithm for cographs, *SIAM J. Computing* 14 (1985) 926–934
17. A. COURNIER, M. HABIB, A new linear algorithm for modular decomposition, *LIRMM, University Montpellier* (1995), Preliminary version in: *Trees in Algebra and Programming – CAAP '94, Lecture Notes in Comp. Sci.* 787 (1994) 68–84
18. V. CHVÁTAL, A semi-strong perfect graph conjecture, *Annals of Discrete Math.* 21 (1984) 279–280
19. E. DAHLHAUS, J. GUSTEDT, R.M. MCCONNELL, Efficient and practical modular decomposition, Tech. Report TU Berlin FB Mathematik, 524/1996 (1996), Conf. Proc. 8th SODA (1997) 26–35
20. G. DING, Recognizing the P_4-structure of a tree, *Graphs and Combinatorics* 10 (1994) 323–328

21. S. FÖLDES, P.L. HAMMER, Split graphs, 8th South-Eastern Conference (F. HOFF-
MAN et al., eds.) Louisiana State Univ. Baton Rouge, Louisiana (1977), *Congressus
Numer.* 19 (1977) 311–315
22. V. GIAKOUMAKIS, P_4-laden graphs: A new class of brittle graphs, *Information
Process. Letters* 60 (1996) 29–36
23. M.C. GOLUMBIC, Algorithmic Graph Theory and Perfect Graphs, Academic Press,
New York (1980)
24. P.L. HAMMER, B. SIMEONE, The splittance of a graph, *Combinatorica* 1 (1981)
275–284
25. C.T. HOÀNG, Perfect graphs, Ph.D. Thesis, McGill University, Montreal, 1985
26. C.T. HOÀNG, Perfectly orderable graphs: a survey, manuscript, 1999
27. C.T. HOÀNG, N. KHOUZAM, On brittle graphs, *J. Graph Theory* 12 (1988) 391–404
28. C.T. HOÀNG, B. REED, Some classes of perfectly orderable graphs, *J. Graph
Theory* 13 (1989) 445–463
29. B. JAMISON, S. OLARIU, P_4-reducible graphs – a class of uniquely tree represen-
table graphs, *Studies in Appl. Math.* 81 (1989) 79–87
30. B. JAMISON, S. OLARIU, A new class of brittle graphs, *Studies in Appl. Math.* 81
(1989) 89–92
31. B. JAMISON, S. OLARIU, A unique tree representation for P_4-sparse graphs, *Discrete Appl. Math.* 35 (1992) 115–129
32. B. JAMISON, S. OLARIU, A linear-time algorithm to recognize P_4-reducible graphs,
Theor. Comp. Sci. 145 (1995) 329–344
33. B. JAMISON, S. OLARIU, Linear time optimization algorithms for P_4-sparse graphs,
Discrete Appl. Math. 61 (1995) 155–175
34. B. JAMISON, S. OLARIU, p-components and the homogeneous decomposition of
graphs, *SIAM J. Discr. Math.* 8 (1995) 448–463
35. V.B. LE, Bipartite-perfect graphs, *submitted.* Extended abstract in *Electronic Notes in Discrete Math.*, http://www.elsevier.nl/locate/endm
36. R. LIN, S. OLARIU, A fast parallel algorithm to recognize P_4-sparse graphs, *Discrete Appl. Math.* 81 (1998) 191–215
37. R.M. MCCONNELL, J. SPINRAD, Linear-time modular decomposition and efficient
transitive orientation of comparability graphs, 5th. Ann. ACM–SIAM Symp. on
Discrete Algorithms, Arlington, Virginia, 1994, pp 536–543
38. R.M. MCCONNELL, J. SPINRAD, Modular decomposition and transitive orientation, *Discrete Math.* 201 (1999) 189–241
39. M. MIDDENDORF, F. PFEIFFER, On the complexity of recognizing perfectly orderable graphs, *Discrete Math.* 80 (1990) 327–333
40. M. PREISSMAN, D. DE WERRA, N.V.R. MAHADEV, A note on superbrittle graphs,
Discrete Math. 61 (1986) 259–267
41. B. REED, A semi-strong perfect graph theorem, *J. Comb. Theory (B)* 43 (1987)
223–240
42. A. SCHÄFFER, Recognizing brittle graphs: Remarks on a paper of Hoàng and
Khouzam, *Discrete Appl. Math.* 31 (1991) 29–35
43. J.P. SPINRAD, J.L. JOHNSON, Brittle, bipolarizable, and P_4-simplicial graph recognition, manuscript, 1999
44. S. SORG, Die P_4-Struktur von Kantengraphen bipartiter Graphen, *Diploma thesis*,
Mathematisches Institut der Universität zu Köln (1997)

Coarse Grained Parallel Algorithms for Detecting Convex Bipartite Graphs

Edson Cáceres[1], Albert Chan[2], Frank Dehne[2], and Giuseppe Prencipe[3]

[1] Departamento de Computação e Estatística, Universidade Federal de Mato Grosso do Sul, Campo Grande, Brasil. edson@dct.ufms.br
[2] School of Computer Science, Carleton University, Ottawa, Canada K1S 5B6. {achan,dehne}@scs.carleton.ca
[3] Dipartimento di Informatica, Corso Italia 40, 56125 Pisa, Italy. prencipe@di.unipi.it

Abstract. In this paper, we present parallel algorithms for the *coarse grained multicomputer* (CGM) and the *bulk synchronous parallel computer* (BSP) for solving two well known graph problems: (1) determining whether a graph G is *bipartite*, and (2) determining whether a bipartite graph G is *convex*.

Our algorithms require $O(\log p)$ and $O(\log^2 p)$ communication rounds, respectively, and linear sequential work per round on a CGM with p processors and N/p local memory per processor, $N=|G|$. The algorithms assume that $\frac{N}{p} \geq p^\epsilon$ for some fixed $\epsilon > 0$, which is true for all commercially available multiprocessors. Our results imply BSP algorithms with $O(\log p)$ and $O(\log^2 p)$ supersteps, respectively, $O(g \log(p)\frac{N}{p})$ communication time, and $O(\log(p)\frac{N}{p})$ local computation time.

Our algorithm for determining whether a bipartite graph is convex includes a novel, coarse grained parallel, version of the *PQ tree* data structure introduced by Booth and Lueker. Hence, our algorithm also solves, with the same time complexity as indicated above, the problem of testing the consecutive-ones property for $(0, 1)$ matrices as well as the chordal graph recognition problem. These, in turn, have numerous applications in graph theory, DNA sequence assembly, database theory, and other areas.

1 Introduction

In this paper, we study the problem of detecting bipartite graphs and convex bipartite graphs. That is, given an arbitrary graph G, determine whether G is a *bipartite* graph and, given a bipartite graph G, determine whether G is a *convex* bipartite graph. Bipartite and convex bipartite graphs are formally defined as follows.

Definition 1. *A graph $G = (V, E)$ is a bipartite graph if V can be partitioned into two sets A and B such that $A \cap B = \emptyset$, $A \cup B = V$ and $E \subseteq ((A \times B) \cup (B \times A))$. A bipartite graph G is also denoted as $G = (A, B, E)$.*

U. Brandes and D. Wagner (Eds.): WG 2000, LNCS 1928, pp. 83–94, 2000.

Definition 2. *A bipartite graph* $G = (A, B, E)$ *is a convex bipartite graph if there exists an ordering* $(b_1, b_2, \cdots, b_{|B|})$ *of* B *such that, for all* $a \in A$ *and* $1 \le i < j \le |B|$, *if* $(a, b_i) \in E$ *and* $(a, b_j) \in E$ *then* $(a, b_k) \in E$ *for all* $i \le k \le j$.

These, and closely related, problems has been extensively studied for the sequential [1,15] and the shared memory (PRAM) parallel [4,5,11,12,13,14] domain. Unfortunately, theoretical results from PRAM algorithms do not necessarily match the speedups observed on *real* parallel machines. In this paper, we present parallel algorithms that are more practical in that the assumptions and cost model used reflects better the reality of commercially available multiprocessors. More precisely, we will use a version of the BSP model, referred to as the *coarse grained multicomputer* (CGM) model. In contrast to the BSP model, the CGM [6,7,8,9] allows only bulk messages in order to minimize message overhead costs. A CGM is comprised of a set of p processors $P_1, \ldots, P_p$ with $O(N/p)$ local memory per processor and an arbitrary communication network (or shared memory). All algorithms consist of alternating local computation and global communication rounds. Each communication round consists of routing a single h-relation with $h = O(N/p)$, i.e. each processor sends $O(N/p)$ data and receives $O(N/p)$ data. We require that all information sent from a given processor to another processor in one communication round is packed into one long message, thereby minimizing the message overhead. A CGM computation/communication round corresponds to a BSP superstep with communication cost $g\frac{N}{p}$ (plus the above "packing requirement"). Finding an optimal algorithm in the coarse grained multicomputer model is equivalent to minimizing the number of communication rounds as well as the total local computation time. The CGM model has the advantage of producing results which correspond much better to the actual performance of implementations on commercially available parallel machines. In addition to minimizing communication and computation volume, it also minimizes important other costs like message overheads and processor synchronization.

In this paper, we present parallel CGM algorithms for detecting bipartite graphs and convex bipartite graphs. The algorithms require $O(\log p)$ and $O(\log^2 p)$ communication rounds, respectively, and linear sequential work per round. They assume that the local memory per processor, N/p, is larger than p^ϵ for some fixed $\epsilon > 0$. This assumption is true for all commercially available multiprocessors. Our results imply BSP algorithms with $O(\log p)$ supersteps, $O(g \log(p)\frac{N}{p})$ communication time, and $O(\log(p)\frac{N}{p})$ local computation time.

The algorithm for detecting bipartite graphs is fairly simple and is essentially a combination of tools developed in [3]. The larger part of this paper deals with the problem of detecting *convex* bipartite graphs. This is clearly a much harder problem. It has been extensively studied in the literature and is closely linked to the *consecutive ones* problem for $(0, 1)$-matrices as well as chordal graph recognition [1,4,5,11,12,13,14,15].

Our algorithm for determining whether a bipartite graph is convex includes a novel, coarse grained parallel, version of the *PQ tree* data structure introduced by Booth and Lueker [1]. Hence, our algorithm also solves, with the same time complexity as indicated above, the problem of testing the consecutive-ones property for $(0, 1)$-matrices as well as the chordal graph recognition problem. These, in turn, have numerous applications in graph theory, DNA sequence assembly, database theory, and other areas. [1,4,5,11,12,13,14,15]

2   Detecting *Bipartite* Graphs

In this section, we present a simple CGM algorithm for detecting bipartite graphs. It is a straight-forward combination of tools developed in [3].

Algorithm 1 Detection of Bipartite Graphs
Input: A Graph $G = (V, E)$ with vertex set V and edge set E, $|G| = N$, stored on a CGM with p processors and $O(N/p)$ memory per processor; $N/p \geq p^\epsilon$ for some fixed $\epsilon > 0$. V and E are arbitrarily distributed over the memories of the CGM. **Output:** A Boolean indicating whether G is a bipartite graph and, if it is, a partition of V into two disjoint set A and B such that $E \subseteq ((A \times B) \cup (B \times A))$.
   (1) Compute a spanning forest of G [3].
   (2) For each tree in the forest, select one arbitrary node as the root. Apply the CGM Euler Tour algorithm in [3] to determine the distance between each node and the root of its tree. Classify the nodes into two groups: the nodes with an odd numbered distance to the root, and the nodes with an even numbered distance to the root.
   (3) Each processor examines the edges stored in its local memory. If any such edge has two vertices that belong to the same group, the result for that processor is "failure"; otherwise, the result is "success".
   (4) By applying CGM sort [10] to all "failure"/"success" values, it is determined whether there was any processor with a "failure" result. If there was any "failure", the graph G is *not* bipartite. Otherwise, G *is* a bipartite graph, and the two groups of vertices identified in Step 2 are the sets A and B.

Theorem 1. *Algorithm 1 detects whether $G = (V, E)$, $|G| = N$, is a bipartite graph and, if so, partitions E into sets A and B such that $E \subseteq ((A \times B) \cup (B \times A))$ in $O(\log p)$ communication rounds and $O(\frac{N}{p})$ local computation per round on a CGM with p processors and $O(\frac{N}{p})$ memory per processor, $\frac{N}{p} \geq p^\epsilon$ for some fixed $\epsilon > 0$.*

Proof. Omitted due to page restrictions. To be included in the full version of this paper. □

3   Detecting *Convex* Bipartite Graphs

We now turn our attention to the problem of testing whether a given bipartite graph is a convex bipartite graph. The sequential solution, presented by Booth and Lueker [1], introduced a data structure called *PQ-tree*. Our coarse grained parallel solution will include a novel coarse grained parallel version of the PQ-tree. We will first review Booth and Lueker's PQ-tree definition.

Definition 3. *A tree T is a PQ-tree if every internal node of T can be classified as either a P-node or a Q-node. A P-node is an internal node that has at least 2 children, and the children can be permuted arbitrarily. A Q-node is an internal node that has at least 3 children, and the children can only be permuted in two ways: the original order or the reverse order. The leaves of the PQ-tree are elements of a universal set $S = \{a_1, \ldots, a_n\}$, usually called the ground set.*

The order of the ground set in the PQ-tree, from left to right, is called its *frontier*. The frontier of a PQ-tree is clearly a permutation of the ground set. Given a PQ-tree T and using only permissible permutations of its internal nodes, we can generate a number of permutations of S. We will denote with $L(T)$ the set of all these permissible permutations. A PQ-tree T' is *equivalent* to T if T' can be transformed into T using only permissible permutations of the internal nodes (if $L(T')$ and $L(T)$ have the same elements).

Given a set $A \subset S$, we say that $\lambda \in L(T)$ *satisfies* A if all elements of A appear consecutively in λ. The main operation on a PQ-tree T is called *reduce*: given a *reduction* set $\mathcal{A} = \{A_1, \ldots, A_k\}$ of subsets of S and a PQ-tree T, we want obtain a PQ-tree T', if it exists, such that each permutation in $L(T')$ satisfies every A_i, $1 \le i \le k$.

Let $m = \Sigma_{i=1}^{k} |A_i|$ and $N = n + m$. In order to store T and $\mathcal{A}$, we require a coarse grained multicomputer with p processors and N/p local memory per processor.

Two particular PQ-trees are the *universal* and the *empty* tree: the first one has only one internal node (the root of T) and that internal node is a P-node; the second one (also called a *null* PQ-tree) is used to represent an impossible reduction, that is when it is impossible to reduce a PQ-tree with respect to a given reduction set.

3.1 Multiple *Disjoint* Reduce Operations on a PQ-Tree

In this section, we will present a coarse grained parallel algorithm for the special case of performing multiple *disjoint* reductions on a PQ-tree. We will then use this solution to develop the general algorithm in the subsequent section. More precisely, given a PQ-tree T we will first study how to perform the *reduce* operation for a set $\mathcal{A} = \{A_1, \ldots, A_k\}$ of subsets of the universal set S where $A_1, \ldots, A_k$ are disjoint. We shall refer to our algorithm as *Algorithm MDReduce*. For ease of discussion, each set A_i is assigned a unique color, and we color the leaves of the PQ-tree accordingly. Some of the PQ-tree definitions used are from [1,12].

We start with a pre-processing phase which extends the coloring δ of the leaves to a coloring Δ of all nodes of the PQ-tree T. For an internal node v of T, we say that a color is *complete* at v if all the leaves with that color are descendants of v. We say a color is *incomplete* at v if some, but not all, of the leaves of that color are descendants of v. We say that a color *covers* v if all the leaves below v are of that color, and that v is *uncovered* if no color covers v. Let $LCA(c)$ be the lowest common ancestor of all leaves with color c. Let

$COLORS(v)$ denote the set of colors assigned to leaves that are descendents of v. Let $INC(v)$ be the set of colors which are incomplete at v. Then $INC(v) = COLORS(v) - \{c: LCA(c)$ is a descendent of $v\}$.

Algorithm 2 Pre-Processing the PQ-Tree
Input: The original PQ-tree T.
Output: The original PQ-tree T in which each node is assigned a "coloring" Δ, or, if failure occurs, a null tree.

(1) Apply the coarse grained parallel Lowest Common Ancestor (LCA) algorithm [3].
(2) Expand T into a binary tree B.
(3) Perform tree contraction on B; see [3]. For each node v_b in B, let v_p be the node in T from which v_b is created. Let w_1 and w_2 be the children of v_b. The operation for the tree contraction is $INC(v_b) = INC(w_1) \cup INC(w_2) - \{c: LCA(c)$ is a descendent of $v_p\}$. If at any point the size of INC is more than two, stop and return a null tree.
(4) Let c_v be a new color unique to node v. Each processor, for all its nodes, v, calculates $\Delta(v) = < c_1, c_2 >$ as follows: If two colors are incomplete at v, then c_1 and c_2 are these colors. If only one color c is incomplete at v but c does not cover v, then $c_1 = c$ and $c_2 = c_v$. If one color c is incomplete at v and covers v, then $c_1 = c_2 = c$. If no color is incomplete at v, then $c_1 = c_2 = c_v$.

Lemma 1. *On a coarse grained multicomputer with p processors and $O(\frac{N}{p})$ storage per processor, Algorithm 2 can be completed in $O(\log p)$ communication rounds with $O(\frac{N}{p})$ local computation per round.*

Proof. Omitted due to page restrictions. To be included in the full version of this paper. □

A node v in a PQ-tree is *orientable* if it is a Q-node and the two colors in its $\Delta(v) = < c_1, c_2 >$ are different, i.e. $c_1 \neq c_2$.

For a color c, define $h_v(c) = \begin{cases} c & \text{if } c \in INC(v) \\ c_v & \text{if } c \notin INC(v) \end{cases}$

For a PQ-tree T where w_1 and w_k are the leftmost and rightmost elements, respectively, of the frontier $fr_T(v)$, let $l_T = h_v(\delta(w_1))$ and $r_T = h_v(\delta(w_k))$. If $lr_T[v] = < l_T[v], r_T[v] >$ then we use the following notation: $< a, b > \sim < a', b' >$ if $\{a, b\} = \{a', b'\}$.

Algorithm 3 Processing P-Nodes
Input: The PQ-Tree output from Algorithm 2.
Output: The original PQ-tree T in which all the P-nodes have been processed, or, if failure occurs, a null tree.

(1) If the input PQ-tree T is a null tree, return T.
(2) Each processor sets variable $FAILURE$ to $FALSE$
(3) Each processor, for each P-node v, reorder the children of v such that for each color c all children covered by c are consecutive.
(4) Each processor, for each P-node v and each color c, if there are at least two children covered by c (and at least one child not covered by c) then insert a new P-node w_c between these c-covered children and v.

(5) Each processor, for each P-node v, constructs an auxiliary graph G_v whose nodes are the children of v and where for each color c there is an edge between children v_i and v_j at which c is incomplete if v_i or v_j is covered by c, or there is no child covered by c. If any node has more than 2 neighbors, set $FAILURE$ to $TRUE$ to indicate a failure condition.

(6) Perform a multi-broadcast of the variable $FAILURE$. If any of the broadcast values is $TRUE$, return a null tree.

(7) Each processor uses list-ranking to identify the connected components of each G_v and verifies that each of these connected components is a simple path. If any of these components is a cycle, set $FAILURE$ to $TRUE$ to indicate a failure condition. We call these paths *color chains*.

(8) Perform a multi-broadcast of the variable $FAILURE$. If any of the broadcast values is $TRUE$, return a null tree.

(9) Each processor, for each color chain χ containing at least 2 nodes, chooses one of the 2 orientations of χ arbitrarily. Reorder the children of v so that the nodes of χ are consecutive, and insert a new Q-node between these nodes of v.

(10) Each processor, for each P-node v, let $S = \{v_i : v_i$ is a child of v, and $INC(v_i)=\emptyset\}$. If every child of v is in S, then return. Otherwise, reorder the children of v to make S consecutive, insert a new P-node v' between v and the subset S (if $|S| > 1$), and rename v to be a Q-node.

Lemma 2. *On a coarse grained multicomputer with p processors and $O(\frac{N}{p})$ storage per processor, Algorithm 3 can be completed in $O(\log p)$ communication rounds with $O(\frac{N}{p})$ local computation per round.*

Proof. Omitted due to page restrictions. To be included in the full version of this paper. □

For each Q-node v, we define an orientation $\overline{LR}(v)$ which is either $\Delta(v_i)$ or $\Delta(v_i)^R$. Note that if $\Delta(v_i) =< c_1, c_2 >$ than $\Delta(v_i)^R =< c_2, c_1 >$ For $< a, b >\sim< a', b' >$ and $a \neq b$, we define $< a, b > swap < a', b' >$ equals $TRUE$ if $< a, b >=< b', a' >$, $FALSE$ if $< a, b >=< a', b' >$. For a Q-node v, *flip* is defined as the operation which re-orders all its children in reverse order.

Algorithm 4 Processing Q-Nodes
Input: The PQ-tree output from Algorithm 3.
Output: The original PQ-tree T in which all the Q-nodes have been processed, or, if failure occurs, a null tree.

(1) If the input PQ-tree T is a null tree, return T.

(2) Each processor sets variable $FAILURE$ to $FALSE$

(3) Each processor, for each Q-node v and children be $v_1, \ldots, v_s$, assign to each $\overline{LR}[v_i]$ either $\Delta(v_i)$ or $\Delta(v_i)^R$ such that every color in the sequence $\overline{LR}[v_1], \ldots, \overline{LR}[v_s]$ occurs consecutively, and such that $h_v(< \overline{L}[v_1], \overline{R}[v_s >]) \sim \Delta(v)$. If this is impossible, set $FAILURE$ to $TRUE$ to indicate a failure condition, otherwise, set $LR[v]$ to $h_v(< \overline{L}[v_1], \overline{R}[v_s] >)$.

(4) Perform a multi-broadcast of the variable $FAILURE$. If any of the broadcast values is $TRUE$, return a null tree.

(5) Each processor for each node v: if v is orientable, then set $OPP[v]$ to $LR[v]$ *swap* $\overline{LR}[v]$, otherwise, set $OPP[v]$ to $FALSE$.

(6) Each processor for each node v: set $REV[v]$ to $\bigoplus_{u \text{ is an ancestor of } v} OPP[u]$ (Note: $\bigoplus$ denotes "exclusive-or").

(7) For each orientable node v, if $REV[v]$ is $TRUE$, then flip v.

Lemma 3. *On a coarse grained multicomputer with p processors and $O(\frac{N}{p})$ storage per processor, Algorithm 4 can be completed in $O(\log p)$ communication rounds with $O(\frac{N}{p})$ local computation per round.*

Proof. Omitted due to page restrictions. To be included in the full version of this paper. □

Algorithm 5 Post-Processing the PQ-Tree
Input: The PQ-tree output from Algorithm 4, with all R-nodes renamed.
Output: Result of Algorithm *MDReduce*.
(1) If T is a null tree, return.
(2) Each processor temporarily cuts the links of its Q-nodes to their parents.
(3) Each processor performs pointer jumping for all its nodes that are children of R-nodes to determine their lowest Q-node ancestor.
(4) Each processor restores the links cut in Step 2.
(5) Each processor eliminates its R-nodes by setting the parents of their children to their lowest Q-node ancestors.

Lemma 4. *On a coarse grained multicomputer with p processors and $O(\frac{N}{p})$ storage per processor, Algoritm 5 can be completed using in $O(\log p)$ communication rounds with $O(\frac{N}{p})$ local computation per round.*

Proof. Omitted due to page restrictions. To be included in the full version of this paper. □

Theorem 2. *On a coarse grained multicomputer with p processors and $O(\frac{N}{p})$ storage per processor, Algorithm MDReduce performs a multiple disjoint reduce for a PQ-tree T in $O(\log p)$ communication rounds with $O(\frac{N}{p})$ local computation per round.*

Proof. Omitted due to page restrictions. To be included in the full version of this paper. □

3.2 Multiple *General* Reduce Operations on a PQ-Tree

Using the coarse grained parallel *MDreduce* algorithm presented in the previous section, we will now develop coarse grained parallel algorithm for the general *MReduce* operation: given a PQ-tree T over the ground set S with n elements, perform the reduce operation for an arbitrary reduction sets $\mathcal{A} = \{A_1, \ldots, A_k\}$

Our CGM algorithm for the general *MReduce* operation consists of two phases. In the first phase, we execute $3 \log p$ times an algorithm which is a CGM

implementation of a PRAM algorithm proposed by Klein [12]. We call this operation $\mathrm{MREDUCE}_1(T, \{A_1, \ldots, A_k\}, 0)$. Our contribution here is the implementation of the various shared memory PRAM steps on a distributed memory CGM,
which is non trivial. After this first phase, we have reduced the problem to one in
which we are left with a set of smaller PQ-trees over ground sets whose size is at
most n/p. Hence, each tree can be stored in the local memory of one processor.
However, we can not guarantee that all the *reduction sets* of these PQ-trees do
also fit in the local memory of one processor. In the second phase of our algorithm, we use a merging strategy to complete the algorithm. We will refer to
this phase as the *Merging Phase*.

First Phase: For a node v of a PQ-tree, $leaves_T(v)$ denotes the set of pendant
leaves of v, i.e. leaves of T having v as ancestor. Let $lca_T(A)$ denote the least
common ancestor in T of the leaves belonging to A. Suppose that $v = lca_T(A)$
has children $v_1, \ldots v_s$ in order. We say A is *contiguous* in T if either (1) v is a
Q-node, and for some consecutive subsequence $v_p, \ldots, v_q$ of the children of v,
$A = \bigcup_{p \leq i \leq q} leaves_T(v_i)$, or (2) v is a P-node or a leaf, and $A = leaves_T(v)$.

Suppose that E is contiguous in T. $T|E$ denotes the subtree consisting of
$lca_T(E)$ and those children of $lca_T(E)$ whose descendents are in E (it is still a
PQ-tree whose ground set is E). For a set A, define

$$A_i|E = \begin{cases} A_i \cap E & \text{if } A_i \cap E \neq E \\ \emptyset & \text{if } A_i \cap E = E \end{cases}$$

Let $\star_E$ denote $lca_T(E)$. T/E denotes the subtree of T obtained by omitting
all the proper descendents of $lca_T(E)$ that are ancestors of elements of E (it is
still a PQ-tree whose ground set is $S - E \cup \{\star_E\}$). For a set A, define

$$A_i/E = \begin{cases} A_i - E \cup \{\star_E\} & \text{if } A_i \supseteq E \\ A_i - E & \text{otherwise} \end{cases}$$

Algorithm 6 $\mathrm{MREDUCE}_1(T, \{A_1, \ldots, A_k\}, i)$:
(1) If $i = 3 \log p$, return.
(2) Purge the collection of input sets A_i of empty sets. If no sets remain, return.
(3) Let n be the size of the ground set of T. If $n \leq 4$, carry out the reduction one by one.
    If the size of the input is smaller than the size of the local memory of the processors,
    than solve the problem sequentially using the Booth and Lueker's algorithm.
(4) Otherwise, let $\mathcal{A}$ be the family of (nonempty) sets A_i. Let $\mathcal{S}$ consist of the sets A_i
    such that $|A_i| \leq n/2$. We call such sets "small". Let $\mathcal{L}$ be the remaining, "large",
    sets in $\mathcal{A}$. Find the connected components of the intersection graph of $\mathcal{A}$, find a
    spanning forest of the intersection graph of $\mathcal{S}$, and find the intersection $\cap \mathcal{L}$ of the
    large sets.
(5) Proceed according to one of the following cases:
    (a) The intersection graph of $\mathcal{A}$ is disconnected. In this case, let $\mathcal{C}_1, \ldots, \mathcal{C}_r$ be the
        connected components of $\mathcal{A}$. For $i = 1, \ldots, r$, let E_i be the union of sets in
        the connected component $\mathcal{C}_i$. Call MDREDUCE to reduce T with respect to
        the disjoint sets $E_1, \ldots, E_r$. Next, for each $i = 1, \ldots, r$ in parallel, recursively
        call $\mathrm{MREDUCE}_1(T|E_i, \mathcal{C}_i, i + 1)$.

(b) The union of sets in some connected component of S has cardinality at least $n/4$. In this case, from the small sets making up this large connected component, select a subset whose union has cardinality between $n/4$ and $3n/4$. Let E be this union, and call SUBREDUCE($T, E, \{A_1, \ldots, A_k\}, i$).

(c) The cardinality of the intersection of the large sets is at most $3n/4$. In this case, from the large sets choose a subset whose intersection has cardinality between $n/4$ and $3n/4$. Let E be this intersection, and call SUBREDUCE($T, E, \{A_1, \ldots, A_k\}, i$).

(d) The other case do not hold. In this case, let E be the intersection of the large sets, and call SUBREDUCE($T, E, \{A_1, \ldots, A_k\}, i$).

In the full version of this paper, we show how to implement the above on a coarse grained multicomputer with p processors and $O(\frac{n}{p})$ storage per processor in $O(\log p)$ communication rounds. The non trivial parts are Step 4, Step 5b, the computation of E, T/E, and $T|E$, as well as the SUBREDUCE operation. The latter involves another operation called GLUE. Due to page restrictions, we can not present this part of our result in the extended abstract. Instead, we give one example which shows the coarse grained parallel computation of the set E in Step 5(b) of Algorithm 6.

Algorithm 7 Computation of E.
Input: The set S and the spanning forest of its intersection graph.

(1) In order to find a connected component C in the spanning forest of S, such that the union of its sets has cardinality at least $n/4$, order all the components according to the labeling given by the coarse grained parallel *spanning forest* algorithm [3].

(2) Sort each component with respect to the values of its elements and mark as "valid" only one element per distinct value.

(3) Sort again with respect to the components' labels. Compute the cardinality of the union of the elements of each component (that is the *size* of each component), with a prefix-sum computation, counting only the "valid" elements. (Hence, we do not count twice the elements with same values and compute correctly the cardinality of the union.)

(4) If a processor finds a component whose size is $\geq n/4$, then it broadcasts the label of this component. Otherwise it broadcast a "not-found" message.

(5) If everybody sent "not-found", go to step 4(c) of MREDUCE algorithm. Otherwise, among all the labels received in the previous step, choose as C the component with the smallest label.

(6) For each of the sets comprising C, compute the distance in the spanning tree (from the root) using the coarse grained parallel Euler-tour technique [3].

(7) Sort the sets according to distance, and let $B_1, \ldots, B_s$ be the sorted sequence. Sort each sets with respect to the values of its elements and mark as "valid" only one element per distinct value. Sort the sets again, according to distance, and let $\hat{i}$ be the minimum i such that $|\bigcup_{j=1}^{i} B_j| \geq n/4$. ($\hat{i}$ can be found with a prefix-sum computation on the "valid" elements.) Broadcast $\hat{i}$.

(8) Mark all "valid" elements in $B_1, \ldots, B_{\hat{i}}$ as elements of E.

Second Phase: Consider the tree R of recursive calls in MREDUCE$_1$. We observe that, after $l = 3\log_{4/3} p$ levels of R (when the first part of our algorithm stops), the sizes of the ground sets associated with the nodes in R at level l are at most n/p. This is due to the fact that the descendants of a node u in R that are 3 levels below u are *smaller* than u by approximately a factor $3/4$. More precisely, if $n(u)$ denotes the size of the ground set of $T(u)$ (the subtree rooted at u) then, for every node w three levels below u, $n(u) \leq 3n(w)/4 + 1$. Hence, each PQ-tree obtained at the end of the first phase fits completely into the local memory of one processor.

Unfortunately, the same argument does not hold for the reduction sets. Recall that $m = \Sigma_{i=1}^{k}|A_i|$. Let u be an internal node of R, $A_{u_1}, \ldots, A_{u_j}$ its reduction sets, and $m_u = \Sigma_{i=1}^{j}|A_{u_i}|$. Since the sizes of the reduction sets of the children of u depend strictly on the A_{u_i} and on how they intersect with the set E computed for u, it is possible that the A_{u_i} are split in an unbalanced way. That is, we can have $\Sigma_{i=1}^{j}|A_{u_i}|E| = O(m_u)$ and $\Sigma_{i=1}^{j}|A_{u_i}/E| = O(1)$ (or vice versa). If this continues up to level $3 \log p$ of R, it is possible that for a recursive call associated with a node v at level l, $\Sigma_{i=1}^{f}|A_v i| > m/p$.

Therefore, while the ground set of $T(v)$, and hence $T(v)$, can fit in one processor, the reduction sets could possibly not. Thus, at this point of the computation, we can not simply use the sequential algorithm of Booth and Lueker [1] for completing the reduction.

Our idea for solving this problem is the following. Let us consider a node v at level l in R that has $m_u > m/p$. Since, at any level of recursion, the sum of the sizes of all reduction sets is at most $2m$, we can create α_v copies of $T(v)$, with $\alpha_v = \lfloor \frac{m_v}{m/p} \rfloor$. We observe that

$$\Sigma_{v \in l}\alpha_v = \Sigma_{v \in l}\lfloor \frac{m_v}{m/p} \rfloor \leq \Sigma_{v \in l}\frac{m_v}{m/p} \leq \frac{p}{m}\Sigma_{v \in l}m_v \leq \frac{p}{m} \cdot 2m = 2p.$$

Hence, we require at most two copies per processor. The reduction problem of each node v at level l of R will be solved by the α_v processors that have copies of $T(v)$. The next step is the *distribution* of the reduction sets associated to v among these α_v processors. Each of these α_v processors can solve locally the problem of reducing $T(v)$ with respect to the reduction sets that it has stored, using Booth and Lueker's algorithm [1]. For each processor, let $T'(v)$ refer to this reduced tree. Now, we need to merge these α_v trees, $T'(v)$. More precisely, we need to compute a PQ-tree $\widehat{T}(v)$ such that $L(\widehat{T}(v)) = L(\overline{T}(v))$, where $\overline{T}(v)$ is the PQ-tree that we would have obtained by reducing $T(v)$ directly with respect its reduction sets. For the construction of $\widehat{T}(v)$, we merge the $T'(v)$ trees in a binary tree fashion.

Algorithm 8 Merging Phase

Input: h PQ-trees $T(i)$, with $|T(i)| \leq n/p$ and $\Sigma_i|T(i)| \leq n$, and their reduction sets.
Output: The $T(i)$ reduced with respect their reduction sets.
   (1) Let m_i be the sum of the sizes of the reduction sets of T_i. Make $\alpha_i = \lfloor \frac{m_i}{m/p} \rfloor$ copies of each $T(i)$. Distribute the reduction sets of each T_i between the processors that have the copies of $T(i)$.
   (2) Each processor executes the sequential algorithm [1] for its PQ-trees with the reduction sets that it has stored. Let $T'(i)$ refer to the trees obtained.

(3) The α_i processors associated with each $T(i)$ merge the $T''(v)$ trees in a binary tree fashion. More details are outlined below.

The following Theorem 3 shows that the merge operation in Step 3 of Algorithm 8 reduces to a tree intersection operation. We have designed a CGM algorithm for tree intersection which implements Step 3 of Algorithm 8. Due to page restrictions, we can not include a description of our tree intersection algorithm in this extended abstract. It will be included in the full version of this paper.

Theorem 3. *Let T be a PQ-tree over the ground set S and let T' be a copy of T. Let T^* and $T'{*}$ be the result of the reduction of T with respect to $\{A_1, \ldots, A_r\}$ and of T' with respect to $\{B_1, \ldots, B_t\}$, respectively. Let $\overline{T}$ be the PQ-tree obtained by reducing T with respect to $\{A_1, \ldots, A_r, B_1, \ldots, B_t\}$. Then,*

$$\lambda \in L(\overline{T}) \Leftrightarrow \lambda \in L(T^*) \cap L(T'{*}).$$

Proof. $L(\overline{T})$ is the intersection of the sets of all orderings that satisfy $A_1, \ldots, A_r$, $B_1, \ldots, B_t$, and $L(T)$. $L(\overline{T})$ is always the same, independently of the order in which we reduce T. If $\lambda \in L(\overline{T})$, then λ must belong to the intersection between the set of all orderings that satisfy $A_1, \ldots, A_r$ and L(T) and it must also belong to the intersection between the set of all orderings that satisfy $B_1, \ldots, B_t$ and L(T), that is $\lambda \in L(T^*)$, $\lambda \in L(T'{*})$ and $\lambda \in L(T)$. Hence λ belongs to $L(T^*) \cap L(T'{*})$. The reverse can be shown analogously. □

In summary, we obtain

Theorem 4. *On a coarse grained multicomputer with p processors and $O(\frac{N}{p})$ storage per processor, Algorithm MReduce performs a reduce operation for a PQ-tree T in $O(\log^2 p)$ communication rounds with $O(\frac{N}{p})$ local computation per round.*

3.3   Convex Bipartite Graphs

Recall the definition of convex bipartite graphs (Definition 2). Given a bipartite graph $G = (A, B, E)$ with $A = \{a_1, a_2, \cdots, a_k\}$ and $B = \{b_1, b_2, \cdots, b_n\}$. Let $\mathcal{A} = \{A_1, \ldots, A_k\}$ where $A_i = \{b \in B : (a_i, b) \in E\}$, and let T be a PQ-tree over the ground set B consisting of a root with children $b_1, b_2, \cdots, b_n$. The problem of determining whether G is convex and, if this is the case, computing the correct ordering of the elements in B is equivalent to the *MReduce* operation on T with respect to $\mathcal{A}$.

Theorem 5. *On a coarse grained multicomputer with p processors and $O(\frac{N}{p})$ storage per processor, the problem of determining whether G is convex (and computing the correct ordering of the elements in B) can be solved in $O(\log^2 p)$ communication rounds with $O(\frac{N}{p})$ local computation per round.*

References

1. K.S. Booth and G.S. Lueker, "Testing for the Consecutive Ones Property, Interval Graphs, and Graph Planarity Using PQ-Tree Algorithms," in *Journal of Computer and System Sciences*, vol. 13, pages 335-379, 1976.
2. P. Bose, A. Chan, F. Dehne, and M. Latzel, "Coarse grained parallel maximum matching in convex bipartite graphs," in Proc. *13th International Parallel Processing Symposium* (IPPS'99), 1999.
3. E. Caceres, F. Dehne, A. Ferreira, P. Flocchini, I. Rieping, A. Roncato, N. Santoro, and S.W. Song, "Efficient parallel graph algorithms for coarse grained multicomputers and BSP," in Proc. 24th International Colloquium on Automata, Languages and Programming (ICALP'97), Bologna, Italy, 1997, Springer Verlag Lecture Notes in Computer Science, vol. 1256, pages 390-400.
4. L. Chen and Y. Yesha, "Parallel recognition of the consecutive ones property with applications," J. Algorithms, vol. 12, no. 3, pages 375-392. 1991.
5. L. Chen, "Graph isomorphism and identification matrices: parallel algorithms," IEEE Trans. on Parallel and Distr. Systems, vol. 7, no. 3, March 1996, pages 308 ff.
6. F. Dehne (Ed.), "Coarse grained parallel algorithms," Special Issue of *Algorithmica*, vol. 24, no. 3/4, 1999, pages 173-426.
7. F. Dehne, A. Fabri, and A. Rau-Chaplin, "Scalable Parallel Geometric Algorithms for Coarse Grained Multicomputers," in Proc. *ACM 9th Annual Computational Geometry*, pages 298–307, 1993.
8. F. Dehne, A. Fabri, and C. Kenyon, "Scalable and Architecture Independent Parallel Geometric Algorithms with High Probability Optimal Time," in Proc. *6th IEEE Symposium on Parallel and Distributed Processing*, pages 586–593, 1994.
9. F. Dehne, X. Deng, P. Dymond, A. Fabri, and A.A. Kokhar, "A randomized parallel 3D convex hull algorithm for coarse grained multicomputers," in Proc. *ACM Symposium on Parallel Algorithms and Architectures* (SPAA'95), pages 27–33, 1995.
10. M.T. Goodrich, "Communication efficient parallel sorting," ACM Symposium on Theory of Computing (STOC), 1996.
11. X. He and Y. Yeshua, "Parallel recognition and decomposition of two termninal series parallel graphs," Information and Computation, vol. 75, pages 15-38, 1987
12. P. Klein, *Efficient Parallel Algorithms for Planar, Chordal, and Interval Graphs* PhD. Thesis, MIT, 1988.
13. P. Klein. "Efficient Parallel Algorithms for Chordal Graphs". *Proc. 29th Symp. Found. of Comp. Sci., FOCS* 1989, pages 150–161.
14. P. Klein. "Parallel Algorithms for Chordal Graphs". In *Synthesis of parallel algorithms*, J.H. Reif (editor). Morgan Kaufmann Publishers, 1993, pages 341–407.
15. A.C. Tucker, "Matrix characterization of circular-arc graphs," Pacific J. Mathematics, vol. 39, no. 2, pages 535-545, 1971.
16. L. Valiant, "A bridging model for parallel computation," Communications of the ACM, vol. 33, no. 8, August 1990.

Networks with Small Stretch Number*
(Extended Abstract)

Serafino Cicerone and Gabriele Di Stefano

Dipartimento di Ingegneria Elettrica, Università dell'Aquila,
67040 Monteluco di Roio, L'Aquila, Italy. {cicerone,gabriele}@ing.univaq.it

Abstract. In a previous work, the authors introduced the class of *graphs with bounded induced distance of order* k, (BID(k) for short) to model non-reliable interconnection networks. A network modeled as a graph in BID(k) can be characterized as follows: if some nodes have failed, as long as two nodes remain connected, the distance between these nodes in the faulty graph is at most k times the distance in the non-faulty graph. The smallest k such that $G \in$ BID(k) is called *stretch number* of G. In this paper we give new characterization, algorithmic, and existence results about graphs with small stretch number.

1 Introduction

The main function of a network is to provide connectivity between the sites. In many cases it is crucial that (properties about) connectivity is preserved even in the case of (multiple) faults in sites. Accordingly, a major concern in network design is fault-tolerance and reliability. That means in particular that the network to be constructed shall remain reliable even in the case of site faults.

According to the actual applications and requirements, the term 'reliability' may stand for different features. In this work, it concerns bounded distances, that is our goal is to investigate about networks in which distances between sites remain *small* even in the case of faulty sites. As the underlying model, we use unweighted graphs, and measure the distance in a network in which node faults have occurred by a shortest path in the subnetwork that is induced by the non-faulty components. Using this model, in [7] we have introduced the class BID(k) of *graphs with bounded induced distance* of order k. A network modeled as a graph in BID(k) can be characterized as follows: if some nodes have failed, as long as two nodes remain connected, the distance between these nodes in the faulty graph is at most k times the distance in the non-faulty graph.

Some characterization, complexity, and structural results about BID(k) are given in [7]. In particular, the concept of *stretch number* has been introduced: the stretch number $s(G)$ of a given graph G is the smallest rational number

---

* Work partially supported by the Italian MURST Project "Teoria dei Grafi ed Applicazioni". Part of this work has been done while the first author, supported by the DFG, was visiting the Department of Computer Science, University of Rostock, Germany.

U. Brandes and D. Wagner (Eds.): WG 2000, LNCS 1928, pp. 95–106, 2000.

k such that G belongs to BID(k). Given the relevance of graphs in BID(k) in the area of communication networks, our purpose is to provide characterization, algorithmic, and existence results about graphs having small stretch number.

Results. We first investigate graphs having stretch number at most 2. In this context we show that: (i) there is no graph G with stretch number $s(G)$ such that $2 - \frac{1}{i} < s(G) < 2 - \frac{1}{i+1}$, for each integer $i \geq 1$ (this fact was conjectured in [7]); (ii) there exists a graph G such that $s(G) = 2 - \frac{1}{i}$, for each integer $i \geq 1$. These results give a partial solution to the following more general problem: Given a rational number k, is k an *admissible stretch number*, i.e., is there a graph G such that $s(G) = k$? We complete the solution to this problem by showing that every rational number $k \geq 2$ is an admissible stretch number. Finally, we give a characterization result in term of forbidden subgraphs for the class BID($2 - \frac{1}{i}$), for each integer $i > 1$. This characterization result allows us to design a polynomial time algorithm to solve the recognition problem for the class BID($2 - \frac{1}{i}$), for each $i \geq 1$ (if k is not fixed, this problem is Co-NP-complete for the class BID(k) [7]). We conclude the paper by showing that such an algorithmic approach cannot be used for class BID(k), for each integer $k \geq 2$.

Related works. In literature there are several papers devoted to fault-tolerant network design, mainly starting from a given desired topology and introducing fault-tolerance to it (e.g., see [4,15,19]). Other works follow our approach.

In [14], a study about our concepts is performed: they give characterizations for graphs in which *no delay* occurs in the case that a *single* node fails. These graphs are called *self-repairing*. In [8], authors introduce and characterize new classes of graphs that guarantee constant stretch factors k even when a multiple number of *edges* have failed. In a first step, they do not limit the number of edge faults at all, allowing for *unlimited* edge faults. Secondly, they examine the case where the number of edge faults is *bounded* by a value ℓ. The corresponding graphs are called k–self-spanners and (k, ℓ)–self-spanners, respectively. In both cases, the names are motivated by strong relationships to the concept of k–*spanners* [21]. Related works are also those concerning distance-hereditary graphs [18]. In fact, distance-hereditary graphs correspond to the graphs in BID(1), and graphs with bounded induced distance can be also viewed as a their parametric extension (in fact, BID(k) graphs are mentioned in the survey [2] as k–distance-hereditary graphs). Distance-hereditary graphs have been investigated to design interconnection network topologies [6,11,13], and several papers have been devoted to them (e.g., see [1,3,5,10,12,16,20,23]).

The remainder of this extended abstract is organized as follows. Notations and basic concepts used in this work are given in Section 2. In Section 3 we recall definitions and results from [7]. Section 4 shows the new characterization results, and in Section 5 we answer the question about admissible stretch numbers. In Section 6 we give the complexity result for the recognition problem for the class BID($2 - \frac{1}{i}$), for every integer $i \geq 1$, and in Section 7 we give some final remarks.

Due to space limitations, some proofs and technical details are omitted and will be provided in the full paper.

2   Notation

In this work we consider finite, simple, loopless, undirected and unweighted graphs $G = (V, E)$ with node set V and edge set E. We use standard terminologies from [2,17], some of which are briefly reviewed here.

A *subgraph* of G is a graph having all its nodes and edges in G. Given a subset S of V, the *induced subgraph* $\langle S \rangle$ of G is the maximal subgraph of G with node set S. $|G|$ denotes the cardinality of V. If x is a node of G, by $N_G(x)$ we denote the *neighbors* of x in G, that is, the set of nodes in G that are adjacent to x, and by $N_G[x]$ we denote the *closed neighbors* of x, that is $N_G(x) \cup \{x\}$. $G - S$ is the subgraph of G induced by $V \setminus S$.

A sequence of pairwise distinct nodes $(x_0, \ldots , x_n)$ is a *path* in G if $(x_i, x_{i+1}) \in E$ for $0 \le i < n$, and is an *induced path* if $\langle \{x_0, \ldots , x_n\} \rangle$ has n edges. A graph G is *connected* if for each pair of nodes x and y of G there is a path from x to y in G.

A *cycle* C_n in G is a path $(x_0, \ldots , x_{n-1})$ where also $(x_0, x_{n-1}) \in E$. Two nodes x_i and x_j are *consecutive* in C_n if $j = (i+1) \bmod n$ or $i = (j+1) \bmod n$. A *chord* of a cycle is an edge joining two non-consecutive nodes in the cycle. H_n denotes an *hole*, i.e., a cycle with n nodes and without chords. The *chord distance* of a cycle C_n is denoted by $cd(C_n)$, and it is defined as the minimum number of consecutive nodes in C_n such that every chord of C_n is incident to some of such nodes. We assume $cd(H_n) = 0$.

The length of a shortest path between two nodes x and y in a graph G is called *distance* and is denoted by $d_G(x, y)$. Moreover, the length of a longest induced path between them is denoted by $D_G(x, y)$. We use the symbols $P_G(x, y)$ and $p_G(x, y)$ to denote a longest and a shortest induced path between x and y, respectively. Sometimes, when no ambiguity occurs, we use $P_G(x, y)$ and $p_G(x, y)$ to denote the sets of nodes belonging to the corresponding paths. $I_G(x, y)$ denotes the set containing all the nodes (except x and y) that belong to a shortest path from x to y.

If x and y are two nodes of G such that $d_G(x, y) \ge 2$, then $\{x, y\}$ is a *cycle-pair* if there exist a path $p_G(x, y)$ and a path $P_G(x, y)$ such that $p_G(x, y) \cap P_G(x, y) = \{x, y\}$. In other words, if $\{x, y\}$ is a cycle-pair, then the set $p_G(x, y) \cup P_G(x, y)$ induces a cycle in G.

Let G_1, G_2 be graphs having node sets $V_1 \cup \{m_1\}$, $V_2 \cup \{m_2\}$ and edge sets E_1, E_2, respectively, where $\{V_1, V_2\}$ is a partition of V and $m_1, m_2 \notin V$. The *split composition* [9] of G_1 and G_2 with respect to m_1 and m_2 is the graph $G = G_1 * G_2$ having node set V and edge set $E = E_1' \cup E_2' \cup \{(x, y) \mid x \in N(m_1), y \in N(m_2)\}$, where $E_i' = \{(x, y) \in E_i \mid x, y \in V_i\}$ for $i = 1, 2$.

3   Basic Definitions and Previous Results

In this section we recall from [7] some definitions and results useful in the remainder of the paper.

Definition 1. [7] *Let k be a real number. A graph $G = (V, E)$ is a bounded induced distance graph of order k if for each connected induced subgraph G' of G:*

$$d_{G'}(x, y) \leq k \cdot d_G(x, y), \quad \text{for each } x, y \in G'.$$

The class of all the bounded induced distance graphs of order k is denoted by $\mathrm{BID}(k)$.

From the definition it follows that every class $\mathrm{BID}(k)$ is hereditary, i.e., if $G \in \mathrm{BID}(k)$, then $G' \in \mathrm{BID}(k)$ for every induced subgraph G' of G.

Definition 2. [7] *Let G be a graph, and $\{x, y\}$ be a pair of connected nodes in G. Then:*

1. *the* stretch number $s_G(x, y)$ *of the pair $\{x, y\}$ is given by $s_G(x, y) = \frac{D_G(x,y)}{d_G(x,y)}$;*
2. *the* stretch number $s(G)$ *of G is the maximum stretch number over all possible pairs of connected nodes, that is, $s(G) = \max_{\{x,y\}} s_G(x, y)$;*
3. *$\mathcal{S}(G)$ is the set of all the pairs of nodes inducing the stretch number of G, that is, $\mathcal{S}(G) = \{\{x, y\} \mid s_G(x, y) = s(G)\}$.*

The stretch number of a graph determines the minimum class which a given graph G belongs to. In fact, $s(G) = \min\{t : G \in \mathrm{BID}(t)\}$. As a consequence, $G \in \mathrm{BID}(k)$ if and only if $s(G) \leq k$.

Lemma 1. [7] *Let $G \in \mathrm{BID}(k)$, and $s(G) > 1$. Then, there exists a cycle-pair $\{x, y\}$ that belongs to $\mathcal{S}(G)$.*

Theorem 1. [7] *Let G be a graph and $k \geq 1$ a real number. Then, $G \in \mathrm{BID}(k)$ if and only if $cd(C_n) > \left\lceil \frac{n}{k+1} \right\rceil - 2$ for each cycle C_n, $n > 2k + 2$, of G.*

4   New Characterization Results

Graphs in $\mathrm{BID}(1)$ have been extensively studied and different characterizations have been provided. One of these characterization is based on forbidden induced subgraphs [1], and in [7] this result has been extended to the class $\mathrm{BID}(\frac{3}{2})$. In this section we further extend this characterization to the class $\mathrm{BID}(2 - \frac{1}{i})$, for every integer $i \geq 2$.

Lemma 2. *Let G be a graph with $1 < s(G) < 2$, and let $\{x, y\} \in \mathcal{S}(G)$ be a cycle-pair. If C is the cycle induced by $p_G(x, y) \cup P_G(x, y)$, then every internal node of $p_G(x, y)$ is incident to a chord of C.*

Proof. Omitted.   □

Theorem 2. *Given a graph G and an integer $i \geq 2$, then $G \in \mathrm{BID}(2 - \frac{1}{i})$ if and only if the following graphs are not induced subgraphs of G:*

1. H_n, for each $n \geq 6$;
2. cycles C_6 with $cd(C_6) = 1$;
3. cycles C_7 with $cd(C_7) = 1$;
4. cycles C_8 with $cd(C_8) = 1$;
5. cycles C_{3i+2} with $cd(C_{3i+2}) = i$.

Proof. Only if part. Holes H_n, $n \geq 6$, have stretch number at least 2. Cycles with 6, 7, or 8 nodes and chord distance 1 have stretch number equal to 2, 5/2, and 3, respectively. Cycles C_{3i+2} with chord distance equal to i have stretch number at least $\frac{2i+1}{i+1} = 2 - \frac{1}{i+1}$. Since the considered cycles have stretch number greater than $2 - \frac{1}{i}$, then they are forbidden induced subgraphs for every graph belonging to $\mathrm{BID}(2 - \frac{1}{i})$.

If part. Given an arbitrary integer $i \geq 2$, we prove that every graph $G \notin \mathrm{BID}(2 - \frac{1}{i})$ contains one of the forbidden subgraphs or a proper induced subgraph $G' \notin \mathrm{BID}(2 - \frac{1}{i})$. In the latter case, we can recursively apply to G' the following proof.

If $G \notin \mathrm{BID}(2 - \frac{1}{i})$ then, by Theorem 1, G contains a cycle C_n, $n \geq 6$, as induced subgraph such that $0 \leq cd(C_n) \leq \left\lceil \frac{i \cdot n}{3i-1} \right\rceil - 2$. This means that a cycle-pair $\{x, y\} \in \mathcal{S}(G)$ generates C_n. In particular, we can assume that C_n is induced by the nodes of the two internal node-disjoint paths $P_G(x, y) = (x, u_1, u_2, \ldots, u_p, y)$ and $p_G(x, y) = (x, v_1, v_2, \ldots, v_q, y)$, such that $p + q + 2 = n$ and $cd(C_n) = q$.

If $q = 0$ then we obtain the holes H_n, $n \geq 6$. If $q = \left\lceil \frac{i \cdot n}{3i-1} \right\rceil - 2$ and $n = 6, 7, 8, 3i + 2$, then we obtain the other forbidden subgraphs.

Now, we show that if $n \geq 9$, $n \neq 3i + 2$, and q fulfills $1 \leq q \leq \left\lceil \frac{i \cdot n}{3i-1} \right\rceil - 2$, then C_n contains one of the given forbidden subgraphs or an induced subgraph G' such that $G' \notin \mathrm{BID}(2 - \frac{1}{i})$.

By Lemma 2, every node v_k, $1 \leq k \leq q$, must be incident to a chord of C_n, otherwise C_n has a stretch number greater or equal to 2 and hence it is itself a forbidden subgraph of G. As a consequence, we can denote by r_j the largest index j' such that v_j and $u_{j'}$ are connected by a chord of C_n, i.e. $r_j = \max\{j' \mid (v_j, u_{j'})$ is a chord of $C_n\}$. Informally, r_j gives the *rightmost* chord connecting v_j to some vertex of $P_G(x, y)$.

Notice that, if $r_1 > 3$ then the subgraph of C_n induced by the nodes $v_1, x, u_1, \ldots, u_{r_1}$ is forbidden, since it is a cycle with at least 6 nodes and chord distance at most 1. Hence, in the remainder of this proof we assume that $r_1 \leq 3$.

Let us now analyze two distinguished cases for C_n, according whether the chord distance q of C_n either (*i*) fulfills $1 \leq q < \left\lceil \frac{i \cdot n}{3i-1} \right\rceil - 2$, or (*ii*) is equal to $\left\lceil \frac{i \cdot n}{3i-1} \right\rceil - 2$.

(i) We consider C_n with $n \geq 9$ and chord distance q such that $1 \leq q < \left\lceil \frac{i \cdot n}{3i-1} \right\rceil - 2$. If $C_{n'}$ denotes the subgraph induced by the nodes of C_n except the nodes $x, u_1, \ldots, u_{r_1-1}$, then $C_{n'}$ is a cycle with $n' \geq n-3$ nodes and chord distance at most $q-1$. To prove that $C_{n'}$ is forbidden, we have to show that $\left\lceil \frac{i \cdot n'}{3i-1} \right\rceil - 2 \geq q-1$:

$$\left\lceil \frac{i \cdot n'}{3i-1} \right\rceil - 2 \geq \left\lceil \frac{i \cdot n - 3i}{3i-1} \right\rceil - 2 \geq q-1,$$

$$\left\lceil \frac{i \cdot n - 3i}{3i-1} \right\rceil - 2 > q-2,$$

$$\left\lceil \frac{i \cdot n - 3i}{3i-1} + 2 \right\rceil - 2 > q,$$

$$\left\lceil \frac{i \cdot n + 4i - 2}{3i-1} \right\rceil - 2 > q.$$

The last inequality holds because $4i - 2 \geq 0$ for each integer $i \geq 1$, and $\left\lceil \frac{i \cdot n}{3i-1} \right\rceil - 2 > q$.

(ii) We consider C_n with $n \geq 9$ and chord distance q such that $q = \left\lceil \frac{i \cdot n}{3i-1} \right\rceil - 2$. In this case q is given whenever a fixed value for n is chosen. In general, since $n \geq 9$, it follows that $q \geq 2$.

Let us analyze again the cycle $C_{n'}$. Recalling that $n' \geq n - 3$ and $cd(C_{n'}) \leq q - 1$, then

$$\left\lceil \frac{i \cdot n'}{3i-1} \right\rceil - 2 \geq \left\lceil \frac{i \cdot n - 3i}{3i-1} \right\rceil - 2 \geq q-1$$

is equivalent to

$$\left\lceil \frac{i \cdot n - 1}{3i-1} \right\rceil - 2 \geq q.$$

In the following we show that, for every n such that $9 \leq n \leq 6i$, either this relation holds or n is equal to $3i + 2$. This means that the cycle $C_{n'}$ is forbidden for each cycle C_n, $9 \leq n \leq 6i$.

Since $\left\lceil \frac{i \cdot n}{3i-1} \right\rceil - 2 = q$ holds by hypothesis, we have to study when $\left\lceil \frac{i \cdot n - 1}{3i-1} \right\rceil \geq \left\lceil \frac{i \cdot n}{3i-1} \right\rceil$. This relation does not hold if and only if there exists an integer m such that $\frac{i \cdot n - 1}{3i-1} \leq m < \frac{i \cdot n}{3i-1}$, that is $\frac{i \cdot n - 1}{3i-1} = m$. Then, since this equality is equivalent to $n = 3m - \frac{m-1}{i}$, m can be equal to $\ell \cdot i + 1$ only, for each integer $\ell \geq 0$. As a consequence, $n = 3m - \frac{m-1}{i} = 3(\ell \cdot i + 1) - \ell$, $\ell \geq 0$. For $\ell = 0$ we obtain $n = 3$ (but we are considering $n \geq 9$), for $\ell = 1$ and $\ell = 2$ the value of n is $3i + 2$ and $n = 6i + 1$, respectively. The cycle with $3i + 2$ nodes is one of the forbidden cycles in the statement of the theorem. As a conclusion, the induced cycle $C_{n'}$ shows that C_n contains a forbidden induced subgraph when $9 \leq n \leq 6i$.

It remains to be considered the case when $n \geq 6i + 1$. In this case $q = \left\lceil \frac{i \cdot n}{3i-1} \right\rceil - 2$ implies $q \geq 2i$, and hence we can compute the value r_i. Since $v_i, v_{i-1}, \ldots, v_1, x, u_1, \ldots, u_{r_i}$ induce a cycle with chord distance i, then it has at most $3i+1$ nodes otherwise it is forbidden. In other words, $r_i \leq 2i$. The subgraph $C_{n''}$ induced by the nodes of C_n except the nodes $v_{i-1}, \ldots, v_1, x, u_1, \ldots, u_{r_i-1}$ is a cycle with $n'' \geq n - 3i + 1$ nodes and chord distance at most $q - i$. To prove that $C_{n''}$ is forbidden, let us show that $\left\lceil \frac{i \cdot n''}{3i-1} \right\rceil - 2 \geq q - i$. The inequality

$$\left\lceil \frac{i \cdot n''}{3i-1} \right\rceil - 2 \geq \left\lceil \frac{i \cdot (n - 3i + 1)}{3i-1} \right\rceil - 2 \geq q - i$$

is equivalent to

$$\left\lceil \frac{i \cdot n}{3i-1} \right\rceil - 2 \geq q.$$

The last relation holds by hypothesis, and this concludes the proof   □

5 Admissible Stretch Numbers

In [7], it was conjectured that there exists no graph G such that $2 - \frac{1}{i} < s(G) < 2 - \frac{1}{i+1}$, for each integer $i \geq 1$. In this section we show that the conjecture is true. Moreover, we extend the result by showing that it is possible to answer to the following general question: Given a rational number $t \geq 1$, is there a graph G such that $s(G) = t$?. In other words, we can state when a given positive rational number is an *admissible* stretch number.

Definition 3. *A positive rational number t is called* admissible stretch number *if there exists a graph G such that $s(G) = t$.*

In the remainder of this section we first show that the conjecture recalled above is true, and then we show that each positive rational number greater or equal than 2 is an admissible stretch number.

Lemma 3. *If p and q are two positive integers such that $2 - \frac{1}{i} < \frac{p}{q} < 2 - \frac{1}{i+1}$, for some integer $i \geq 1$, then $q > i$.*

Proof. Omitted.   □

Theorem 3. *If t is a rational number such that $2 - \frac{1}{i} < t < 2 - \frac{1}{i+1}$, for some integer $i \geq 1$, then t is not an admissible stretch number.*

Proof. We have to show that there exists no graph G such that $2 - \frac{1}{i} < s(G) < 2 - \frac{1}{i+1}$, for each integer $i \geq 1$.

By contradiction, let us assume that there exist an integer $i \geq 1$ and a graph G such that $2 - \frac{1}{i} < s(G) < 2 - \frac{1}{i+1}$. By Lemma 1 there exists a cycle-pair $\{x, y\} \in \mathcal{S}(G)$. If we assume that $P_G(x, y) = (x, u_1, u_2, \ldots, u_{p-1}, y)$ and $p_G(x, y) = (x, v_1, v_2, \ldots, v_{q-1}, y)$, then $p_G(x, y) \cup P_G(x, y)$ induces a cycle C, and $s(G) = \frac{p}{q}$. By Lemma 3, the relation $q > i$ holds; then, the node v_i exists in the path $p_G(x, y)$. By Lemma 2, the node v_i is incident to a chord of C, and hence we can define the integer r, $1 \leq r \leq q - 1$, such that

$$r = \max\{j \mid (v_i, u_j) \text{ is a chord of } C\}.$$

Let us now denote by C_L the cycle induced by the nodes $v_i, v_{i-1}, \ldots, v_1, x, u_1,$ $u_2, \ldots, u_r$, and by C_R the cycle induced by the nodes $v_i, v_{i+1}, \ldots, v_{q-1}, y, u_{p-1},$ $u_{p-2}, \ldots, u_r$. In other words, the chord (v_i, u_r) divides C into the *left* cycle C_L, and the *right* cycle C_R.

First of all, let us compute the stretch number of the cycle C_R. Since $p_G(x, y) = (x, v_1, v_2, \ldots, v_{q-1}, y)$ then $p_{C_R}(v_i, y) = (v_i, v_{i+1}, \ldots, v_{q-1}, y)$. Moreover, since the path $(v_i, u_r, u_{r+1}, \ldots, u_{p-1}, y)$ is induced in C, then its length implies $D_{C_R}(v_i, y) \geq p - r + 1$. Then

$$s(C_R) \geq s_{C_R}(v_i, y) \geq \frac{p - r + 1}{q - i}.$$

Since C_R is an induced subgraph of G then

$$\frac{p - r + 1}{q - i} \leq \frac{p}{q}.$$

This inequality is equivalent to

$$\frac{p}{q} \leq \frac{r - 1}{i}.$$

From the relations

$$2 - \frac{1}{i} < \frac{p}{q} \leq \frac{r - 1}{i}$$

we obtain that $r > 2i$, that is $r \geq 2i + 1$.

Let us now compute the stretch number of the cycle C_L when $r \geq 2i + 1$. In this case, $p_{C_L}(x, u_r) = (x, v_1, v_2, \ldots, v_i, u_r)$ and $P_{C_L}(x, u_r) = (x, u_1, u_2, \ldots, u_r)$. Then

$$s(C_L) \geq s_{C_L}(x, u_r) = \frac{r}{i + 1} \geq \frac{2i + 1}{i + 1} \geq 2 - \frac{1}{i + 1}.$$

The obtained relation implies that $s(C_L) > s(G)$. This is a contradiction since C_L is an induced subgraph of G. $\square$

In order to show that each rational number equal or greater than 2 is admissible as stretch number, let us consider the graph $G(n_1, n_2, \ldots, n_t)$ obtained by composing t holes $H_{n_1}, H_{n_2}, \ldots, H_{n_t}$ by split composition, where $n_i \geq 5$ for $1 \leq i \leq t$. The holes correspond to the following chordless cycles (as an example, see Figure 1, where $t = 5$) :

- $H_{n_1} = (l_1, x_0, x_1, m'_1, r_1, \ldots)$;
- $H_{n_i} = (l_i, m_i, x_i, m'_i, r_i, \ldots)$, for each i such that $1 < i < t$;
- $H_{n_t} = (l_t, m_t, x_t, x_{t+1}, r_t, \ldots)$.

The holes are composed by means of the split composition as follows:

$$G(n_1, n_2, \ldots, n_t) = H_{n_1} * H_{n_2} * \cdots * H_{n_t},$$

and the marked nodes between H_{n_i} and $H_{n_{i+1}}$ are m_i' and m_{i+1}, $1 \le i < t$, respectively.

In the following, we denote by V_1 (V_t, resp.) the set containing all the nodes of the hole H_{n_1} (H_{n_t}, resp.) but x_0, x_1, and m_1' (m_t, x_t, and x_{t+1}, resp.); we denote by V_i the set containing all the nodes of the hole H_{n_i} but m_i, x_i, and m_i', $1 \le i \le t$. Finally, we denote by X the set $\{x_0, x_1, \ldots, x_{t+1}\}$.

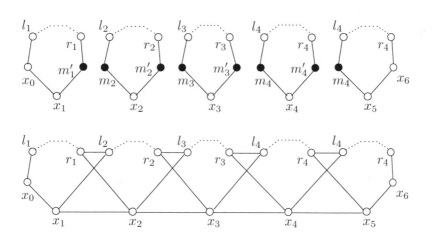

Fig. 1. *The graph $G(n_1, n_2, n_3, n_4, n_5)$ obtained by the split composition of 5 holes. The i-th hole has $n_i \ge 5$ nodes.*

Lemma 4. *Given the graph $G = G(n_1, n_2, \ldots, n_t)$, the following facts hold:*

1. $s_G(x_0, x_{t+1}) = \frac{\sum_{i=1}^{t} n_i - 3t + 1}{t+1}$;
2. *if $i < j$ then $p_G(x_i, x_j) \cup P_G(x_i, x_j)$ induces a subgraph isomorphic to $G(n_i, n_{i+1}, \ldots, n_j)$;*
3. *if $n_t \ge \frac{\sum_{i=1}^{t-1} n_i}{t-1}$ then $s_G(x_0, x_{t+1}) > s_G(x_0, x_t)$;*
4. *there exists a pair $\{u, v\} \in \mathcal{S}(G)$ such that $u \in X$ and $v \in X$;*
5. *if $n_i = n$ for some fixed integer n, $1 \le i \le t$, then $s(G) = s_G(x_0, x_{t+1}) = \frac{nt - 3t + 1}{t+1}$.*

Proof. Omitted. □

Notice that the stretch number of nodes x_0 and x_{t+1} in $G(n_1, n_2, \ldots, n_t)$ does not depend on how many nodes are in each hole; it depends only on the total number of nodes in $G(n_1, n_2, \ldots, n_t)$ and on the number t of used holes.

Theorem 4. *If t is a rationale number such that $t \ge 2$, then t is an admissible stretch number.*

Sketch of the Proof. Let us suppose that $t = p/q$ for two positive integers p and q. If $q = 1$ then $G = H_{2p+2}$, if $q = 2$ then $G = H_{p+2}$. When $q \geq 3$ we show that the graph G is equal to $G(n_1, n_2, \ldots, n_{q-1})$ for suitable integers $n_1, n_2, \ldots, n_{q-1}$.

Let $b = 3 + \left\lfloor \frac{p-1}{q-1} \right\rfloor$ and $r = (p-1) \bmod (q-1)$. Let us choose the sizes of the holes $H_{n_1}, H_{n_2}, \ldots, H_{n_{q-1}}$ according to the following strategy: r holes contain exactly $b+1$ nodes, while the remaining $q - 1 - r$ contain exactly b nodes. By Fact 1 of Lemma 4 it follows that $s_G(x_0, x_q) = \frac{\sum_{i=1}^{q-1} n_i - 3(q-1) + 1}{q} = \frac{p}{q}$. To prove that $S(G) = p/q$, by Fact 4 of Lemma 4, we have to prove that $s_G(x_i, x_j) \leq p/q$, $1 \leq i, j \leq q - 2$. This property holds only if we are able to determine a deterministic method to decide whether the hole H_{n_i}, $1 \leq i \leq q - 1$, contains either b or $b + 1$ nodes. In the full paper we show that such a deterministic method exists.   □

Corollary 1. *For each integer $i \geq 1$, $2 - \frac{1}{i}$ is an admissible stretch number.*

Proof. From Fact 5 of Lemma 4, it follows that $G = G(n_1, n_2, \ldots, n_{i-1})$ such that $n_j = 5$ for each $1 \leq j \leq i - 1$, has stretch equal to $2 - \frac{1}{i}$.   □

The results provided by Corollary 1, Theorem 3, and Theorem 4 can be summarized in the following two corollaries.

Corollary 2. *Let t be an admissible stretch number. Then, either $t \geq 2$ or $t = 2 - \frac{1}{i}$ for some integer $i \geq 1$.*

Corollary 3. *For every admissible stretch number t, split composition can be used to generate a graph G with $s(G) = t$.*

Notice that, by Theorem 4, we can also use every irrational number greater than 2 to define graph classes containing graphs with bounded induced distance. For instance, $\mathrm{BID}(\pi) \neq \mathrm{BID}(k)$ for every rational number k.

6   Recognition Problem

The recognition problem for $\mathrm{BID}(1)$ can be solved in linear time [1,16]. In [7], this problem has been shown Co-NP-complete for the generic case (i.e., when k is not fixed), and the following question has been posed: What is the largest constant k such that the recognition problem for $\mathrm{BID}(k)$ can be solved in polynomial time?

In this section we show that Theorem 2 can be used to devise a polynomial algorithm to solve the recognition problem for the class $\mathrm{BID}(k)$, for every $k < 2$.

Lemma 5. *There exists a polynomial time algorithm to test whether a given graph G contains, as induced subgraph, a cycle C_n with $n \geq 6$ and $cd(C_n) \leq 1$.*

Proof. Omitted.   □

Theorem 5. *For any fixed integer $i \geq 1$, the recognition problem for the class $\mathrm{BID}(2 - \frac{1}{i})$ can be solved in polynomial time.*

Proof. For $i = 1$ the problem can be solved in linear time [1,16]. By Theorem 2, a brute-force, rather naive algorithm for solving the recognition problem for the class $\mathrm{BID}(2 - \frac{1}{i})$, $i > 1$, is: test if G contains, as induced subgraph, (1) a cycle C_n with $n \geq 6$ and $cd(C_n) \leq 1$, or (2) a cycle C_{3i+2} with chord distance equal to i. To perform Test 1 above, we can use the algorithm of Lemma 5, and to perform Test 2 we can check whether any subset of $3i + 2$ nodes of G forms a cycle with chord distance equal to i. The latter test can be implemented in polynomial time since the number of subsets of nodes with $3i + 2$ elements is bounded by n^{3i+2}.   □

7   Conclusions

In this paper we provide new results about graph classes that represent a parametric extension of the class of distance-hereditary graphs. In any graph G belonging to the generic new class $\mathrm{BID}(k)$, the distance between every two connected nodes in every induced subgraph of G is at most k times their distance in G.

The recognition problem for $\mathrm{BID}(2 - \frac{1}{i})$ can be solved in polynomial time (Theorem 5), and the corresponding algorithm is based on Theorem 2. Can the same approach be used in order to solve the same problem for class $\mathrm{BID}(k)$, $k \geq 2$? In other words, if $k \geq 2$ is an integer, is it possible to characterize $\mathrm{BID}(k)$ by listing all its forbidden induced subgraphs? For instance, holes H_n with $n \geq 2k + 3$, and a *finite* number of cycles having different chord distance. Unfortunately, the following theorem states that it is not possible.

Theorem 6. *For each integers $k \geq 2$ and $i \geq 2$, there exists a minimal forbidden induced cycle for the class $\mathrm{BID}(k)$ with chord distance equal to i.*

Proof. Omitted.   □

Many problems are left open. For instance, what is the largest constant k such that the recognition problem for $\mathrm{BID}(k)$ can be solved in polynomial time? Moreover, several algorithmic problems are solvable in polynomial time for $\mathrm{BID}(1)$. Can some of these results be extended to $\mathrm{BID}(k)$, $k > 1$?

References

1. H. J. Bandelt and M. Mulder. Distance-hereditary graphs. *Journal of Combinatorial Theory, Series B*, 41(2):182–208, 1986.
2. A. Brandstädt, V. B. Le and J. P. Spinrad. *Graph classes - a survey*. SIAM Monographs on Discrete Mathematics and Applications, Philadelphia, 1999.
3. A. Brandstädt and F. F. Dragan. A linear time algorithm for connected r-domination and Steiner tree on distance-hereditary graphs. *Networks*, 31:177–182, 1998.
4. J. Bruck, R. Cypher, and C.-T. Ho. Fault-tolerant meshes with small degree. *SIAM J. on Computing*, 26(6):1764–1784, 1997.
5. S. Cicerone and G. Di Stefano. Graph classes between parity and distance-hereditary graphs. *Discrete Applied Mathematics*, 95(1-3): 197–216, August 1999.

6. S. Cicerone, G. Di Stefano, and M. Flammini Compact-Port routing models and applications to distance-hereditary graphs. In *6th Int. Colloquium on Structural Information and Communication Complexity (SIROCCO'99)*, pages 62–77, Carleton Scientific, 1999.

7. S. Cicerone and G. Di Stefano. Graphs with bounded induced distance. In *Proc. 24th International Workshop on Graph-Theoretic Concepts in Computer Science, WG'98*, pages 177–191. Lecture Notes in Computer Science, vol. 1517, Springer-Verlag, 1998. To appear on *Discrete Applied Mathematics*.

8. S. Cicerone, G. Di Stefano, and D. Handke. Survivable networks with bounded delay: The edge failure case (Extended Abstract). In *Proc. 10th Annual International Symp. Algorithms and Computation, ISAAC'99*, pages 205–214. Lecture Notes in Computer Science, vol. 1741, Springer-Verlag, 1999.

9. W. H. Cunningham. Decomposition of directed graphs. *SIAM Jou. on Alg. Disc. Meth.*, 3:214–228, 1982.

10. A. D'Atri and M. Moscarini. Distance-hereditary graphs, steiner trees, and connected domination. *SIAM Journal on Computing*, 17:521–530, 1988.

11. G. Di Stefano. A routing algorithm for networks based on distance-hereditary topologies. In *3rd Int. Coll. on Structural Inform. and Communication Complexity (SIROCCO'96)*, Carleton Scientific, 1996.

12. F. F. Dragan. Dominating cliques in distance-hereditary graphs. In *4th Scandinavian Workshop on Algorithm Theory (SWAT'94)*, volume 824 of *Lecture Notes in Computer Science*, pages 370–381. Springer-Verlag, 1994.

13. A. H. Esfahanian and O. R. Oellermann. Distance-hereditary graphs and multidestination message-routing in multicomputers. *Journal of Comb. Math. and Comb. Computing*, 13:213–222, 1993.

14. A. M. Farley and A. Proskurowski. Self-repairing networks. *Paral. Proces. Letters*, 3(4):381–391, 1993.

15. J. P. Hayes. A graph model for fault-tolerant computing systems. *IEEE Transactions on Computers*, C-25(9):875–884, 1976.

16. P. L. Hammer and F. Maffray. Completely separable graphs. *Discr. Appl. Mathematics*, 27:85–99, 1990.

17. F. Harary. *Graph Theory*. Addison-Wesley, 1969.

18. E. Howorka. Distance hereditary graphs. *Quart. J. Math. Oxford*, 2(28):417–420, 1977.

19. F. T. Leighton, B. M. Maggs, and R. K. Sitaraman. On the fault tolerance of some popular bounded-degree networks. *SIAM J. on Computing*, 27(6):1303–1333, 1998.

20. F. Nicolai. Hamiltonian problems on distance-hereditary graphs. Technical Report SM-DU-264, University Duisburg, 1994.

21. D. Peleg and A. Schaffer. Graph spanners. *Journal of Graph Theory*, 13:99–116, 1989.

22. J. van Leeuwen and R.B. Tan. Interval routing. *The Computer Journal*, 30:298–307, 1987.

23. H. G. Yeh and G. J. Chang. Weighted connected domination and steiner trees in distance-hereditary graphs. In *4th Franco-Japanese and Franco-Chinese Conference on Combinatorics and Computer Science*, volume 1120 of *Lecture Notes in Computer Science*, pages 48–52. Springer-Verlag, 1996.

Efficient Dispersion Algorithms for Geometric Intersection Graphs

Peter Damaschke

FernUniversität, Theoretische Informatik II, 58084 Hagen, Germany.
Peter.Damaschke@fernuni-hagen.de

Abstract. The dispersion problem in a graph requires to find a subset of vertices of prescribed size, so as to maximize the minimum distance between the chosen vertices. We propose efficient algorithms solving the dispersion problem in interval graphs, circular-arc graphs, and trapezoid graphs. Graphs are supposed to be represented geometrically, rather than by their edge sets.

1 Introduction

Given a graph and a number q, the dispersion problem requires to find a subset of q vertices so as to maximize the minimum distance between the chosen vertices. The problem is motivated e.g. by the location of undesirable facilities. It can be naturally generalized to weighted graphs, i.e. with arbitrary positive vertex weights and edge lengths. (The subset is demanded to have total weight at least q.) The problem is NP-hard for general graphs, since it includes the maximum independent set problem, and it can be approximated in polynomial time within ratio 2, see e.g. [14]. For $q = 2$, it is nothing else than the problem of computing the diameter.

At least in the unweighted case, there is an obvious relationship between dispersion, k-independent sets and k-th powers of graphs. If maximum independent sets can be found in polynomial time in the class of powers of graphs from a class $\mathcal{G}$ then the dispersion problem in $\mathcal{G}$ can be solved in polynomial time, too. Fortunately, a number of well-known graph classes are closed under graph powers. However, the straightforward use of this observation yields $\Omega(n^2)$ time dispersion algorithms in such classes. In order to obtain more efficient algorithms, we have to avoid explicit insertion of edges to get the k-th powers of the given graph where we look for maximum independent sets.

To the best of our knowledge, only few subquadratic dipersion algorithms have been provided recently: The dispersion problem can be solved in $O(n)$ time for unweighted trees [1], in $O(n \log n)$ time for weighted paths [14], and in $O(n \log^4 n)$ time for weighted trees [2]. In the present note we give likewise efficient algorithms for interval graphs, circular-arc graphs, and trapezoid graphs.

U. Brandes and D. Wagner (Eds.): WG 2000, LNCS 1928, pp. 107–115, 2000.
© Springer-Verlag Berlin Heidelberg 2000

2   Preliminaries

We introduce our notion and report some very well-known facts.

Let $G = (V, E)$ be an undirected graph with n vertices and m edges. We will always assume that G is connected, without explicit notice; this is no loss of generality. The distance $d(u, v)$ of two vertices u, v is the length, i.e. number of edges, of a shortest path connecting u and v. The diameter $diam(G)$ is the maximum distance among vertices of G. A distance query in a graph takes a pair u, v as its input and outputs $d(u, v)$. The edges of the k-th power G^k of G are all pairs uv such that $d(u, v) \leq k$ in G. We have $G^{ab} = (G^a)^b$.

A q-dispersion set in G is a subset $S \subseteq V$ with $|S| = q$ such that $md(S) = \min\{d(u, v) \mid u, v \in S, u \neq v\}$ is maximized, among all subsets of V with q vertices. The dispersion problem requires to find a q-dispersion set, given G and q. A set S is called k-independent if $md(S) > k$. A 1-independent set is simply called independent. Note that k-independent sets in G are exactly the independent sets in G^k.

An interval family I is a set of intervals $[l_i, r_i]$, $i = 1, \ldots, n$ on the real line. The interval graph G associated with I has the intervals of I as its vertices, and two vertices are adjacent iff the corresponding intervals have a nonempty intersection. G is called the intersection graph of I. Similarly, as circular-arc graph is the intersection graph of a family of arcs on a circle. A trapezoid graph is the intersection graph of a family of trapezoids two opposite sides of which lie on a pair of parallel straight lines.

In an interval family I, we may w.l.o.g. assume that all l_i, r_i are mutually distinct. The vertex represented by $[l_i, r_i]$ is named v_i. We sometimes informally identify vertices and intervals. We say that I is sorted (with respect to right interval endpoints) if $r_1 \leq \ldots \leq r_n$. Instead of $u = v_i$, $v = v_j$, $i < j$ we simply write $u < v$.

One easily sees that a sorted interval family has the following properties:

(1) If $u < v < w$ and uv, uw are edges then vw is also an edge.

(2) If $u < v < w$ and uw is an edge then at least one of uv, vw is also an edge.

An ordering of the vertices of a graph satisfying (1) is called a perfect elimination ordering (PEO). A graph admits a PEO if and only if it is chordal, i.e. contains no chordless cycles of more than three vertices. All interval graphs are chordal, but not vice versa.

A maximum independent set in a PEO can be found in $O(n + m)$ time: Scan the PEO from left to right. Starting with $S = \emptyset$, add a vertex to S if it is not adjacent to one of the previous vertices of S. (The correctness proof is a simple exercise.) If a PEO additionally satisfies (2) then we may put any new vertex into S if it is not adjacent to the last vertex added to S. This reduces the time bound to $O(n)$. We refer to this algorithm as the greedy algorithm.

3   Efficient Dispersion in Interval Graphs

As a remarkable result of [6], one can answer distance queries in an interval graph in constant time, after an $O(n)$ time preprocessing. Actually, the result holds on circular-arc graphs, and preprocessing may be done in parallel, on $O(n/\log n)$ CREW PRAM processors in $\log n$ time. (We also mention that a weaker version with $O(n \log n)$ preprocessing time and $O(\log n)$ answer time is easily obtained by standard doubling techniques.)

Our second prerequisite addresses common PEOs of interval graphs. Common PEOs of chordal powers of graphs have been studied in [3,4], and the following lemma can be derived from these papers, however it is more convenient to give a short self-contained proof.

Lemma 1. *A sorted interval family is a common PEO of all powers of the represented interval graph.*

Proof. Consider vertices $u < v$, $u < w$ with $d(u, v) \leq k$ and $d(u, w) \leq k$. We have to show $d(v, w) \leq k$. So let P be a shortest path from u to v, and Q a shortest path from u to w. These paths are 1-shaped.

Let v' and w' be the first vertex on P and Q, respectively, after u. Due to (1), $v'w'$ is an edge. If u' denotes the leftmost vertex of v', w' then we have 1-shaped paths from u' to v and w, of length $k - 1$ and k, or vice versa. By an inductive argument we eventually obtain 1-shaped paths from some vertex x to v and w, such that one of them, w.l.o.g. to v, has length 1 or 0. More precisely, we either have $v < x$ and vx is an edge, or $x = v$.

In case $x = v$ we are done, since $d(x, w) \leq k$. If $d(v, x) = 1$ and $d(x, w) < k$, the assertion is also proven. Assume that $d(v, x) = 1$ and still $d(x, w) = k$. By construction, this implies $x < w'$. But since $u < v < x < w'$, property (0) yields that vw' is an edge. Thus $d(v, w) \leq k$.   □

In the following we assume that the interval graph is already given as a sorted interval family I. If I is not sorted yet, we have to spend $O(n \log n)$ time in general, but note that there are natural cases where faster sorting is possible. For example, if all endpoints are integers within a range of s, then I can be sorted in $O(n + s)$ time by bucketing, which is $O(n)$ if $s = O(n)$.

Lemma 1 immediately yields a simple and efficient dispersion algorithm:

Theorem 1. *A q-dispersion set in an interval graph can be found in time $O(n \log(n/q^2))$, if the graph is given as a sorted interval family.*

Proof. In $O(n)$ time, set up the data structure supporting constant-time distance queries. Let s denote the distance between the first vertex u and the last vertex v in the given PEO of G. We may obtain s by a distance query. One may also compute a shortest path in $O(n)$ time in a straightforward way, but the algorithm only needs the number s.

Any shortest path $(u = x_0, x_1, \ldots, x_s = v)$ is monotone in the PEO, that is $x_0 < x_1 < \ldots x_s$. It partitions the PEO into s consecutive blocks of vertices, where the ith block contains the vertices succeeding x_{i-1} up to x_i. (Vertex x_0 is added to the first block.) By (1) and (2), every vertex in the ith block is adjacent to x_i. Hence, two vertices in the ith and jth block have distance at most $k+2$, where $k = |j - i|$. On the other hand, their distance it at least $k - 2$, otherwise there would be a path from x_i to x_j shorter than k.

We easily conclude that the cardinality q_k of a maximum k-independent set in G satisfies

$$\frac{s-1}{k+3} \leq \left\lfloor \frac{s-1}{k+3} \right\rfloor + 1 \leq q_k \leq \frac{s-1}{k-1} + 1.$$

This can be written as

$$\frac{s-1}{q_k} - 3 \leq k \leq \frac{s-1}{q_k - 1} + 1.$$

Remember that we search for the maximum k with $q_k \geq q$. By maximality of k we have $q_{k+1} < q$. Since the left-hand inequality holds similarly for $k + 1$, we finally get

$$\frac{s-1}{q} - 4 < k \leq \frac{s-1}{q-1} + 1.$$

The difference is $O(s/q^2) = O(n/q^2)$. Thus it suffices to search for k among $O(n/q^2)$ candidates. Since q_k decreases monotone with k, we can use binary search.

Due to Lemma 1, for computing a maximum k-independent set in G we may apply the $O(n)$ time greedy algorithm to the given PEO, with the only modification that adjacency queries to G^k are replaced with distance queries to G.   □

In particular, the time bound is $O(n)$ if $q = \Omega(\sqrt{n})$, but only $O(n \log n)$ for smaller q. For an acceleration we have to exploit more structure of interval graphs.

First note that the k-independent sets computed by the greedy algorithm (in the k-the power) consist of (inclusion-)minimal intervals only: Assume that the greedy algorithm puts u and next some non-minimal interval v into S. Since v includes some minimal interval w, and w precedes v in the ordering, w is considered before v. But $d(u, v) > k$ obviously implies $d(u, w) > k$, hence w would be chosen, contradiction. Similarly, the first interval in S is a minimal one.

We conclude that the following modified greedy algorithm outputs a maximum k-independent set in $O(n)$ time: Mark all minimal intervals. Scan the PEO from left to right. Starting with $S = \emptyset$, add a vertex to S if it is not adjacent to the last vertex added to S AND it is a minimal interval.

We can also formulate:

Lemma 2. *If $u < v < w$, w is a minimal interval, and $d(u, w) \leq k$ then $d(u, v) \leq k$.*

Proof. Let P be a shortest path from u to w. Since some edge of P bridges v, by (2), v is adjacent to some endvertex of that edge. This gives immediately $d(u, v) \leq k$, unless the bridging edge is the last one of P, and v is adjacent to w only. But then interval w includes v, contradiction. $\square$

A block in an ordered interval family I is a set of consecutive elements, with respect to the ordering.

Lemma 3. *The vertices in a block of size b have mutual distances at most $2b$.*

Proof. It suffices to show that two consecutive vertices $u < v$ have distance at most 2. As we know, any shortest path P from u to v is 1-shaped. Let w be the second vertex on P. Since no vertex is between u and w, we have $v \leq w$. By (2), uv or vw is an edge, or $w = v$. $\square$

Now we are ready to present our fast dispersion algorithm.

Theorem 2. *A q-dispersion set in an interval graph can be found in time $O(n \log \log n)$, if the graph is given as a sorted interval family.*

Proof. Mark the minimal intervals. This can be done in $O(n)$ time in a straightforward way. Set up the data structure for constant time distance queries, in $O(n)$ time. Partition I into blocks of approximately $\log n$ vertices each. For fixed k we determine a maximum k-independent set in the following way.

If $k \leq 2 \log n$ then apply the simple $O(n)$ time greedy algorithm.

If $k > 2 \log n$ then build a set S as the greedy algorithm does, but consider only the minimal intervals, and check only the first minimal interval v in every block. If the distance of v to the currently last element $u \in S$ is at most k then this is also true for all vertices in the preceding block, by Lemma 2. If $d(u, v) > k$ then search the preceding block plus v for the leftmost minimal interval having distance larger than k to u. Put this vertex into S to be the new u. By Lemma 2 again, it suffices to perform binary search in the block, rather than testing all candidates. Since, by Lemma 3, the distances within the block are bounded by $2 \log n < k$, no further vertex from the same block can be added to S. Thus we need no more than $O(\log \log n)$ distance queries in each of the $n/\log n$ blocks.

Finally, we have to search for the maximum k which admits a k-independent set of size q. For $k \leq 2 \log n$ we need $O(\log \log n)$ steps of binary search, each of duration $O(n)$. The $O(\log n)$ steps with $k > 2 \log n$ need $O(n \log \log n/ \log n)$ time each. $\square$

It seems that we cannot extend the result to strongly chordal graphs, a class including both trees and interval graphs and being closed under graph powers. Even if a strong elimination ordering is already given, it is unlikely that a common PEO of all powers can be efficiently constructed. (The results of [4] discourage such a conjecture.) In particular, a strong elimination ordering itself does not have this property in general; this is wrong even for trees. Moreover, a subquadratic preprocessing that supports fast distance queries in strongly chordal graphs is missing.

4   Extension to Circular-Arc Graphs

Similarly, the results of [6], [9], and [11] can be combined to an $o(n^2)$ dispersion algorithm for circular-arc graphs, provided that the graph is given as a circular-arc family, sorted by their right endpoints. The "left" and "right" endpoint of an arc is understood with respect to the clockwise orientation. In this section we outline such an algorithm.

As shown in [11], a certain extension of the greedy algorithm on interval graphs computes a maximum independent set in a circular-arc graph: First start on an arbitrary vertex u. Scan the circle clockwise and find the next right endpoint of an arc v being not adjacent to u. Continue with $u := v$, and so on. Mark every visited vertex. As soon as some vertex u is visited the second time, the process runs into a "cycle" from which it is quite easy to select a maximum independent set. (The principle was independently discovered also in [12,15].)

A suitable implementation of this algorithm via some auxiliary directed graph runs in $O(n)$ time. It is important to notice that the greedy algorithm uses two things only: the circular ordering of vertices with respect to right endpoints of the representing arcs, and adjacency queries.

In [9], the following has been proven:

Lemma 4. *If G^k is a circular-arc graph, and $k = 1$ or $diam(G^k) \geq 4$, then G^{k+1} is also a circular-arc graph. Moreover, an arc representation of G^{k+1} is obtained from one of G^k by extending the arcs in one direction only, w.l.o.g. counterclockwise.*

It follows that all powers G^k of a circular-arc graph G with $k = 2$ or $k < diam(G)/3 + 1$ have a common circular ordering of the right endpoints. Hence we can find, for any such k, a maximum k-independent set in $O(n)$ time: Apply the greedy algorithm to the given circular ordering, while replacing adjacency queries to G^k with distance queries to G.

Altogether we obtain:

Theorem 3. *A q-dispersion set in a circular-arc graph can be found in time $O(n \log n)$, if the graph is given as an arc family, and $q \geq 7$.*

Proof. Let A be the given arc family with intersection graph G. Check in $O(n)$ time whether A covers the entire circle (otherwise G is an interval graph, and we are done). Let c be the minimum number of arcs from A which cover the whole circle. It is not hard to see that $diam(G) \geq \lfloor c/2 \rfloor$.

Search for the largest k that admits a k-independent set of size at least q. If we consider some $k \geq \max\{3, diam(G)/3 + 1\}$ during our search, we have $k \geq (c-1)/6 + 1$. Since there exists a circle cover of size c, any k-independent set in G has at most $c/(k-1)$ vertices. Trivially, this is at most $c/2$, and for $c \geq 13$, we have $c/(k-1) \leq 6c/(c-1) \leq 6.5$. In either case, these are less than q vertices. Therefore, also the greedy algorithm applied to G^k (which is not guaranteed to work correctly for such k!) cannot supply a k-independent set of size q. On the other hand, it works correctly for $k < diam(G)/3 + 1$, as discussed above. Hence we can use binary search to find the best k, which implies the time bound.   □

Thus we have a very short proof of the $O(n \log n)$ bound, but annoyingly, this approach fails for $q \leq 6$. In order to close this gap, we have to make an additional effort and generalize the maximum independent set algorithm itself. We follow the lines of [11].

Given an arc family A, an arc in A is called a min-arc if it includes no other arc from A. The min-arcs in A form a cyclic ordering in an obvious way, w.l.o.g. clockwise. If A is already sorted, one can recognize the min-arcs and construct this ordering in $O(n)$ time. For any positive integer k, trivially, there exists a maximum k-independent set in the intersection graph G of A that consists of min-arcs only. Hence it suffices to search for such k-independent sets.

For a min-arc v we define $NEXT_k(v)$ to be the first min-arc w following v in the cyclic ordering, such that $d(v, w) > k$. Note that w does not always exists, in that case $NEXT_k(v)$ is undefined. Define a directed graph D_k whose vertices are the min-arcs of A, with directed edges $(v, NEXT_k(v))$. Note that any vertex of D_k has at most one outgoing edge. A moment of thinking reveals that any vertex having an ingoing edge must also have an outgoing edge. It follows that any path starting in a non-isolated vertex leads to a cycle (not in a dead end).

For a min-arc x let $C(x)$ denote the maximal path in D_k starting with x, such that all y in $C(x)$ satisfy $d(x, y) > k$.

Lemma 5. *The sequence $C(x)$ is embedded into the cyclic ordering of min-arcs, and the vertices of $C(x)$ form a k-independent set,*

Proof. For the first assertion, note that no directed edge (of D_k) in $C(x)$ can jump over x in the circle, otherwise we had a contradiction to the definition of D_k.

Assume that $C(x)$ is not k-independent. Then there exist vertices u, v being non-neighbors in $C(x)$ with $d(u, v) \leq k$. Removal of arcs u and v divides the circle in two parts. (In the following assume that u and v do not intersect, that is $d(u, v) > 1$. However case $d(u, v) = 1$ can be settled similarly.) Consider a shortest path P connecting u and v. The union of arcs in P must cover at least one of the two mentioned parts completely. Since u, v are non-neighbors in $C(x)$, w.l.o.g. $w = NEXT_k(u)$ is between u and v in this part and is therefore adjacent to some vertex of P. This contradicts $d(u, w) > k$, unless v is the only vertex of P adjacent to w. But this is also impossible, as v is a min-arc, and w is between u and v on a segment covered by P. Thus we have shown that $C(x)$ is k-independent.   □

Lemma 6. *$C(x)$ is a largest k-independent set containing x.*

Proof. $C(x)$ uniquely partitions the cyclic ordering of min-arcs into segments, each beginning with the next member of $C(x)$. Assume that there is a larger k-independent set $C' \ni x$, consisting of min-arcs only. Some of the mentioned segments beginning with, say, $u \in C(x)$ must contain two vertices $v, w \in C'$, with v before w, thus $u \neq w$. By construction of $C(x)$ we have $d(u, w) \leq k$. If a shortest path from u to w covers v then we get a contradiction to $d(v, w) > k$, as in the previous proof. Hence a shortest path P from u to w must cover x (including case $u = x$ at the moment). Since $x, w \in C'$ we have $d(x, w) > k$ which implies $u \neq x$, and u is the only vertex of P adjacent to x. Again this contradicts the fact that u is a min-arc.   □

If x has an outgoing edge (x, y) then $C(y)$ has at least as many vertices as $C(x)$: Note that $C(x)$ contains y, and $C(y)$ is a largest k-independent set containing y. In particular, for all x within a cycle of D_k, the $C(x)$ have equal size, and some cycle must consist of vertices x such that every $C(x)$ is a maximum k-independent set. Using constant-time distance queries in G, it is now rather straightforward to conclude:

Theorem 4. *A maximum k-independent set in a circular-arc graph can be found in $O(n)$ time, provided that the graph is given as a sorted arc family. Consequently, a q-dispersion set can be found in $O(n \log n)$ time.*

Similarly as in the interval graph case, we can improve this time bound for large enough q.

Theorem 5. *A q-dispersion set in a circular-arc graph can be found in time $O(n \log(n/q^2))$, if the graph is given as a sorted arc family.*

Proof. Compute a greedy covering of the circle in $O(n)$ time. That is, start in an arbitrary point and extend the covered segment counterclockwise, as much as possible, by adding a new arc, until the whole circle is covered. Let c denote the number of arcs in this greedy covering. Any k-independent set has at least $\lfloor c/(k+1) \rfloor$ and at most $c/(k-1)$ vertices. Given q, it suffices to search for the maximum k in a range of $O(c/q^2)$ values.   □

Here it is left open whether the $O(n \log \log n)$ technique for interval graphs can be extended to circular-arc graphs. The difficulty is to find, in sublinear time, some x in a cycle of D_k which yields maximum $C(x)$.

5   Dispersion in Weighted Trapezoid Graphs

For permutation graphs, an $O(n)$ time preprocessing that supports answering distance queries in constant time is known [5]. In trapezoid graphs one has $O(n \log n)$ preprocessing time and $O(\log k)$ answer time if the distance is k [10]. Permutation graphs are contained in the class of trapezoid graphs which is closed under powers [8], and maximum weight independent sets in trapezoid graphs can be found in $O(n \log n)$ time [7]. (The algorithm in [13] is not useful for our purpose, as it is linear in the number of vertices and edges which may be $O(n^2)$.)

However this does not lead to an $o(n^2)$ dispersion algorithm in the same way as in the previous sections, because things do not go well together. Remember that in previous cases (i) powers preserve some common vertex ordering, and (ii) the independent set algorithm used relies on this ordering and adjacency queries in G^k only (which are realized by distance queries in G). The algorithm in [7] violates (ii), since it explicitly uses the coordinates in a trapezoid representation (or equivalently, in a box representation) of the graph.

Nevertheless we obtain an $O(n \log^2 n)$ time dispersion algorithm in weighted trapezoid graphs in another way, without using fast distance queries. The key is the following lemma being implicit in [10].

Lemma 7. *A trapezoid representation of G^k can be constructed in $O(n \log n)$ time from a trapezoid representation of G.*

Together with an $O(n \log n)$ time algorithm for maximum weight independent sets in trapezoid graphs [7], we get immediately:

Theorem 6. *The dispersion problem for trapezoid graphs with positive vertex weights can be solved in $O(n \log^2 n)$ time, if the graph is given as a trapezoid family.*

References

1. B.K. Bhattacharya, M.E. Houle: Generalized maximum independent sets for trees, *Computing - The 3rd Australasian Theory Symposium CATS'97*
2. B.K. Bhattacharya, M.E. Houle: Generalized maximum independent sets for trees in subquadratic time, *10th ISAAC'99, LNCS* 1741, 435-445
3. A. Brandstädt, V.D. Chepoi, F.F. Dragan: Perfect elimination orderings of chordal powers of graphs, *Discrete Math.* 158 (1996), 273-278
4. A. Brandstädt, F.F. Dragan, F. Nicolai: LexBFS-orderings and powers of chordal graphs, *Discrete Math.* 171 (1997), 27-42
5. H.S. Chao, F.R. Hsu, R.C.T. Lee: On the shortest length queries for permutation graphs, *Int. Computer Symp., Workshop on Algorithms 1998*, NCKU Taiwan, 132-138
6. D.Z. Chen, D.T. Lee, R. Sridhar, C.N. Sekharan: Solving the all-pair shortest path query problem on interval and circular-arc graphs, *Networks* (1998), 249-257
7. S. Felsner, R. Müller, L. Wernisch: Trapezoid graphs and generalizations, geometry and algorithms, *Discrete Applied Math.* 74 (1997), 13-32
8. C. Flotow: On powers of m-trapezoid graphs, *Discrete Applied Math.* 63 (1995), 187-192
9. C. Flotow: On powers of circular arc graphs and proper circular arc graphs, *Discrete Applied Math.* 69 (1996), 199-207
10. F.R. Hsu, Y.L. Lin, Y.T. Tsai: Parallel algorithms for shortest paths and related problems on trapezoid graphs, *10th ISAAC'99, LNCS* 1741, 173-182
11. W.L. Hsu, K.H. Tsai: Linear-time algorithms on circular arc graphs, *Info. Proc. Letters* 40 (1991), 123-129
12. D.T. Lee, M. Sarrafzadeh, Y.F. Wu: Minimum cuts for circular-arc graphs, *SIAM J. Computing* 19 (1990), 1041-1050
13. Y.D. Liang: Comparative studies of algorithmic techniques for cocomparability graph families, *Congr. Num.* 122 (1996), 109-118
14. D.J. Rosenkrantz, G.K. Tayi, S.S. Ravi: Capacitated facility dispersion problems, to appear in *J. of Comb. Optimization*
15. S.Q. Zheng: Maximum independent sets of circular-arc graphs: simplified algorithms and proofs, *Networks* 28 (1996), 15-19

Optimizing Cost Flows
by Modifying Arc Costs and Capacities*

Ingo Demgensky, Hartmut Noltemeier, and Hans-Christoph Wirth

University of Würzburg, Department of Computer Science, Am Hubland,
97074 Würzburg, Germany.
{demgensky,noltemei,wirth}@informatik.uni-wuerzburg.de

Abstract. We examine a network upgrade problem for cost flows.
A budget can be distributed among the arcs of the network. An investment on a single arc can be used either to decrease the arc flow cost,
or to increase the arc capacity, or both. The goal is to maximize the flow
through the network while not exceeding bounds on the budget and on
the total flow cost.

The problems are NP-hard even on series-parallel graphs. We provide an
approximation algorithm on series-parallel graphs which, for arbitrary
$\delta, \varepsilon > 0$, produces a solution which exceeds the bounds on the budget
and the flow cost by factors $1+\delta$ and $1+\varepsilon$, respectively, while the amount
of flow is at least that of an optimum solution. The running time of the
algorithm is polynomial in the input size and $1/(\delta\varepsilon)$.

1 Introduction and Related Work

Weighted graphs can be used to model a wide range of problems, e. g. in the
area of transportation, logistics, and telecommunication. While the underlying
instances are static, realistic applications often admit local improvements instead of re-implementing the whole network. This leads to the area of *network
upgrade* problems. All these problems have in common that investing a budget
on parts of the graph allows to change weightings on the graph elements while
the topological structure of the graph remains untouched.

Popular models in the area of network upgrade problems are the *node upgrade*
model, where upgrading a node decreases the length of all incident edges [11,4,5],
and the *edge upgrade* model, where upgrading an edge decreases its length [10,6].
In [7] the authors investigate a cost flow problem where a budget invested on the
arcs can be used to increase the capacity of that arcs. Given a flow value F, the
goal is to compute an upgrade strategy of minimum cost such that the resulting
network admits a flow of at least F. The authors give an FPAS on series-parallel
graphs.

In this paper we investigate a cost flow problem where the budget can be
used to decrease the unit flow costs or to increase the capacity of arcs independently. This model is motivated e. g. in the case of a transportation company

* Supported by the Deutsche Forschungsgemeinschaft (DFG), Grant NO 88/15–3.

U. Brandes and D. Wagner (Eds.): WG 2000, LNCS 1928, pp. 116–126, 2000.

specialized on perishable food. Upgrading the capacity, i. e., using larger transportation containers, involves even the use of larger cooling aggregates such that
the transportation costs per unit of food are not affected. On the other hand,
by investing in the infrastructure one can lower the unit costs without affecting
capacities.

We provide hardness results on series-parallel graphs (Section 3) and give an
approximation algorithm in the spirit of an FPAS (Section 4,5). In Section 6,
we evolve a combined problem—decreasing flow costs and increasing capacities
simultaneously—which turns out to be solvable within essentially the same running time.

2   Preliminaries and Problem Formulation

A flow cost problem is defined by a directed graph $G = (V, R)$ with arc capacities u and arc costs c. An arc function x is called a *(feasible) flow* from s to t
$(s, t \in V)$, if $0 \leq x \leq u$ for each arc, and for each node $v \in V \setminus \{s, t\}$ the inflow
equals the outflow. The *flow value* $F(x)$ is given by the net outflow of node s.
The *cost* of a flow x is given by $\sum_{r \in R} c(r)x(r)$.

We model a scenario where the flow costs can be decreased by investing a
budget on the arcs. To this end, we are given a *price* function p and a *discount*
function d on the set of arcs. An *investment* on the graph is described by an
arc function y with the following meaning: An investment of $y(r)$ units on arc r
costs $y(r)p(r)$, while the flow cost reduce from $c(r)x(r)$ to $\big(c(r) - y(r)d(r)\big)x(r)$.
An additional arc function $c_{\min}$ specifies a lower bound on the flow costs.

Naturally, we have the following restrictions on the functions: $u, c > 0$, $c >
c_{\min}$, and $0 < d \leq c - c_{\min}$. These restrictions are without loss of generality:
The case $c(r) = 0$ can be modeled by setting $c(r) = d(r) = 1$ and $p(r) = 0$.
If $c(r) = c_{\min}(r)$, this means that the flow costs can not be lowered; this can
be modeled by setting $p(r)$ to some large constant which exceeds the available
budget. The same applies to the case $d(r) = 0$.

Definition 1 (FCLP). *An instance of the* flow cost lowering problem *(FCLP)
is given by a directed graph $G = (V, R)$ with terminals $s, t \in V$ and functions
$u, c, c_{\min}, p, d \colon R \to \mathbb{N}_0$ restricted as noted above. Further, we are given constants
$B, C \in \mathbb{N}$ as limits on the upgrade budget and the flow cost, respectively.*

*The goal is to find a solution $\sigma = (x, y)$, specified by a feasible flow x and an
upgrade strategy y on the arcs, such that*

1. *(upgrade investment)* $B(\sigma) := \sum_{r \in R} y(r)p(r) \leq B$,
2. *(flow cost)* $\qquad C(\sigma) := \sum_{r \in R} \big(c(r) - y(r)d(r)\big)x(r) \leq C$,
3. *(flow value)* $\qquad F(\sigma) := F(x)$ *is maximized.*

Using a canonical notation for multi-criteria optimization problems [8], FCLP
is denoted by *(flow value: max, upgrade investment: upper bound, flow cost:
upper bound)*. A solution σ is called *feasible*, if $B(\sigma) \leq B$ and $C(\sigma) \leq C$. An
approximation algorithm with *performance* $(\alpha_1, \alpha_2, \alpha_3)$ is a polynomial running
time algorithm, which returns an *almost feasible* solution (i. e., a solution with

$B(\sigma) \leq \alpha_2 \cdot B$ and $C(\sigma) \leq \alpha_3 \cdot C)$ with $F(\sigma) \geq 1/\alpha_1 \cdot F(\sigma^*)$, where σ^* is an optimal feasible solution. If there is no feasible solution, the algorithm is free to provide this information or to return any almost feasible solution.

For multi-criteria optimization problems, by interchanging bounds and optimization objective one gets a *family of related problems*. Results on the hardness carry over to the related problems. Moreover, for integral objectives as in the case of FCLP, even approximability factors carry over to related problems [12,9, 8,1]. For bicriteria problems, related problems are also known as *dual* problems.

A *series-parallel graph* is defined by the following rules: The single arc (s,t) is a series-parallel graph with two *terminals*, the *source* s and the *target* t. For series-parallel graphs G_1, G_2 with sources s_1, s_2 and targets t_1, t_2, respectively, a new series-parallel graph with source s_1 and target t_2 can be obtained by either identifying $t_1 \equiv s_2$ (series composition), or identifying $s_1 \equiv s_2$ and $t_1 \equiv t_2$ (parallel composition).

For graph $G = (V, R)$, we use the abbreviations $n := |V|$ and $m := |R|$ where appropriate.

3   Hardness Results

In this section we show the hardness of FCLP and provide lower bounds on the approximability.

Theorem 1 (Non-approximability of FCLP). *For any (polynomial time computable) function $\alpha(n)$, the existence of an $(1, 1, \alpha(n))$-approximation algorithm for FCLP on series-parallel graphs with n nodes implies $P = NP$.*

Proof. We show the hardness by a reduction from KNAPSACK [3, Problem MP9]. An instance of KNAPSACK is given by a finite set $A = \{a_1, \ldots, a_k\}$ of items, each of weight $w(a_i) \geq 0$ and value $v(a_i) \geq 0$, and two numbers $W, V \in \mathbb{N}$. It is NP-complete to decide whether there is a subset $A' \subseteq A$ such that $w(A') \leq W$ and $v(A') \geq V$.

Assume there is a $(1, 1, \alpha)$-approximation algorithm for FCLP. Given an instance of KNAPSACK, we construct a graph with vertex set $\{s, t\}$ joined by k parallel arcs. For item a_i, arc r_i has capacity $u(r_i) := v(a_i)$, and price $p(r_i) := w(a_i)$. Further, for all arcs set the initial flow cost $c(r) := \alpha \cdot V + 1$, the minimal flow cost $c_{\min}(r) := 1$, and the discount $d(r) := \alpha \cdot V$. Finally, set the flow cost constraint $C := V$, and the budget constraint $B := W$.

We claim that there is a solution of the FCLP instance with flow value at least V if and only if there is a solution to the KNAPSACK instance.

Assume that there is a solution of FCLP with flow value $F \geq V$ obeying both constraints. Let R' be the set of upgraded arcs, choose A' to be the corresponding set of items. Due to the flow cost constraint it is easy to see that the solution of FCLP uses only upgraded arcs. Then $\sum_{a' \in A'} v(a') = \sum_{r' \in R'} u(r') \geq F \geq V$ and $\sum_{a' \in A'} w(a') = \sum_{r' \in R'} p(r') \leq B = W$, therefore A' is a solution for KNAPSACK. Conversely, constructing a solution for FCLP out of a solution of

KNAPSACK is easily achieved by upgrading the arcs corresponding to the items of the solution.

Since the flow cost of an arc r which is not upgraded exceeds $\alpha \cdot V$, any $(1, 1, \alpha)$-approximation algorithm for FCLP must in fact solve the underlying KNAPSACK problem exactly. □

Theorem 2 (Non-approximability of FCLP). *For any $\alpha < \ln(n/2 - 1)$, the existence of a $(1, \alpha, 1)$-approximation algorithm for FCLP on bipartite graphs with n nodes implies $NP \subseteq DTIME(N^{O(\log \log N)})$.*

Proof. The proof uses a reduction from DOMINATING SET (DS) [3, Problem GT2]. An instance of DS is given by a graph $G = (V, E)$ and a number $K \in \mathbb{N}$. A subset $V' \subseteq V$ is called *dominating*, if each node in V is either contained in V' or incident to a node from V'. It is NP-complete to decide whether G admits a dominating set of size at most K.

Given an instance $G = (V, E)$ of DS, construct a graph G' with node set $V_1' \cup V_2' \cup \{s, t\}$, where each V_i' is a copy of V. Insert an arc between $v \in V_1'$ and $w \in V_2'$ if and only if w is dominated by v in the original graph. For each $v \in V_1'$ insert an arc (s, v) of cost $c := 2$, discount $d := 1$, minimum cost $c_{\min} := 1$ and price $p := 1$. For each $w \in V_2'$ add an arc (w, t) of capacity $u := 1$. If not specified yet, set the capacity on the remaining arcs to $u := n$, and the costs and prices to zero. Notice that the resulting graph is bipartite, and only the arcs incident to s cause flow costs.

It is easy to see that a solution for FCLP with budget constraint B, flow cost constraint n, and flow value n implies a dominating set in the original graph of size at most B. Conversely, a dominating set of size B implies the existence of a solution of FCLP for budget constraint B and flow cost constraint n which has flow value n.

Assume that there is a $(1, \alpha, 1)$-approximation algorithm for FCLP. By trying all n instances with budget constraint $B = 1, \ldots, n$ and flow cost constraint n and taking the best solution, we get an α-approximation algorithm for Minimum Dominating Set. Using the result of Feige [2] and the fact that the number of nodes in the constructed graph satisfies $n' = 2n + 2$, we can conclude that a $(1, \ln(n/2 - 1), 1)$-approximation algorithm for FCLP on bipartite graphs with n nodes implies $NP \subset \text{DTIME}(N^{O(\log \log N)})$. □

4   Algorithm for FCLP on Series-Parallel Graphs

In this section we provide a pseudopolynomial time algorithm which solves FCLP on series-parallel graphs exactly. The main idea is to find a distribution of the given budget and flow cost on the arcs such that the resulting flow value is maximized.

To this end, we define

$$\text{OPT}_G(\beta, \gamma)$$

to be the flow value $F(\sigma^*)$ of an optimal solution σ^* for FCLP with budget limit β and cost limit γ on graph G. We give an algorithm

$$A(G, u, c, d, p; B, C)$$

which computes the value $\mathrm{OPT}_G(B, C)$ and the corresponding upgrade strategy and flow.

First we compute a decomposition tree of the series-parallel graph G. Schoenmakers has shown that in time $O(n+m)$ one can compute such a decomposition tree or give the information that the graph is not series-parallel [13]. Then, an optimal solution can be built by traversing the tree bottom-up. Notice that the size of the decomposition tree is in $O(m) \subseteq O(n)$.

For a single arc r, the numbers $\mathrm{OPT}_r(\beta, \gamma)$ for all values $0 \le \beta \le B$ and $0 \le \gamma \le C$ can be determined in total running time $O(BC)$: Since

$$y(r) = \min \left\{ \left\lfloor \frac{\beta}{p(r)} \right\rfloor, \left\lfloor \frac{c(r) - c_{\min}(r)}{d(r)} \right\rfloor \right\},$$

it follows that

$$\mathrm{OPT}_r(\beta, \gamma) = x(r) = \min \left\{ \left\lfloor \frac{\gamma}{c(r) - d(r)y(r)} \right\rfloor, u(r) \right\}.$$

Notice that the above expressions yield finite numbers even if one of the denominators equals zero.

If G is the series composition of two subgraphs G_1 and G_2, then

$$\mathrm{OPT}_G(\beta, \gamma) = \max_{\substack{0 \le i \le \beta \\ 0 \le j \le \gamma}} \min\{\mathrm{OPT}_{G_1}(i, j), \mathrm{OPT}_{G_2}(\beta - i, \gamma - j)\}.$$

If G is the parallel composition of G_1 and G_2, then

$$\mathrm{OPT}_G(\beta, \gamma) = \max_{\substack{0 \le i \le \beta \\ 0 \le j \le \gamma}} \mathrm{OPT}_{G_1}(i, j) + \mathrm{OPT}_{G_2}(\beta - i, \gamma - j).$$

Assuming that the numbers $\mathrm{OPT}_{G_\nu}(i, j)$ on the subgraphs are already computed for all values $0 \le i \le \beta$ and $0 \le j \le \gamma$, the running time for calculating $\mathrm{OPT}_G(\beta, \gamma)$ is in $O(\beta\gamma)$ in both cases. Since the decomposition tree of G consists of $O(m)$ nodes, the total running time for computing number $\mathrm{OPT}_G(B, C)$ (and simultaneously keeping track of the corresponding flow and upgrade strategy) is in $O(mB^2C^2)$.

5   Polynomial Time Approximation Algorithm

In this section we use a scaling technique derived from [7] to achieve a polynomial time approximation algorithm for FCLP with performance guarantee $(1, 1+\delta, 1+\varepsilon)$ for arbitrary $\delta, \varepsilon > 0$.

The running time of the algorithm presented in the previous section is in $O(mB^2C^2)$ when called with bounds B and C on the investment budget and total flow cost. In order to achieve a solution $\sigma = (x, y)$ in polynomial time we scale the bounds to polynomial size. To this end it is also necessary to scale the costs and prices on the arcs by appropriate factors. The budget and cost value of σ on the original (unscaled) instance exceed the given bounds by small factors due to the rounding errors arising from the scaling process. In exchange to this inaccuracy the resulting flow value is at least the optimum flow value.

As a preprocessing step, we first set up

$$\hat{x}(r) := u(r) \quad \text{and} \quad \hat{y}(r) := \left\lfloor \left| \frac{c(r) - c_{\min}(r)}{d(r)} \right| \right\rfloor,$$

and $c_r(r) := c(r)\hat{x}(r)\hat{y}(r)$, $d_r(r) := d(r)\hat{x}(r)\hat{y}(r)$, and $p_r(r) := p(r)\hat{y}(r)$. As we will see later, this modification of the scaling technique from [7] helps to make the rounding error on each arc independent of the size of investment and flow on an arc.

Algorithm A is modified as follows. To compute the value $\text{OPT}_r(\beta, \gamma)$ for a single arc r, it computes

$$y(r) = \min \left\{ \left\lfloor \left| \frac{\beta \hat{y}(r)}{p_r(r)} \right| \right\rfloor, \hat{y}(r) \right\} \tag{1}$$

and hence

$$\text{OPT}_r(\beta, \gamma) = x(r) = \min \left\{ \left\lfloor \frac{\gamma \hat{x}(r)\hat{y}(r)}{c_r(r) - d_r(r)y(r)} \right\rfloor, \hat{x}(r) \right\}. \tag{2}$$

The scaling is performed as follows. Given $\delta, \varepsilon > 0$, choose $\varepsilon' := 3/4 \cdot \varepsilon$ and set

$$c_r^C = \begin{cases} \left\lceil \frac{c_r}{\frac{C\varepsilon'}{3m}} \right\rceil = \left\lceil \frac{3mc_r}{C\varepsilon'} \right\rceil & \text{if } C \neq 0, \\ c_r & \text{otherwise,} \end{cases}$$

$$d_r^C = \begin{cases} \left\lceil \frac{d_r}{\frac{C\varepsilon'}{3m}} \right\rceil = \left\lceil \frac{3md_r}{M\varepsilon'} \right\rceil & \text{if } C \neq 0, \\ d_r & \text{otherwise,} \end{cases}$$

$$p_r^B = \begin{cases} p_r & \text{if } B = 0, \\ \left\lfloor \left(1 + \frac{1}{\delta}\right) 3m \right\rfloor \hat{y} + 1 & \text{if } p > B \wedge B \neq 0, \\ \left\lceil \frac{3mp_r}{B\delta} \right\rceil & \text{otherwise.} \end{cases}$$

Notice that $c_r^C, d_r^C > 0$, since $m, c_r, d_r, \varepsilon' > 0$. Moreover, $p_r^B = 0 \iff p_r = 0$, since $m, \delta > 0$. Algorithm A is then called with parameter list

$$A(G, \hat{x}, \hat{y}, c_r^C, d_r^C, p_r^B; \lfloor (1 + 1/\delta)\, 3m \rfloor, \lfloor (1 + 1/\varepsilon')\, 3m \rfloor).$$

Notice that the running time of the algorithm is now in $O(m^5(1 + \frac{1}{\delta})^2(1 + \frac{4}{3\varepsilon})^2)$.

The following lemma shows that the solution exceeds the given bounds on the investment budget and the flow cost only by small factors.

Lemma 1. *For any solution σ returned by the algorithm A, we have $B(\sigma) \leq (1 + \delta)B$ and $C(\sigma) \leq (1 + \varepsilon)C$.*

Proof. We have

$$B(\sigma) = \sum_{r \in R} yp = \sum_{r \in R} y\frac{p_r}{\hat{y}} \leq \frac{B\delta}{3m} \sum_{r \in R} y\frac{p_r^B}{\hat{y}}$$

$$\leq \frac{B\delta}{3m} \left(1 + \frac{1}{\delta}\right) 3m = (1 + \delta)B \,.$$

On the other hand,

$$C(\sigma) = \sum_{r \in R} x\frac{c_r - d_r y}{\hat{x}\hat{y}} \leq \sum_{r \in R} x\frac{\frac{C\varepsilon'}{3m}c_r^C - \frac{C\varepsilon'}{3m}(d_r^C - 1)y}{\hat{x}\hat{y}}$$

$$= \frac{C\varepsilon'}{3m} \sum_{r \in R} \left(\frac{c_r^C x - d_r^C xy}{\hat{x}\hat{y}} + \frac{xy}{\hat{x}\hat{y}}\right)$$

$$\leq \frac{C\varepsilon'}{3m} \left(\left(1 + \frac{1}{\varepsilon'}\right) 3m + m\right) = \left(1 + \frac{4}{3}\varepsilon'\right) C = (1 + \varepsilon)C \,.$$

At this point, we profit from the preprocessing step, since the rounding error per arc, $xy/\hat{x}\hat{y} \leq 1$, is independent of y. □

It remains to show that the overall flow value is at least the flow value of an optimum solution.

Lemma 2. *For any solution σ returned by the algorithm A, we have $F(\sigma) \geq OPT_G(B, C)$.*

Proof. Let $\sigma^* = (x^*, y^*)$ be an optimal solution of the original (unscaled) instance. (Notice that FCLP always has a feasible solution, e.g. $(x \equiv 0, y \equiv 0)$.) We show that there is a solution $\sigma = (x, y)$ on the scaled instance with $x \geq x^*$ and $y \geq y^*$. Since the algorithm computes all possible distributions of the budget and the flow costs, it suffices to give one appropriate distribution (β, γ) of budget and flow cost. The detailed proof is of rather technical nature and can be found in the appendix. □

The following theorem summarizes the results of the current section.

Theorem 3 (Approximability of FCLP). *For any $\delta, \varepsilon > 0$, there is an approximation algorithm for FCLP with performance $(1, 1 + \delta, 1 + \varepsilon)$. Its running time is in $O(m^5(1 + \frac{1}{\delta})^2(1 + \frac{4}{3\varepsilon})^2)$.* □

6   Extension to Augmentable Arc Capacities

In this section we extend FCLP to cover a situation where the invested budget can also be used to augment the capacities of the arcs. The resulting problem is called *flow cost lowering capacity augmenting problem (FCLCAP)*.

An instance of FCLCAP is given by a directed graph $G = (V, R)$ with terminals $s, t \in V$ and functions $u, u_{max}, c, c_{min}, p_d, d, p_a, a \colon R \to \mathbb{N}_0$. The meaning of the parameter functions is similar to those given in Definition 1. u_{max} specifies an upper bound beyond which the capacity of an arc can not be increased. Analogously to the function pair p_d/d, the function p_a specifies the price for augmenting the arc capacity by an amount of a. Note that the available budget can be split arbitrarily to decrease the flow cost or increase the capacity.

The goal is to find a solution $\sigma = (x, y_a, y_d)$, specifying a feasible flow x and investment strategies y_a, y_d for increasing capacities and lowering flow costs, respectively, such that

1. (investment) $B(\sigma) := \sum_{r \in R} y_a(r) p_a(r) + y_d(r) p_d(r) \leq B$,
2. (flow cost) $C(\sigma) := \sum_{r \in R} \big(c(r) - y_d(r) d(r) \big) x(r) \leq C$,
3. (flow value) $F(\sigma) := F(x)$ is maximized.

Notice that the condition on the feasibility of x is changed to $x(r) \leq u(r) + y_a(r) \cdot a(r)$. The canonical notation of FCLCAP is *(flow value: max, upgrade and augmentation investment: upper bound, flow cost: upper bound)*.

It is easy to see that FCLCAP contains FCLP as a special case. Therefore all hardness and non-approximability results immediately carry over to FCLCAP.

Theorem 4 (Approximability of FCLCAP). *For any $\delta, \varepsilon > 0$, there is an approximation algorithm for FCLCAP with performance $(1, 1 + \delta, 1 + \varepsilon)$. Its running time is in $O(m^5 (1 + \frac{1}{\delta})^2 (1 + \frac{6}{5\varepsilon})^2)$.*

Proof. The proof is similar to the proof given in Section 5. The main difference consists in the behavior of the underlying pseudopolynomial algorithm A. The computation of $\mathrm{OPT}_r(\beta, \gamma)$ for a leaf r of the decomposition tree is modified as follows: With

$$y_a(r) = \min \left\{ \left\lfloor \frac{\beta_a}{p_a(r)} \right\rfloor, \left\lfloor \frac{u_{max}(r) - u(r)}{a(r)} \right\rfloor \right\} \quad \text{and}$$

$$y_d(r) = \min \left\{ \left\lfloor \frac{\beta_d}{p_d(r)} \right\rfloor, \left\lfloor \frac{c(r) - c_{min}(r)}{d(r)} \right\rfloor \right\},$$

we have

$$\mathrm{OPT}_r(\beta, \gamma) = x(r) = \max_{\beta_a + \beta_d = \beta} \min \left\{ \left\lfloor \frac{\gamma}{c(r) - d(r) y_d(r)} \right\rfloor, u(r) + a(r) y_a(r) \right\}.$$

The computations for interior nodes of the decomposition tree are exactly the same as described above.

After changing the definition of $\hat{x}$ to

$$\hat{x}(r) := \left\lfloor \frac{u_{max}(r) - u(r)}{a(r)} \right\rfloor a(r) + u(r),$$

the scaling technique outlined in the previous section can be applied to FCLCAP straightforward. This leads to the stated result. $\qquad \square$

7   Conclusions and Further Remarks

We have given approximation algorithms for FCLP and FCLCAP with performance $(1, 1 + \delta, 1 + \varepsilon)$ for arbitrary $\delta, \varepsilon > 0$ and running time polynomial both in the input size and in $1/(\delta\varepsilon)$. Analog results we could show for the related problems with interchanged optimization objectives.

On the other hand, we have shown that $(1, 1, \alpha(n))$ is a lower bound on the approximability of FCLP and FCLCAP on series-parallel graphs for any (polynomial time computable) function $\alpha(n)$. A lower bound of $(1, \ln(n/2 - 1), 1)$ could be shown to hold on bipartite graphs, where n denotes the number of nodes in the graph. All results rely on the common conditions $P \neq NP$ or $NP \not\subseteq DTIME(N^{O(\log \log N)})$, respectively.

In realistic applications often problems arise where upgrading of an arc is not performed continuously but in one single step. There are examples where this is true for arc capacities as well as for arc flow costs. In our model, these cases are handled by restricting y_a or y_d to take on values only from $\{0, 1\}$ by setting $u_{\max}$ and $c_{\min}$ appropriately. Therefore the approximation results also carry over to those 0/1-cases.

References

1. I. Demgensky, *Netzwerkoptimierung – Flussausbauprobleme, Flusskostensenkungsprobleme (network optimization by augmenting capacities and decreasing flow costs, in German)*, Diploma Thesis, University of Würzburg, January 2000.
2. U. Feige, *A threshold of* ln *n for approximating set cover*, Proceedings of the 28th Annual ACM Symposium on the Theory of Computing (STOC'96), 1996, pp. 314–318.
3. M. R. Garey and D. S. Johnson, *Computers and intractability (a guide to the theory of NP-completeness)*, W.H. Freeman and Company, New York, 1979.
4. S. O. Krumke, M. V. Marathe, H. Noltemeier, R. Ravi, S. S. Ravi, R. Sundaram, and H. C. Wirth, *Improving spanning trees by upgrading nodes*, Theoretical Computer Science **221** (1999), no. 1–2, 139–156.
5. S. O. Krumke, M. V. Marathe, H. Noltemeier, S. S. Ravi, and H.-C. Wirth, *Upgrading bottleneck constrained forests*, Discrete Applied Mathematics, to appear, 2000.
6. S. O. Krumke, H. Noltemeier, S. S. Ravi, M. V. Marathe, and K. U. Drangmeister, *Modifying networks to obtain low cost subgraphs*, Theoretical Computer Science **203** (1998), no. 1, 91–121.
7. S. O. Krumke, H. Noltemeier, R. Ravi, S. Schwarz, and H.-C. Wirth, *Flow improvement and flows with fixed costs*, Proceedings of the International Conference of Operations Research Zürich (OR'98), Editors: H.-J. Lüthi and P. Kall, Operations Research Proceedings, Springer, 1999.
8. S. O. Krumke, *On the approximability of location and network design problems*, Ph.D. thesis, Lehrstuhl für Informatik I, Universität Würzburg, December 1996.
9. M. V. Marathe, R. Ravi, R. Sundaram, S. S. Ravi, D. J. Rosenkrantz, and H. B. Hunt III, *Bicriteria network design problems*, Proceedings of the 22nd International Colloquium on Automata, Languages and Programming (ICALP'95), Lecture Notes in Computer Science, vol. 944, 1995, pp. 487–498.

10. C. Phillips, *The network inhibition problem*, Proceedings of the 25th Annual ACM Symposium on the Theory of Computing (STOC'93), May 1993, pp. 776–785.
11. D. Paik and S. Sahni, *Network upgrading problems*, Networks **26** (1995), 45–58.
12. R. Ravi, M. V. Marathe, S. S. Ravi, D. J. Rosenkrantz, and H. B. Hunt III, *Many birds with one stone: Multi-objective approximation algorithms*, Proceedings of the 25th Annual ACM Symposium on the Theory of Computing (STOC'93), May 1993, pp. 438–447.
13. B. Schoenmakers, *A new algorithm for the recognition of series parallel graphs*, Tech. Report CS-R9504, Centrum voor Wiskunde en Informatica (CWI), Amsterdam, 1995.

8 Appendix

Proof (of Lemma 2). For shorter notation we omit parameter r, i. e. we write f instead of $f(r)$ for all functions f where there are no ambiguities. Assume

$$\beta(r) := \left\lceil \frac{p_r^B}{\hat{y}} y^* \right\rceil$$

for each arc $r \in R$. Clearly, $\hat{y} \geq y^*$. On the other hand, if $p_r^B > 0$,

$$\left\lfloor \frac{\beta \hat{y}}{p_r^B} \right\rfloor = \left\lfloor \frac{\left\lceil \frac{p_r^B}{\hat{y}} y^* \right\rceil}{p_r^B} \hat{y} \right\rfloor \geq \left\lfloor \frac{\frac{p_r^B}{\hat{y}} y^*}{p_r^B} \hat{y} \right\rfloor = y^*.$$

Hence, by (1) it follows $y \geq y^*$.

Assume

$$\gamma(r) := \left\lceil \frac{x^* \left(c_r^C - d_r^C y^* \right)}{\hat{x} \hat{y}} \right\rceil$$

for each $r \in R$. Clearly, $\hat{x} \geq x^*$. Moreover,

$$\left\lfloor \frac{\gamma \hat{x} \hat{y}}{c_r^C - d_r^C y} \right\rfloor = \left\lfloor \frac{\left\lceil \frac{x^* (c_r^C - d_r^C y^*)}{\hat{x} \hat{y}} \right\rceil \hat{x} \hat{y}}{c_r^C - d_r^C y} \right\rfloor \overset{y \geq y^*}{\geq} \left\lfloor \frac{\frac{x^* (c_r^C - d_r^C y^*)}{\hat{x} \hat{y}} \hat{x} \hat{y}}{c_r^C - d_r^C y^*} \right\rfloor = x^*.$$

By (2), we have $x \geq x^*$. Since the source s has no incoming arcs, this implies $F(\sigma) \geq F(\sigma^*) = \mathrm{OPT}_G(B, C)$.

It remains to show that σ is a feasible solution on the scaled instance.

$$B(\sigma) = \sum_{r \in R} \beta = \sum_{\substack{r \in R \\ \beta \neq 0}} \beta = \sum_{\substack{r \in R \\ \beta \neq 0}} \left\lceil \frac{p_r^B}{\hat{y}} y^* \right\rceil$$

$$\leq \sum_{\substack{r \in R \\ \beta \neq 0}} \left(\frac{\frac{3 m p_r}{B \delta} + 1}{\hat{y}} y^* + 1 \right) = \sum_{\substack{r \in R \\ \beta \neq 0}} \left(\frac{3 m p_r y^*}{B \delta \hat{y}} + \frac{y^*}{\hat{y}} + 1 \right)$$

$$\leq \frac{3m}{\delta} \left(\sum_{\substack{r \in R \\ \beta \neq 0}} \frac{p_r y^*}{B \hat{y}} \right) + 2m \leq \frac{3m}{\delta} + 2m \leq \left\lfloor \frac{3m}{\delta} + 3m \right\rfloor = \left\lfloor (1 + \frac{1}{\delta}) 3m \right\rfloor.$$

On the other hand,

$$
\begin{aligned}
C(\sigma) = \sum_{r \in R} \gamma &= \sum_{r \in R} \left\lceil \frac{x^* c_r^C - d_r^C x^* y^*}{\hat{x}\hat{y}} \right\rceil \\
&\leq \sum_{r \in R} \left(\frac{\left(\frac{3mc_r}{C\varepsilon'} + 1\right) x^*}{\hat{x}\hat{y}} - \frac{\frac{3mc_r}{C\varepsilon'} x^* y^*}{\hat{x}\hat{y}} + 1 \right) \\
&= \sum_{r \in R} \left(\frac{3mc_r x^*}{C\varepsilon'\hat{x}\hat{y}} + \frac{x^*}{\hat{x}\hat{y}} - \frac{3md_r x^* y^*}{C\varepsilon'\hat{x}\hat{y}} + 1 \right) \\
&\leq \frac{3m}{\varepsilon'} \left(\sum_{r \in R} \frac{c_r x^* - d_r x^* y^*}{C\hat{x}\hat{y}} \right) + 2m \\
&= \frac{3m}{\varepsilon'} + 2m \leq \left\lfloor \frac{3m}{\varepsilon'} + 3m \right\rfloor = \left\lfloor (1 + \frac{1}{\varepsilon'})3m \right\rfloor .
\end{aligned}
$$

This shows the claim. $\square$

Update Networks and Their Routing Strategies

Michael J. Dinneen and Bakhadyr Khoussainov

Department of Computer Science, University of Auckland, Auckland, New Zealand.
{mjd,bmk}@cs.auckland.ac.nz

Abstract. We introduce the notion of update networks to model communication networks with infinite duration. In our formalization we use bipartite finite graphs and game-theoretic terminology as an underlying structure. For these networks we exhibit a simple routing procedure to update information throughout the nodes of the network. We also introduce an hierarchy for the class of all update networks and discuss the complexity of some natural problems.

1 Introduction

A network can be viewed as a finite directed graph whose nodes represent a type of system (e.g., processors, data repositories, and servers) and edges represent communication channels between the nodes. The process of communication in a network is essentially an infinite duration process. The interacting network has to be robust (e.g., be fault-tolerant and free of dead-locks). We are interested in those networks that possess liveness and fairness properties. Liveness means that all nodes are actively participating in the collective system. Fairness means that critical information of the dynamic network is distributed to all of the nodes.

To give an example, suppose we have data stored on each node of a network and we want to continuously update all nodes with consistent data. For instance, we are interested in addressing redundancy issues in distributed databases. Often one requirement is to share key information between all nodes of the distributed database. We can do this by having a data packet of current information continuously go through all nodes of the network. We call a network of this type an update network.

The operation of an update packet over time enters only a finite number of nodes and produces an infinite sequence $v_1, v_2, \ldots$, called a *run-time sequence*. Since the number of nodes is finite, some of the nodes, called *persistent nodes*, appear infinitely often in the run-time sequence. The success of a run-time sequence is determined by whether or not the sequence of nodes satisfies the liveness and fairness properties. That is, the set of persistent nodes coincides with the set of all nodes of the network. We can regard the run-time sequences as plays of a two-player game where one player, called *Survivor*, tries to ensure that persistent states span the network and the other player, called *Adversary*, does not.

One possible model for an update network is thus based on a finite directed graph as the underlying structure for games between Survivor and Adversary. We formalize this in the following definition, where for this paper we restrict our attention to the bipartite case.

U. Brandes and D. Wagner (Eds.): WG 2000, LNCS 1928, pp. 127–136, 2000.

Definition 1. *An* update (bipartite) game *is a bipartite graph* $G = (V, E)$ *with partite sets* S *and* A. *There are two players, called Survivor and Adversary, which control vertices* S *and* A, *respectively. We stipulate that the out-degree of every vertex is nonzero.*

The game rules allow moves along edges of G. Each *play* of an infinite duration game is a sequence of vertices $v_0, v_1, \ldots, v_i, \ldots$ such that the game rules are followed. We call a finite prefix sequence of a play a *history*. We say that a vertex v is *visited* if it occurs in the play. Note that either Survivor or Adversary may begin the play. Survivor wins a play if the persistent vertices of the play is the set V, otherwise Adversary wins. A *strategy* for Survivor is a function H from play histories $v_0, \ldots, v_i$, where $v_i \in S$, to the set A such that $(v_i, H(v_0, \ldots, v_i))$ is an edge. A strategy for Advisory is similarly defined.

A given strategy for a player may either win or lose the game when starting at an *initial vertex* v_0, where $v_0 \in V$. A player's *winning strategy* for an initial vertex is one that wins no matter what the other player does.

Example 1. In Figure 1 we present a game $\mathcal{G}$, where $S = \{1, 3, 4\}$, and $A = \{0, 2, 5\}$. Here Adversary has a winning strategy from any starting node. If whenever node 0 is reached, Adversary moves to node 4. Thus, node 1 is only reached at most once during any play.

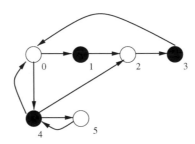

Fig. 1. Example of an update game.

We now formally define update networks.

Definition 2. *An* update game is an update network *if Survivor has a winning strategy for every initial vertex.*

Update networks can also be viewed as message passing networks where only Survivor's nodes actively participate in the updating process, while Adversary's nodes only need to passively forward a message.

Example 2. The graph displayed below in Figure 2 is an update network. Here, no matter what Adversary does node 0 is visited. Survivor then can alternate moving between nodes 2 and 3 to span the network.

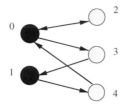

Fig. 2. A simple example of an update game which is an update network.

We end this section with a few related references. Previous work on two-player infinite duration games on finite bipartite graphs is presented in the paper by McNaughton [2] and extended by Nerode *et al.* [3]. Our work focuses on a subclass of the games considered by these authors. Nerode *et al.* provide an algorithm for deciding McNaughton games. Their algorithm runs in exponential time of the graph size for certain inputs. For our games we provide two simple polynomial time algorithms for deciding update networks, partially based on the structural properties of the underlying graphs. We also note that several earlier papers have dealt with finite duration games on automata and graphs (e.g., see [4,5]).

2   Constructing Update Networks

We now want to present several construction techniques for building update networks from smaller update networks. We introduce three primitive operations below for update networks $G_1 = (S_1, A_1, E_1)$ and $G_2 = (S_2, A_2, E_2)$.

Enrichment: Take G_1 and add edges E' to E_1 where all new edges are directed from S_1 to A_1. The resulting graph is called an *enrichment* of G_1, denoted $G_1 \cup E'$.

Identification: Take G_1 and G_2 and subsets T_1 of S_1 and T_2 of S_2 with the same cardinality. Let f be a bijection between T_1 and T_2. Build the graph $(S_1 \cup S_2 \setminus T_2, A_1 \cup A_2, E_1 \cup E)$ where E is E_2 with all vertices in T_2 replaced with their image under f. This graph is called the *identification of G_1 and G_2 under f*, denoted by $G_1 \oplus_f G_2$.

Extension: Take G_1 and G_2 then construct the graph $(S = S_1 \cup S_2, A = A_1 \cup A_2, E_1 \cup E_2 \cup E)$ where E is a new set of edges directed from S to A with at least one edge from S_1 to A_2 and at least one edge from S_2 to A_1. This graph is called the *extension of G_1 by G_2 via E*, denoted $G_1 \cup G_2 \cup E$.

It is easy to observe the following.

Proposition 1. *The enrichment, identification, extension operators applied to update networks yield update networks.*

Example 3. The graphs displayed below in Figure 3 are obtained by applying the operations of enrichment, join and extension to the update networks B_1 and B_2. The black and white vertices are Survivor and Adversary nodes, respectively.

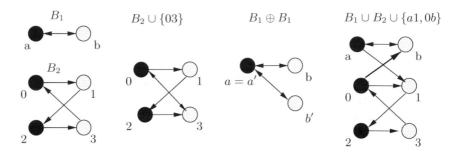

Fig. 3. Illustrating the operations of enrichment, join and extension.

3   Deciding Update Networks

In this section we provide an important operation, called the contraction operation, that applied to update games produces update networks if and only if the original update game is an update network. The operation reduces the size of the underlying bipartite graph, and hence produces a decision procedure for checking if a given update game is an update network.

We can easily characterize those bipartite update networks with only one Survivor vertex. These are the bipartite graphs where out-degree$(s) = |A|$ for the single Survivor vertex s. We now mention several properties for all bipartite update networks [1].

Lemma 1. *If $(V = A \cup S, E)$ is a bipartite update network then for every vertex $s \in S$ there exists at least one $a \in A$ such that $(a, s) \in E$ and out-degree$(a) = 1$*

Proof. Let s be a vertex that does not satisfy the statement of the lemma. Let $A_s = \{a \mid (a, s) \in E\}$. Then out-degree$(a) > 1$ for all $a \in A_s$. Adversary has the following winning strategy. If the play history ends in a then since out-degree$(a) > 1$, Adversary moves to s', where $s' \neq s$ and $(a, s') \in E$. This contradicts the assumption of lemma that we have an update network.   □

For any Survivor vertex s define Forced$(s) = \{a \mid$ out-degree$(a) = 1$ and $(a, s) \in E\}$, which denotes the set of Adversary vertices that are 'forced' to move to s.

Lemma 2. *If B is a bipartite update network such that $|S| > 1$ then for every $s \in S$ there exists an $s' \neq s$ and an $a \in$ Forced(s), such that (s', a, s) is a directed path.*

Proof. If B has more than one vertex in S then there must be a strategy for Survivor to create a play history to visit vertex s from some other vertex $s' \in S$. To do this we need a forced Adversary vertex a (of A) in the neighborhood of s'. There exists such a vertex a by Lemma 1.   □

Definition 3. *Given a bipartite graph $(S \cup A, E)$ a forced cycle is a (simple) cycle $(a_k, s_k, \ldots, a_2, s_2, a_1, s_1)$ for $a_i \in Forced(s_i)$ and $s_i \in S$.*

The following lemma is the main ingredient in characterizing bipartite update networks.

Lemma 3. *If B is a bipartite update network such that $|S| > 1$ then there exists a forced cycle of length at least 4.*

Proof. Take $s_1 \in S$. From Lemma 2 there exists a path (s_2, a_1, s_1) in B such that $s_2 \neq s_1$ and $a_1 \in Forced(s_1)$. Now for s_2 we apply the lemma again to get a path (s_3, a_2, s_2) in B such that $s_3 \neq s_2$ and $a_2 \in Forced(s_2)$. If $s_3 = s_1$ we are done. Otherwise repeat Lemma 2 for vertex s_3. If $s_4 \in \{s_1, s_2\}$ we are done. Otherwise repeat the lemma for s_4. Eventually $s_i \in \{s_1, s_2, \ldots, s_{i-2}\}$ since B is finite.   □

Note if B does not have a forced cycle of length at least 4 then either $|S| = 1$ or B is not a bipartite update network. That is, if $|S| > 1$ then Adversary has a strategy to not visit a vertex s_2 of S whenever the play begins at some different vertex s_1 of S.

We now define a contraction operator for reducing the size of an update game. Let $B = (S \cup A, E)$ be a bipartite update game with a forced cycle $C = (a_k, s_k, \ldots, a_2, s_2, a_1, s_1)$ of length at least 4. The contraction operator applied to B via C, denoted B/C, produces an update game $B' = (S' \cup A', E')$ with $|S'| < |S|$. We construct B' as follows. For new vertices a and s,

$$S' = (S \setminus \{s_1, s_2, \ldots, s_k\}) \cup \{s\} \quad \text{and} \quad A' = (A \setminus \{a_1, a_2, \ldots, a_k\}) \cup \{a\}$$

and

$$\begin{aligned} E' = {}& E(B \setminus \{s_1, a_1, \ldots, s_k, a_k\}) \cup \\ & \{(s, a') \mid a' \in A' \text{ and } (s_i, a') \in E, \text{ for some } i \leq k\} \cup \\ & \{(a', s) \mid a' \in A' \text{ and } (a', s_i) \in E, \text{ for some } i \leq k\} \cup \\ & \{(s', a) \mid s' \subset S' \text{ and } (s', a_i) \in E, \text{ for some } i \leq k\} \cup \{(a, s), (s, a)\}. \end{aligned}$$

We now present a method that helps us decide if a bipartite game is a bipartite update network.

Lemma 4. *If $B = (S \cup A, E)$ is a bipartite update game with a forced cycle C of length at least 4 then B is a bipartite update network if and only if $B' = B/C$ is one.*

Proof (sketch). We show that if B' is an update network then B is also an update network. We first define the natural mapping p from vertices of B onto vertices of B' by

$$p(v) = v \quad \text{if } v \notin C$$
$$p(v) = a \quad \text{if } v \in C \cap S$$
$$p(v) = s \quad \text{if } v \in C \cap A.$$

Then any play history of B is mapped, via the function $p(v) = v'$, onto a play history of B'. Consider a play history $v_0, v_1, \ldots, v_n$ of B that starts at vertex v_0. Let f' be a winning strategy for Survivor when the game begins at vertex v_0'. We use the mapping p to construct Survivor's strategy f in game B by considering the two cases $v_n' = s$ and $v_n' \neq s$. It is not hard to construct a winning strategy f for Survivor in game B whenever f' is a winning strategy in B'.

By a similar case study one can show that if B is an update network then B' is also an update network (see [1] for details).   □

With respect to the above proof, Figure 4 shows how a forced cycle of B is reduced to a smaller forced cycle (of length 2) in B'.

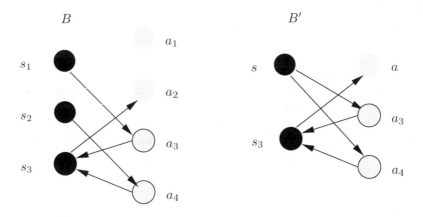

Fig. 4. Showing the bipartite update game reduction of Lemma 4.

Theorem 1. *There exists an algorithm that decides whether a bipartite update game B is a bipartite update network in time $O(n \cdot m)$, where n and m are the order and size of the underlying graph.*

Proof. We show that finding a cycle that is guaranteed to exist by Lemma 3 takes time at most $O(m)$ and that producing B' from B in Lemma 4 takes time at most $O(n + m)$. Since we need to recursively do this at most n times the overall running time is shown to be $O(n \cdot m)$.

The algorithm terminates whenever a forced cycle of length at least four is not found. It decides whether the current bipartite graph is a update network by simply checking that $S = \{s\}$ and out-degree$(s) = |A|$. That is, the singleton Survivor vertex is connected to all Adversary vertices.

Let us analyze the running time for finding a forced cycle C. Recall the algorithm begins at any vertex s_1 and finds an in-neighbor a_1 (of s_1) of out-degree 1 with $(s_2, a_1) \in E$ where $s_2 \neq s_1$. This takes time proportional to the number of edges incident to s_1 to find such a vertex a_1. Repeating with s_2 we find an a_2 in time proportional to the number of edges into s_2, etc. We keep a boolean array to indicate which s_i are in the partially constructed forced path (i.e., the look-up time will be constant time to detect a forced cycle of length at least 4). The total number of steps to find the cycle is at most a constant factor time the number of edges in the graph.

Finally, we can observe that building B' from B and C of Lemma 4 runs in linear time by the definition of S', A' and E'. Note that if the data structure for graphs is taken to be adjacency lists then E' is constructed by copying the lists of E and replacing one or more vertices s_i's or a_j's with one s or a, respectively.

$\square$

The above result indicates the structure of bipartite update networks. These are basically connected forced cycles, with possibly other legal moves for some of Survivor and Adversary vertices.

4 Level Structures of Update Networks

In this section we introduce a hierarchy of update networks which can be used to extract 'efficient' winning strategies for Survivor. The class H_0 contains all update networks with exactly one Survivor node. Assume that the class H_k has been defined. Then an update network G belongs to H_{k+1} if and only if G can be contracted (via some forced cycle) to a G' such that $G' \in H_k$. Thus, by Lemma 4, G is an update network if and only if there exists a sequence $G = G_0, G_1, \ldots, G_k$ of update networks and forced cycles $C_0, C_1, \ldots C_{k-1}$ such $G_k \in H_0$ and G_i is obtained by the contraction operator applied to G_{i-1} via C_i.

Definition 4. *The* level number *k of an update network G is the minimum number of contractions of forced cycles needed to produce an update network with one Survivor node, that is $G \in H_k$ and $G \notin H_{k-1}$.*

Intuitively, the level number represents a complexity of update networks. Thus, given an update network it is quite natural to compute the level number. Computing the level number, in turn, is related to finding a large forced cycles in the network. One approach for estimating this level number can be based on a greedy algorithm that repeatedly finds a largest forced cycle and applies the contraction operator. However the following proposition shows that this approach is not efficient.

Proposition 2. *To find the longest forced cycle in an update game is* **NP-complete.**

Proof. Let LONGFORCED be the decision problem to decide if an update game has a forced cycle of length at least k. We know that deciding, for input a digraph G and integer k, whether G has a cycle of length at least k, called LONGCYCLE, is **NP**-complete. We can polynomial-time many-one reduce the input (G, k) of LONGCYCLE to an input $(G', 2k)$ for LONGFORCED as follows. Let G' be G with every edge subdivided, that is for every edge (u, v) in G is replaced with edges (u, a) and (a, v), where a is a new vertex. Clearly G' is a bipartite graph where the new vertices correspond to the Adversary set A. If G' has a forced cycle of length $2k$ then G has a cycle of length at least k. Obviously, the other direction holds. $\square$

It turns out that, in general, the greedy algorithm does not compute the level number of update networks, as illustrated with the networks shown in Figure 5. Here, the graph has three forced cycles, one small one of length 4 and two large ones of length at least 5. The greedy algorithm would use 4 contractions, while the level number is actually 2.

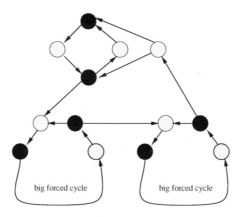

Fig. 5. Why the greedy algorithm does not find the smallest level number.

5   Simple Update Strategies

We now want to consider how to design Survivor winning strategies when we have an update network.

From the previous section we see that one way for constructing winning strategies is to explicitly apply the contraction operation repeatedly while remembering the sequence of forced cycles. From Lemma 4 a sequence $G, G_1, \ldots,$ G_k of update networks is obtained from the original network G. Here the final network G_k with one survival node has a trivial winning strategy, where it moves repeatedly in a cyclic fashion to all of its neighbors. The proof of the lemma indicates how to compute a winning strategy for G_i from the winning strategy for G_{i+1}. (Recall, a forced cycle of length 2 in G_{i+1} is uncontracted to a forced cycle of length at least 4 in G_i.) We call strategies of this type *contraction-based strategies*.

Example 4. Consider the update network G' of Figure 4. If we apply the contraction algorithm again on the forced cycle (a_3, s_3, a, s) we get an update network with one Survivor node s' that has a winning strategy to alternate sending to the Adversary nodes a' and a_4. Doing an uncontraction yields a winning strategy for G' where node s alternates sending to a_3 and a_4 and node s_3 always sends to a.

It turns out there is another simple local winning strategy (which may not be as efficient at updating all the nodes) that only requires each Survivor node to systematically update their neighbors. This strategy needs only to follow simple local rules with out any knowledge of the topology of the network. We now formally present this *cyclic neighbor strategy*. Each Survivor node s keeps an index i_s between 0 and d_s=out-degree(u)-1, initially set to 0. Survivor's next move from node s is to its i_s-th out-neighbor and i_s is incremented modulo d_s. We now show this is a winning strategy.

Theorem 2. *The cyclic neighbor strategy is a winning strategy for any update network.*

Proof. Consider any play consistent with the cyclic neighbor strategy for an update network G. There will be a set of persistent vertices called P. If $P = A \cup S$ then we are done. Now assume otherwise and consider the set $N = (A \cup S) \setminus P$. There is at least one Adversary node in N by Lemma 1 (an adversary vertex of out-degree 1 can not be in P if it's out-neighbor is in N). If we follow the cyclic neighbor strategy there must not be any edges from $P \cap S$ to $N \cap A$, since all vertices in P were reached infinitely often. Thus Adversary has a winning strategy to keep out of N forever (no matter what strategy Survivor uses). This contradicts the fact that G was an update network. □

The above proof leads to the following more general winning strategy.

Corollary 1. *Any Survivor strategy for an update network that from any $s \in S$ moves to each neighbor of s infinitely often is a winning strategy.*

One can compare the cyclic neighbor strategies with the contraction-based strategies. In general the contraction-based strategies are more efficient at updating all nodes of a network more frequently. However, there is a quadratic cost of having to compute these. However, like the cyclic neighbor strategies, there is very little operational cost in using a contraction-based strategy. The only requirement is that a sequence of neighbors (with possible repetitions) has to be kept at each node instead of just a single counter.

We end this section with a couple comments about Adversary winning strategies for update games that are not update networks. The decision algorithm fails because there are no forced cycles of length at least 4 when there remains at least two Survivor nodes. In any case, Adversary knows which Survivor nodes to avoid. This strategy is a counter strategy to Survivor's contraction-based strategy mentioned above.

6   Conclusion

In this paper we have presented a game-theoretic model for studying update networks. We have shown that it is algorithmically feasible to recognize update networks. That is, we have provided an algorithm which solves the update game problem in $O(n \cdot m)$ time. Moreover, our algorithm for the case of bipartite update games is used to give a characterization of bipartite update networks. We have presented two simple routing strategies for updating these networks. We have also discussed the complexity of update networks in terms of the level number invariant. In general, update networks with the lower level numbers have simpler and more efficient updating strategies.

References

1. M.J. DINNEEN AND B. KHOUSSAINOV. Update Games and Update Networks. Proceedings of the Tenth Australasian Workshop on Combinatorial Algorithms, AWOCA'99. R. Raman and J. Simpson, Eds. Pages 7–18, August, 1999.
2. R. MCNAUGHTON. Infinite games played on finite graphs. *Annals of Pure and Applied Logic* 65 (1993), 149–184.
3. A. NERODE, J. REMMEL AND A. YAKHNIS. McNaughton games and extracting strategies for concurrent programs. *Annals of Pure and Applied Logic* 78 (1996), no. 1-3, 203–242.
4. R.J. BÜCHI AND L.H LANDWEBER. Solving sequential conditions by finite-state strategies. *Trans. Amer. Math. Soc.* 138 (1969), 295–311.
5. T.J. SCHÄFER. Complexity of some two-person perfect-information games. *J. Computer & System Sciences* 16 (1978), 185–225.

Computing Input Multiplicity in Anonymous Synchronous Networks with Dynamic Faults

Stefan Dobrev*

Institute of Mathematics, Slovak Academy of Sciences, Department of Informatics,
P.O. Box 56, 840 00 Bratislava, Slovak Republic. stefan@ifi.savba.sk

Abstract. We consider the following problem: Each processor of the network has assigned a (not necessarily unique) input value. Determine multiplicity of each input value. Solving this problem means any input-symmetric function (i.e. function not sensitive to permutations of its input values) can be computed. We consider anonymous synchronous networks of arbitrary topology, in which dynamic link faults [3,6] may occur.

An instance of this problem has been stated as an open problem by N. Santoro at SIROCCO'98: Is it possible to distributively compute parity function (XOR) on anonymous hypercubes with dynamic faults?

We show that if the network size N (the number of processors) is known, the multiplicity of inputs (and thus any input-symmetric function) can be computed on any connected network. The time complexity depends on the details of the model and the amount of topological information, but it is always a low polynomial in N.

1 Introduction

The problem of fault tolerance is very important in the field of distributed computing. Indeed, as more and more processors are being interconnected, the probability of faulty links or nodes increases. According to classical model of fault tolerance – static model – the fault status of any component is fixed during the whole computation. Results based on this model cover fault tolerant distributed agreement as well as broadcasting, gossiping and routing. Consult a survey by Pelc [5] for more details.

The communication model we consider is synchronous point–to–point message passing with dynamic faults, introduced in [3,6]. The computation network is represented by simple non-directed graph $G = (V, E)$, where V and E represent processors and bidirectional communication links, respectively. In the rest of the paper we use N for the number of processors. The links at a processor v are labeled $1, 2, \ldots, deg_v$, where deg_v is the degree of v. These labels are assigned in an arbitrary way, with no relevance to the structure of the network.

The network is synchronous, in the sense that the communication and computation steps alternate. Each communication step takes 1 time unit, the time for

---

* Supported by VEGA grant No. 2/7007/20.

U. Brandes and D. Wagner (Eds.): WG 2000, LNCS 1928, pp. 137–148, 2000.
© Springer-Verlag Berlin Heidelberg 2000

local computation is negligible. At one communication step, a processor can simultaneously communicate with all its neighbours. This communication scheme is called *shouting* or *all port* model. We allow messages of arbitrary size. This unrealistic assumption will be discussed later on.

Only link failures are considered. At each step at most k links may be faulty. The nature of the faults is dynamic – the set of faulty links may vary from step to step. All faults are of the bidirectional crash type, i.e. messages sent in any direction over a faulty link are lost. We suppose k is smaller than the edge connectivity of the network, otherwise it would not be possible to perform broadcast, neither to compute any global function.

The network is anonymous, all processors are identical, equipped with the same algorithm. Computation is started simultaneously by all processors and terminates when all of them reach the termination state. The time complexity is expressed as the total number of communication steps.

The most investigated problem in this model is the time complexity of broadcasting, studied on special topologies [2,3,4] as well as on general networks [1]. In fact, the broadcasting algorithm is always the same: Shout all you know to all your neighbours. The intricacy lies in showing how much time it takes to reach everyone when the adversary dynamically chooses the links to fail. Clearly, $N-1$ steps are enough – at each step at least one new processor is informed. This bound can be improved to $O(N/k)$ [1], where k is the number of links that may fail at one step. Better results can be obtained for special topologies, sometimes almost matching the broadcasting time in the absence of faults (e.g. diameter+2 for hypercubes [2], see also [4]).

The problem of distributed evaluation of an N-ary function f can be stated as follows. At the beginning of the computation each processor is given an input value from a set of input values I, upon termination each processor should know the result of the application of f to the whole input. Since the network is anonymous and the output value is unique, f should be invariant to the permutations of input values imposed by the symmetries of the underlying network. As we consider general graphs, for the rest of the paper we restrict ourselves only to functions which are invariant to all permutations of inputs. (Boolean AND/OR/XOR and the sum of input values are examples of such functions.)

The problem has been studied by Santoro and Widmayer in [6] for non-anonymous complete graphs, with several fault modes (crash, corrupting/adding messages) being considered.

If the network size N is known to processors, a broadcasting algorithm can be used to compute the boolean AND/OR of the inputs [6]: Each processor broadcasts its value and collects the values it sees. After $N-1$ steps it has received all values and can output their boolean AND resp. OR. This approach works well for AND/OR, because these functions depend only on the values present in the input, not on their multiplicity. It is not immediately clear how to compute more input sensitive functions, e.g. parity function (XOR) or sum of the input values, in the presence of dynamic faults. In fact, at SIROCCO'98 rump session N. Santoro stated as an open problem even more restricted question: Is it possible to compute the XOR on anonymous hypercubes with dynamic faults?

In this paper we present a deterministic algorithm *InputMulti*, which, given the network size N, computes the multiplicity of each input value in the presence of dynamic faults. This means that any input-symmetric function can be computed, giving a positive answer to the open problem mentioned above. The knowledge of N is necessary. It is known that in general it is not possible to compute the network size on anonymous networks [7] deterministically. However, computing the multiplicity of inputs implies knowledge of the network size (consider all input values being equal).

Let B be an upper bound on time needed for broadcasting in a given network in the presence of dynamic faults, known to all processors. (Clearly $B \leq N - 1$, however, it can be much less if some topological information is given to processors. For example if processors know that the communication topology is hypercube network, they can work with $B = \log N + 2$.) Then the time complexity of the algorithm *InputMulti* is $2BN$.

Some of the model requirements are unnecessarily strong, we show how to adapt *InputMulti* to the following relaxed variants of the model in section 4:

- **Asynchronous start up:** The computation is started by some non-empty set of initiators, instead of all processors. The cost for initial synchronization is additional B steps.
- **Unidirectional faults:** The fact that a link always fails in both directions allows us to modify any algorithm in such a way that sender can detect failure of his message transmission: Add dummy messages, if necessary, to make sure that at each step, a message is sent via each link in both directions. Then, the fact that no message has been received on link h indicates that the message sent via h has been lost too. Unless stated otherwise, from now on we assume that this technique is implicitly used and sender can detect failure of his message transmission.

  If a link may fail in one direction only, sender can not straightforwardly detect call failure. The algorithm *InputMulti* assumes bidirectional faults, however this assumption is not essential. We present a work around this problem in cost of $2k + 3$ multiplicative factor.
- **Short messages:** Transmission of messages of arbitrary size in one time step is unrealistic. *InputMulti* can be adapted to the case when only messages of limited size (e.g. $O(\log N)$) are allowed, with a slowdown factor of a low polynomial in N (depending on the details of the model).

The paper is organized as follows. In section 2 we present the algorithm *InputMulti* for computing the multiplicity of input values in the presence of dynamic faults and prove its correctness and time complexity. Section 3 contains a technique which, for some symmetric topologies, allows further reduction of the time complexity. Application to hypercubes yields an $O(\log^2 N)$ solution. In section 4 adaptations of the algorithm *InputMulti* to above mentioned relaxations of the model are presented. Finally, concluding remarks can be found in section 5.

2   Computing the Multiplicity of Input Values

The algorithm *InputMulti* works in stages. The first stage is started by all processors. During each stage an asymmetry of the network, input data or occurrence of faults is used to reduce the number of initiators of the next stage. If no asymmetry is detected, the symmetry of the configuration allows computation of the result. In fact, *InputMulti* is a leader election algorithm, which either finishes prematurely by being able to compute the result, or elects a leader. If a leader is elected, computing the result is straightforward, since the network is no longer anonymous.

Each stage consists of a building phase, a broadcasting phase and a resolving step at the end of the stage. Since the network is synchronous, all processors start computation synchronously and they know B, each processor knows when each phase/stage begins and when it ends.

During the building phase flooding of the network from initiators is performed and each vertex is assigned a (not necessarily unique) name. In the course of this phase several broadcast processes are launched to inform other processors of what has been built. The building phase takes B steps - enough for the flooding to reach the whole network. The broadcasting phase takes additional B steps and it allows the broadcasts initiated during the building phase to reach all processors. This ensures that at the time of the resolving step all processors have the same information about the building phase. In the resolving step any asymmetry captured in this information is used to either identify the initiators of the next stage or compute the multiplicity of each input value. In the latter case the algorithm *InputMulti* announces result and terminates.

2.1   Building and Broadcasting Phase

During the building phase a forest of trees rooted at the stage initiators is built. The roots start with a name ϵ, while the name of a son u of a processor v with name x is $x \circ i$, where i is the label of the link (v, u) at v. The names are assigned by the flooding protocols initiated at roots: A processor v with an already assigned name x offers names $x \circ 1, x \circ 2, \ldots, x \circ deg_v$ to its neighbours (neighbour to which link i leads to is offered the name $x \circ i$). If a processor has not been assigned a name yet, it accepts one of the name offers it has received and proceeds by offering names to its neighbours. In addition, acceptance or rejection of each offer is announced by broadcasting *alive/dead* messages. The idea is to have for each offer generated either an *alive* or a *dead* message, but not both (or neither). The broadcasting phase is just a time buffer of B steps to guarantee all *alive* and *dead* messages arrive at all processors. During the both phases each processor collects all *alive* and *dead* messages it has seen during the current stage.

A processor v has a local variable $Name_v$ containing its name, valid only for the current stage. The type of $Name_v$ is an ordered list of integers (ϵ being the empty list). $list \circ x$ means adding element x to the end of the list $list$. val_v is the input value of v and deg_v is its degree.

Broadcast(*message*) denotes a process which broadcasts message *message* to all other processors.

KeepSending((*You are: x*), *link*) is a process which sends message (*You are: x*) via the link *link* in each step until the end of the building phase. If during all these sending steps the link *link* was faulty, a message (*dead, x*) is broadcasted. (The offer never came to the receiver, act as if it was rejected.)

The algorithm for an initiator v:

$Name_v := \epsilon$;
Broadcast((*alive*, ϵ, val_v, deg_v);
KeepSending((*You are: i*), i) for $i = 1, 2, \ldots, deg_v$;{*For all incident links.*}

At a vertex v upon receiving (*You are: m_1*), $\ldots$, (*You are: m_l*) via links $h_1, h_2, \ldots, h_l$[1]:

if v already has a name then
    Broadcast((*dead, m_i*)) for $i = 1, 2, \ldots l$;
else
    $Name_v :=$ lexicographically minimal value m_j among m_i;
    Broadcast((*alive, $Name_v$, val_v, deg_v*));   {*One offer is accepted,*}
    if message (*You are: $Name_v$*) has been received via at least two links then
        Broadcast((*dead, $Name_v$*));   {*others are rejected.*}
    Broadcast((*dead, m_i*)) for all m_i such that $Name_v \neq m_i$;
    KeepSending((*You are: $Name_v \circ i$*), i) for the links i over which you have
        not received (during the current stage) a *You are:* message;

The main process, as well as the child processes KeepSending and Broadcast execute simultaneously and interfere only in an explicitly stated way. This is possible because the link contention is resolved using messages of unlimited size: If several messages are sent simultaneously over a link, they are packed into one big message which is upon arrival unpacked back into original messages.

As the above algorithm is in fact extension of the shout-to-all broadcasting algorithm, each vertex computes its name in time at most B.

2.2 Resolving Step

The resolving step is performed at the time $2B$ since the start of the current stage – after the termination of the broadcasting phase. At that time all processors

---

[1] Only the first (*You are:*) message on a given link (during this stage) is considered here, its subsequent arrivals (due to KeepSending()) are ignored.

have received the same set of *alive/dead* messages and the following processing is performed by each of them:

1. If there is a name for which both *alive* and *dead* messages were received, the processors with lexicographically minimal name among such names become initiators of the next stage.
2. Otherwise, only *alive* or only *dead* messages were received for each name. Consider now the names for which only *alive* messages were received. If there is a name for which different values of val_v or deg_v were received, the processors with minimal such name and with lexically minimal pair (val_v, deg_v) become initiators of the next stage.
3. Otherwise identical *alive* messages were received for each name. In this case the multiplicity of value *val* is given by $|val| \cdot N/|Name|$, where *Name* is the set of all names for which an *alive* message was received during the current stage and $|val|$ is the number of different *alive* messages in which *val* occurs.

As this processing is performed at each processor, after the resolving step each processor knows whether it is an initiator of the next phase, passive element, or the result is computed. Note that if the current stage had a single initiator, all the assigned names are different and the last case applies.

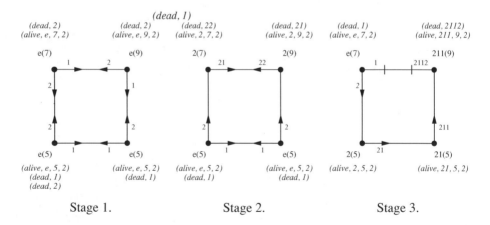

Fig. 1.

Example. An example execution of *InputMulti* on a 4-node ring is shown in Figure 1. Each processor is marked by its name and its input value in the brackets. The arrows represent the *You are:* messages, the messages broadcasted by each processor are shown above/below it.

In the first stage all vertices are initiators, thus starting with name ϵ. The leftmost figure depicts the situation at the end of the first stage. At the time of the resolving step each vertex has collected the following set of *alive/dead* messages: $\{(alive,\ \epsilon,\ 5,\ 2),\ (alive,\ \epsilon,\ 7,\ 2),\ (alive,\ \epsilon,\ 9,\ 2),\ (dead,\ 1),(dead,\ 2)\}$. In this case the rule 2 applies: The processors with the name ϵ and the input value 5 become the initiators of the next stage.

The situation at the end of the stage 2 is depicted in the middle figure. The set of collected *alive/dead* messages is the union of the sets listed for each processor. Again, the rule 2. applies, this time for the processors with name 2.

Finally, the situation at the end of the stage 3 is shown in the rightmost figure. (The top link has been blocked during the whole building phase.) Now rule 3 applies and the multiplicity of input values is computed: $Name = \{\epsilon, 2, 21, 211\}$, $N/|Name| = 4/4 = 1$, $|5| = 2$ (2, 21), $|7| = 1$ (ϵ) and $|9| = 1$ (211).

Note that if the input values of the two top processors were the same (e.g. 7), the rule 3. would apply in the second stage.

2.3 Complexity and Correctness

For a given stage, denote by $\#(x)$ the number of processors with the name x. Note that $\#(x \circ i) \leq \#(x)$, because a processor with the name x sends message (*You are: i*) to at most one of its neighbours.

The analysis of the algorithm *InputMulti* is based on the following lemma:

Lemma 1. *Let the algorithm* InputMulti *reached stage* $r + 1$, $r > 0$. *Then the number of initiators of stage* $r + 1$ *is strictly less then the number of initiators of stage* r.

Proof. The initiators of stage $r + 1$ are determined at the resolving step of stage r by rules 1. and 2.. Suppose the initiators were determined by the rule 1. applied to name x. In that case both $(alive,\ x,\ \dots)$ and $(dead,\ x)$ messages were generated. Clearly $x \neq \epsilon$, because no *dead* message for ϵ can be generated. Let $x = y \circ i$ for some y and i. $(alive,\ x,\ \dots)$ message means there is a processor v with name y whose offer (*You are: $y \circ i$*) has been accepted, while the $(dead,\ x)$ means there is a processor u with the name y whose offer (*You are: $y \circ i$*) has been rejected or never delivered. It follows: The number of initiators of stage $r + 1 = \#(x) < \#(y) \leq \#(\epsilon) =$ the number of initiators of stage r.

Consider now the case when the initiators were determined by the rule 2. applied to name x. Since $\#(x) \leq \#(\epsilon)$ and a proper subset of processors with the name x is chosen as the set of initiators for the next stage, the lemma holds also in this case. □

Since the original number of initiators is N, *InputMulti* terminates at last at the end of a stage with single initiator, and each stage takes $2B$ steps, we get the theorem:

Theorem 1. *The algorithm* InputMulti *terminates in time at most* $2BN$.

The following theorem completes the analysis of the algorithm *InputMulti*:

Theorem 2. *The algorithm* InputMulti *correctly computes the multiplicity of each input value.*

Proof. Since the values at processors with the same name are equal (otherwise rule 2. would apply), the multiplicity of the value *val* is given by a $\sum \#(x)$ over names x, for which the processors with name x have input value *val*. If we show $\forall x, y \in Name : \#(x) = \#(y)$, this sum can be simplified to $|val| \cdot \#(x)$. Moreover, in such a case $\#(x)$ can be computed as $N/|Name|$. The resulting expression for multiplicity of the value *val* is $|val| \cdot N/|Name|$, which is exactly the term used in the resolving phase.

$\forall x, y \in Name : \#(x) = \#(y)$ is equivalent to $\forall x \in Name : \#(x) = \#(\epsilon)$. We prove this assertion by induction on the length of x. The first step is trivial: $\#(\epsilon) = \#(\epsilon)$. The induction step is $\forall x, x \circ i \in Name, \#(x \circ i) = \#(x)$. Since the inequality $\#(x \circ i) \leq \#(x)$ has already been shown and the algorithm terminates only when no dead message was received for all $x \circ i \in Name$, the only case left for discussion is whether the inequality $\#(x \circ i) \leq \#(x)$ might be proper. The negative answer would finish the proof.

Let us assume $\#(x \circ i) < \#(x)$ without generating message (*dead, x \circ i*). That means all $x \circ i$ offers were accepted. Since $\#(x \circ i) < \#(x)$, there was a processor v with name x, which did not offer name $x \circ i$ to any of its neighbours. That means $deg_v < i$. However, $x \circ i \in Name$ implies the existence of a processors v' with the name x and degree at least i. Since the algorithm terminates only if all processors with the same name have the same value and degree, we get a contradiction. □

3   Special Topologies

The previous result shows that the multiplicity of input values can be computed within the time $2BN$. The factor N (which on many topologies (e.g. hypercubes, star graphs and tori) is significantly higher than B) is determined by the number of initiators in the first stage. For a given graph G, let S be the time needed to compute the multiplicity of input values in the absence of faults. The following preprocessing reduces the number of initiators to $2k$ (recall that k is the maximal number of faults in one step):

- Run the computation of multiplicity of input values assuming no faults occur.
- Run concurrently the following fault detection algorithm for S steps: At every step each processor sends dummy message to all its neighbours and tests whether it has received messages on all its links. Missing message indicates a fault. When a processor detects a fault, it broadcasts the time when the fault has occurred.
- Wait B steps – the time sufficient to finish broadcasts. If no fault has been detected, the multiplicities were correctly computed. Otherwise, the processors with minimal time of an incident fault are initiators of the first stage. Obviously, there are at most $2k$ initiators (when all first faults were duplex and vertex-disjoint).

Substituting $2k$ for N (the number of stages) in Theorem 1 and taking into account the cost $S + B$ of this preprocessing we get:

Theorem 3. *Let S be the time needed to compute the multiplicity of each input value on G in the absence of faults. Then the multiplicity of input values can be computed in the presence of dynamic faults in time $S + (4k + 1)B$.*

The following lemma allows the Theorem 3 to be successfully applied to many frequently used symmetric topologies (e.g. hypercubes, rings and cliques). The main idea is to broadcast all values and to use the number of paths of a given length between processors, the multiplicity of received messages for a value x and the time of the arrival of these messages, to compute the multiplicity of x.

Lemma 2. *Let G be a topology for which the number of different paths P_d^l of length l between two vertices of distance d is the same for all such pairs of vertices and for all d and l. Suppose this topology is known to all processors. Then in the absence of faults the multiplicity of input values can be computed in time D, where D is the diameter of G.*

Due to space limitations the proof (as well as the proofs in the remainder of this paper) is left to the reader.

Lemma 2 in combination with Theorem 3 can be applied to hypercubes. Since for hypercubes $D = \log N$, $k = \log N - 1$ and $B = \log N + 2$, we get:

Corollary 1. $O(\log^2 N)$ *steps are sufficient to compute the multiplicity of input values on an anonymous D-dimensional hypercube with dynamic faults.*

4   Relaxing the Model

4.1   Asynchronous Start Up

In the previous discussion all processors have been supposed to start the execution of the algorithm simultaneously. If that is not the case, B steps are sufficient to synchronize them using the following protocol.

- **Upon spontaneous wake up:** Send clock value 1 to all neighbours including yourself.
- **Upon receiving messages** (t_i) **for** $i = 1, \ldots, l$: Set your local clock to the maximal value received. If it is equal to B, terminate. Otherwise send your local clock+1 to all neighbours including yourself.

Since the maximal clock value grows each step by 1 and broadcasting takes at most B steps, each processor receives message containing B at the time B after the wake up of the first processor and no such message was received before.

4.2   Unidirectional Faults

Up to now, the sender was supposed to detect failure of its call using the fact that links always fail bidirectionally. This requirement seems to be quite strong. Fortunately, it is not necessary: The only situation in which bidirectionality of faults is used is when KeepSending detects that it failed to transmit *You are:* message during the whole building phase. In the case of such permanent faults the receiver knows that something is wrong – it has not received any message on the faulty link during this stage. What the receiver does not know is the name that has not come. So, it is the sender who should send the corresponding *dead* message and, for that reason, it needs the mentioned feedback. A possible way around is to make the receiver know the name:

The simple sender(father) – receiver(possible son) communication from the *InputMulti* algorithm is replaced by a three step protocol (see Figure 2.):

- Sender keeps trying to send receiver message (*You are:* y).
- After receiving a (*You are:* y) message, processor keeps trying to send reply (*OK*, y).
- When a processor receives (*OK*, y) message, it keeps trying to send reply (*Proceed*, y).

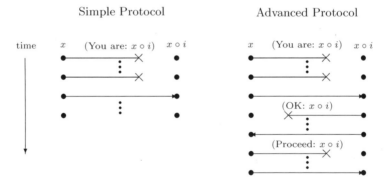

Fig. 2.

A processor computes its name based on *Proceed* messages in the same way as in the simple protocol from *You are:* messages. The *dead* and *alive* messages are generated in the same way, except for:

- At the end of the building phase sender broadcasts (*dead*, y) message if it has not received corresponding (*OK*, y) message to his (*Your are:* y) message.
- At the end of the building phase receiver broadcasts (*dead*, y) message if it has not received corresponding (*Proceed*, y) message to his (*OK*, y) message.

Note, that although both – sender and receiver – might broadcast *dead* message, it is never the case that one of them broadcasts *alive* and the other *dead* message.

Thus, the crucial property which the resolving step and correctness arguments are based on is maintained.

To make *InputMulti* work using this three-step protocol, enough time must be reserved for the building phase to let each processor compute its name. One may naively hope for a constant slowdown factor of 3 for the resulting algorithm, compared to the simple one. However, during 3 consecutive steps the faults can be adversely placed on up to $3k$ links, blocking the completion of the protocol on all of them. Nevertheless, the time for the building phase can be reasonably bounded:

Lemma 3. *In the presence of unidirectional dynamic faults, each processor computes its name within $(2k+3)B$ time steps since the start of the building phase.*

Since the broadcasting phase takes B steps also in the case of unidirectional faults, we get:

Theorem 4. *Let N be the network size, B be an upper bound on time needed for broadcasting in the presence of up to k dynamic faults and let B and k be known to all processors. Suppose that links may fail unidirectionally. Then the multiplicity of input values can be computed in time $(2k+4)BN$.*

4.3   Short Messages

Up to now we have supposed that messages of arbitrary size can be sent in one step. Straightforward simulation of long messages by a sequence of short ones is not possible, because it gives faults time to block much more than k messages. However, as the following theorem shows, a clever simulation is still possible.

Let a *unit* message be a message that can be transmitted in one step, e.g. of size $O(\log N)$.

Theorem 5. *Let s be an upper bound on size (in unit messages) of the maximal message an algorithm A uses on graph G and k be an upper bound on the number of faults in one step, both known to all processors. Let T be the time complexity of A on G. Then there exists a simulation A' of the algorithm A using only unit messages, with time complexity $T(1 + 2(s - 1)(k + 1))$.*

The idea of the proof is to split a long message into short blocks and try sequentially sending these blocks. The transmission of the next block will start only after receiving acknowledgement of arrival of the current one.

Since the size of processor names constructed by *InputMulti* is at most $O(B \log N)$ bits and there are at most $|E|$ broadcasts of *dead* and *alive* messages running concurrently, the maximal message size in unit message can be bounded by $O(N^3)$. As both k and B are less then N, substituting $T = 2BN$ in the Theorem 5 yields:

Corollary 2. *Let N be the network size, known to all processors. Then the multiplicity of input values can be computed in time $O(N^5)$.*

For the case of hypercubes we have $k, B \in O(\log N)$, $T \in O(\log^2 N)$ and $s \in O(B|E|) = O(N \log^2 N)$, which yields much better bound of $O(Tsk) = O(\log^2 N \cdot N \log^2 N \cdot \log N) = O(N \log^5 N)$.

5   Conclusions

We have shown that the multiplicity of input values, and thus any function invariant to permutations of input values, can be computed on anonymous networks with dynamic faults. The time required is small polynomial in the network size N and can be further reduced by additional topological information. The algorithms presented here were designed with simplicity in mind, there is a room for further improvements:

- The resolving step can probably be improved to significantly reduce the number of initiators, eventually reducing the number of stages from N to $\log N$. Although the multiplicity of receiving a given *alive* message is not reliable and can be altered by link faults, it still gives some estimate of relative density of the processors with the corresponding name.
- The $(2k+3)$ estimate for the slowdown of the *InputMulti* using the tree-step protocol, compared to the simple one, is rather rough. It might be possible to improve it, although a different technique is needed. The same applies to the simulation of long messages by short ones.
- The reduction of size and number of messages would be interesting, with most significant implications in the model allowing only short messages.

Nothing is known for the case of more malicious faults (corrupting/adding messages).

Acknowledgement. I would like to thank Dana Pardubská for many comments that helped to improve the presentation of the paper.

References

1. Chlebus, B. S. – Diks, K. – Pelc, A.: *Broadcasting in Synchronous Networks with Dynamic Faults*, Networks, 27 (1996), pp. 309–318.
2. Dobrev, S. – Vrťo, I.: *Optimal broadcasting in hypercubeswith dynamic faults*, Information Processing Letters 71 (1999), pp 81–85.
3. Fraigniaud, P. – Peyrat, C.: *Broadcasting in a hypercube when some calls fail*, Information Processing Letters 39 (1991), pp. 115–119.
4. De Marco, G. – Vaccaro, U.: *Broadcasting in hypercubes and star graphs with dynamic faults*, Information Processing Letters 66 (1998). pp. 321–326.
5. Pelc, A,: *Fault-tolerant broadcasting and gossiping in communication networks*, Networks 28 (1996), pp. 143–156.
6. Santoro, N. – Widmayer, P.: *Distributed function evaluation in presence of transmission faults*, in Proc. of SIGAL'90, Tokyo, 1990; LNCS 450, Springer Verlag, 1990, pp. 358–369.
7. Yamashita, M. – Kameda, T. *Computing on an anonymous networks*, in Proc. of PODC'88, 1988, pp. 117-130.

Diameter of the Knödel Graph

Guillaume Fertin[1], André Raspaud[1], Heiko Schröder[2],
Ondrej Sýkora[3]*, and Imrich Vrťo[4]**

[1] LaBRI, Université Bordeaux I, 351 Cours de la Libération, 33405 Talence, France.
[2] School of Applied Science, Nanyang Technological University, Nanyang Avenue,
Singapore 639798, Republic of Singapore.
[3] Department of Computer Science, Loughborough University, Loughborough,
Leicestershire, LE11 3TU, United Kingdom.
[4] Department of Informatics, Institute of Mathematics, Slovak Academy of Sciences,
P.O. Box 56, 840 00 Bratislava, Slovak Republic.

Abstract. Diameter of the 2^k-node Knödel graph is $\lceil (k+2)/2 \rceil$.

1 Introduction

Recently, diverse properties and invariants of interconnection networks, not only those of parallel machines, have been studied and a number of interesting results has been found, see e.g. [1]. One of the important features of an interconnection network is its message passing ability. The quality of an interconnection network depends mainly on the time delay of the communication between the nodes, which can be either processors or computers or other type of terminals. Let us suppose that a node knows a piece of information and needs to transmit it to every other node in the network. This task is usually called broadcasting. When each node knows a piece of information that has to be transmitted to every other node, we are speaking about gossiping. We refer to [6,8,9] for surveys on broadcasting and gossiping. The Knödel graph was introduced 25 years ago by Knödel [11], as an interconnection network where gossiping can be performed in the minimum time if the 1-port, telephone model is supposed. More precisely, Knödel gave an algorithm in a complete graph on even number of nodes, so that it allows to complete gossiping in the minimum possible time and Fraigniaud and Peters [7] formally defined the Knödel graph, which is the graph underlying the Knödel's communication pattern.

The Knödel graph on 2^k nodes is of special interest. It contains the minimum possible number of edges while the time needed to achieve gossiping is the same as it would be in the complete graph. In other words, this graph is a *minimum gossip graph*. It can also be shown that the Knödel graph on 2^k nodes is a minimum broadcast graph [12] and a minimum linear gossip graph [7]. In [5], it was shown that many of the graphs given as examples of minimum broadcast

* On leave from Institute of Mathematics, Slovak Academy of Sciences, Bratislava.
** This research was supported by the EC grant ALTEC-KIT, The British Council grant to the project LORA-TAIN and the VEGA grant No. 02/7007/20.

U. Brandes and D. Wagner (Eds.): WG 2000, LNCS 1928, pp. 149–160, 2000.

graphs in [4,10] or of minimum gossip graphs [12], are in fact isomorphic to the
Knödel graph. It should be noted that if the number of vertices is not a power
of 2, then the Knödel graph need not be a minimum gossip or broadcast graph.

Knödel graphs were also studied in [3] in connection with the load balancing
problem in parallel systems.

Let us define the Knödel graph in general: The Knödel graph on $n \geq 4$ nodes,
n is even, denoted W_n, is composed of nodes (i,j) where $i = 1,2$ and $0 \leq j \leq n/2 - 1$ and of edges between any two nodes: $(1,j)$ and $(2, j + 2^l - 1 \pmod{n/2})$
for any $j, 0 \leq j \leq n/2 - 1$ and $l, 0 \leq l \leq \lfloor \log_2 n \rfloor - 1$; the edges connecting the
nodes: $(1,j)$ and $(2, j + 2^l - 1 \pmod{n/2})$, for any $j, 0 \leq j \leq n/2 - 1$ are said
to be in dimension l.

Diameter is one of the main invariants of an interconnection network giving
the number of hops in a shortest path that a message has to perform on its way
from the source to the destination node in the worst case. Although the Knödel
graph was introduced 25 years ago, the diameter of the Knödel graph has not
been determined yet.

In this paper we prove:

Theorem 1. *For $k \geq 2$*

$$\text{diameter}(W_{2^k}) = \left\lfloor \frac{k+2}{2} \right\rfloor.$$

In the following section we introduce an alternative definition of the Knödel
graph. In sections 3 and 4 we present the upper and lower bound parts of the
proof of the Theorem, respectively. The last section is devoted to some open
questions and conclusions.

2 Alternative Definition

We introduce an alternative and useful definition of the Knödel graph W_n, where
$n = 2^k$. Let us take the Hamiltonian cycle of the Knödel graph composed of
edges of dimensions 0 and 1 and label the nodes of the Knödel graph as follows:
$(1,j) \equiv 2j, (2,j) \equiv 2j - 1, 0 \leq j \leq n/2 - 1$. It means that a node of the Knödel
graph, labelled by an even number x, is incident to the edges connecting it to
the nodes: $x + 2^l - 3 \pmod{n}$, for each $1 \leq l \leq k$. Similarly, all neighbours of a
node labelled by an odd number x have labels of the form $x - (2^l - 3) \pmod{n}$.
The edges corresponding to l are the edges of the $(l-1)$st dimension of the
general definition of the Knödel graph. See Figure 1 with the Knödel graph W_{16}.

As the Knödel graph is vertex symmetric [5], to determine the diameter, it
is sufficient to study the distance from the node 0 to any node x. If there is a
path from the node 0 to a node x using f edges in dimensions: $l_1, l_2, l_3, ..., l_f$,
then there exists a representation of x in the following form:

$$x = \sum_{i=1}^{f} (-1)^{i-1} (2^{l_i+1} - 3) \pmod{n}.$$

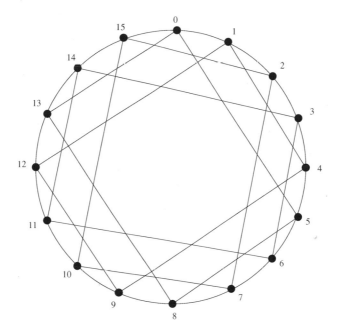

Fig. 1. Knödel graph W_{16}

Conversely, any such representation of x corresponds to a path between 0 and x. Therefore, determining the shortest path between 0 and x is equivalent to finding the above representation of x with minimal possible f. The representation will be called ±representation and we will say that the length of the ±representation is f. Note that for even x the ±3 terms disappear from the above representation.

3   Upper Bound

In this section we show that the diameter of W_{2^k} is at most $\lceil (k+2)/2 \rceil$.

3.1   Even Case

Let x be an even k–bit number. Then

$$x = \sum_{t=1}^{m} (2^{i_t} - 2^{j_t}),$$

such that $k \geq i_1 > j_1 > i_2 > j_2 > ... > i_m > j_m \geq 1$, where each pair $2^{i_t} - 2^{j_t}$ represents the block of ones in the binary representation of x i.e. $2^{i_t} - 2^{j_t} = 2^{i_t-1} + 2^{i_t-2} + ... + 2^{j_t}$. In the Knödel graph each pair $2^{i_t} - 2^{j_t}$ represents two edges: one edge from the i_t–th dimension and the other from the j_t–th dimension which follows after the first one and the ±representation gives a path how the node x can be reached from the node 0. This representation will be called the basic ±representation.

Let $I = \{i_1, i_2, ..., i_m\}$, $J = \{j_1, j_2, ..., j_m\}$. Let $\#(A)$ denote the cardinality of a set A. A maximal sequence of consecutive ones in the binary representation of x is called a block. A maximal set of consecutive numbers from $I \cup J$ is called a ±block. A number a such that $1 \leq a \leq k$ and $a \notin I \cup J$ we will call a gap or a gap number if more suitable. Notice that $0 \notin I \cup J$ and we will not count it to gap numbers. An important fact is that each ±block is accompanied with one gap number which is a number by one greater than the maximum of the block. The only exception is a (possible) block with k.

Let B_e^+ be the set of ±blocks of integers from $I \cup J$ such that the maximums of the ±blocks are from I and the minimums of the ±blocks are from J i.e. the length of a ±block from B_e^+ is even. Let B_e^- be the set of even length ±blocks of integers from $I \cup J$ with maximums from J. Similarly, we define the set of ±blocks B_o^+ (B_o^-) which is the set of the ±blocks with maximums from I (J) and the lengths of the blocks are odd and strictly greater than 1. For the sets of blocks of length 1 we will use notation B_1^+ and B_1^-.

The sum of lengths of all ±blocks in a set of ±blocks B_p^*, where $p \in \{e, o\}$, and $* \in \{+, -\}$ will be called total length of the set and denoted $|B_p^*|$.

In what follows, we show that either the basic ±representation of x satisfies $\#(I \cup J) \leq \lceil (k+2)/2 \rceil$ or we can transform the basic ±representation into another ±representation, where new I and J are sets s.t. $\#(I \cup J) \leq \lceil (k + 2)/2 \rceil$.

First we introduce four reduction rules and then we show some useful lemmas.

Reduction rule R1:
if $a \neq b \in I$ s.t. $a - 1, b + 1 \in J$, exist **then**
do $I = I \cup \{a - 1\} - \{a, b\}, J = J \cup \{b\} - \{a - 1, b + 1\}$ **od**

Reduction rule R2:
if $a, b, c \in I$ are pairwise not equal s.t. $a - 1, b - 1, c + 2 \in J, a - 1 \neq c + 2 \neq b - 1$, exist and $c + 1 \notin I \cup J$ **then**
do $I = I \cup \{a - 1, b - 1\} - \{a, b, c\}, J = J \cup \{c, c + 1\} - \{a - 1, b - 1, c + 2\}$ **od**
The reduction rule R3 is symmetrical to the rule R2.

Reduction rule R3:
if $a, b, c \in I$ are pairwise not equal s.t. $a + 1, b + 1, c - 2 \in J, a + 1 \neq c - 2 \neq b + 1$, exist and $c - 1 \notin I \cup J$ **then**
do $I = I \cup \{c - 1, c - 2\} - \{a, b, c\}, J = J \cup \{a, b\} - \{a + 1, b + 1, c - 2\}$ **od**

Reduction rule R4:
if $k, a \in I, k \neq a, a - 1 \in J$ **then**
do $I = I \cup \{a - 1\} - \{k, a\}, J = J - \{a - 1\}$

The correctness of the rules follows from the equivalences:
R1 : $2^a + 2^b - 2^{a-1} - 2^{b+1} = 2^{a-1} - 2^b$
R2 : $2^a + 2^b + 2^c - 2^{a-1} - 2^{b-1} - 2^{c+2} = 2^{a-1} + 2^{b-1} - 2^{c+1} - 2^c$
R3 : $2^a + 2^b + 2^c - 2^{a+1} - 2^{b+1} - 2^{c-2} = 2^{c-1} + 2^{c-2} - 2^a - 2^b$
R4 : $2^k + 2^a - 2^{a-1} \pmod{2^k} = 2^{a-1}$.

Therefore the rules R1-R4, if applied to a ±representation of x, produce a new ±representation of x.

Note that any application of the reduction rules R1 and R4 creates two new gap numbers $a, b+1$ and k, a, respectively. Any application of the reduction rule R2 introduces three new gap numbers $a, b, c+2$ while the original gap number $c+1$ is changed to a number from J i.e. by applying the rule R2 the number of gaps in the new $\pm$representation is increased by 2. Similar situation holds for the rule R3, where three new gap numbers $a+1, b+1, c$ are created and the original gap number $c-1$ is changed to the number from I. Again, the number of gap numbers in the new $\pm$representation is increased by 2.

Now, we describe the application of the above rules on a basic representation. The following lemma shows a procedure how to get a new $\pm$representation where only one of the set of blocks $B_e^+, B_e^-, B_o^+, B_o^-$ remains non empty and possibly $B_1^+ \neq \emptyset$, $B_1^- \neq \emptyset$.

Lemma 1. *Let $B_e^+ \neq \emptyset$ and $B_e^- \neq \emptyset$. We can reduce both sets of $\pm$blocks to a new set of $\pm$blocks of the same type as it was the type of the set of greater length. Moreover we get a set of $\pm$blocks of length 1 each accompanied with a gap number. A similar proposition holds for all following combinations: $B_o^+ \neq \emptyset$ and $B_o^- \neq \emptyset$; $B_e^+ \neq \emptyset$ and $B_o^+ \neq \emptyset$; $B_e^+ \neq \emptyset$ and $B_o^- \neq \emptyset$; $B_e^- \neq \emptyset$ and $B_o^+ \neq \emptyset$; $B_e^- \neq \emptyset$ and $B_o^- \neq \emptyset$.*

Proof. Let $B_e^+ \neq \emptyset$ and $B_e^- \neq \emptyset$. Let us apply the reduction rule R1, as many times as possible so that we take a from a $\pm$block of B_e^+ and b from a $\pm$block of B_e^- and a and b are the maximal not yet in R1 used numbers of I and J, respectively. We get new blocks of length 1. Furthermore, if in the original $\pm$representation $|B_e^+| = |B_e^-|$ then in the new $\pm$representation $B_e^+ = B_e^- = \emptyset$ and if $|B_e^+| > |B_e^-|$ $(|B_e^+| < |B_e^-|)$ we get a new $\pm$representation where $B_e^+ \neq \emptyset$ and $B_e^- = \emptyset$ $(B_e^+ = \emptyset$ and $B_e^- \neq \emptyset)$. There may be created at most one new $\pm$block which is of odd length, greater than 1. Let the maximum of the block be d. Then $d, d-1 \in I$ $(d, d-1 \in J$ if $|B_e^+| < |B_e^-|)$ and the part of the $\pm$block except d is a part of the original $\pm$block from B_e^+. This new $\pm$block is accompanied with two gap numbers - the gap number accompanying the original $\pm$block and the gap number $d+1$. Therefore, we can consider the new $\pm$block to be an even length $\pm$block with maximum $d-1$ and accompanied with one gap number. Formally we can consider d as a $\pm$block from B_1^+ accompanied with a gap number too. □

One can see that the application of rule R1 affects each element of $I \cup J$ of the basic $\pm$representation at most once and the gap numbers of the basic $\pm$representation are not affected at all. For the rest we need to process the still not affected part of the $\pm$representation. We discuss two cases $B_o^+ \neq \emptyset$ and $B_e^+ \neq \emptyset$ in the following two lemmata.

Lemma 2. *If in a $\perp$ representation $B_o^+ \neq \emptyset$, $B_e^+ = B_e^- = B_o^- = \emptyset$ then we can create a new $\pm$representation where the number of gaps is at least $\#(I \cup J) - 3$.*

Proof. Let $B_o^+ \neq \emptyset$, $B_e^+ = B_e^- = B_o^- = \emptyset$. We can suppose that B_o^+ does not contain any $\pm$block of length $4i+1, i \geq 1$. Otherwise, we can apply the reduction rule R1 so that the $a, a-1, b, b+1$ (see the rule R1) are taken from the same $\pm$block of the length $4i+1$. If we do it to maximal possible extent, we get for each such $\pm$block (except the one with k as a maximum) as many gaps as the number of elements in new I and J. We can also suppose that B_o^+ does not contain more than one $\pm$block of length 3. Otherwise we can efficiently use the reduction R1 on the set of all the $\pm$blocks of length 3. If the number of the $\pm$blocks is odd, then one such $\pm$block remains.

It remains to analyse the case when all $\pm$blocks of B_o^+ are of length $4i+3, i \geq 1$. On each $\pm$block of B_o^+ we can use the reduction rule R1. We can do it in a way that after reduction we get for each such $\pm$block $u, u-1, u-3 \in I$ and $u-2 \in J$. With each such new $\pm$block two gap numbers are accompanied: $u+1 \notin I \cup J$ and the one which was accompanied with the original $\pm$block (except the case when the maximum of the original $\pm$block is equal to k.) If we get two such $\pm$blocks i.e. u, v exist such that: $u, u-1, u-3 \in I, u-2 \in J, v, v-1, v-3 \in I, v-2 \in J$ (it holds: $\{u, u-1, u-2, u-3\} \cap \{v, v-1, v-2, v-3\} = \emptyset$), then we can apply the reduction rule R1 taking for example $u-1, u-2$ from one $\pm$block and $v-2, v-3$ from the other $\pm$block. Create as many as possible pairs of type u, v and apply R1 on all the pairs. After the latest reduction we get at least $\#(I \cup J) - 1$ gaps except in the case where the number of $\pm$blocks in B_o^+ was odd.

Now, let us analyze the odd case. Let $k \notin I$. By applying the same procedure, as it was used for the even case, we get $\#(I \cup J) - 2$ gaps. Let $k \in I$. If $k \in I$ and the number of $\pm$blocks is more than 1, then apply R4 so that a is a maximum of another $\pm$block. Now we may have an even length $\pm$block with maximum $k-1 \in J$ and one even length $\pm$block of form: $a-1, a-2, a-3, a-4, ..., v+1, v$ s.t. $a-1, a-2, a-4, ..., v \in I, a-3, a-5, ..., v+1 \in J$ and $a+1, a, v-1$ are gap numbers. If the length of $\pm$block with maximum $k-1$ is shorter than the total length of the other $\pm$blocks-1, then we can apply the reduction rule R1 so that in place of the $\pm$block with maximum $k-1$ we will get as many gaps as the numbers from J, and this case is reduced to the previous one. We will get at least $\#(I \cup J) - 2$ gaps again.

Let the length of the $\pm$block with maximum $k-1$ be greater than the total length of the other $\pm$blocks-1. Then we can use the reduction rule R1 again so that one couple (in the rule shown as $b+1, b$) is taken from the $\pm$block with maximum $k-1$ and the other one is taken from other $\pm$blocks. In the place of other $\pm$blocks we get the same number of gaps as $\#(I)$ and in the place of the original $\pm$block with maximum $k-1$ we get a $\pm$block of form $u, u-1, u-3, ..., v-3, v-1 \in J, u-2, u-4, u-6, ..., v-2, v \in I$ and we apply the rule R1 to it.

Finally, two $\pm$blocks may rest: either $d, d-1 \in J$ and $g, g-1 \in I$, then the number of gaps is at most by one less than $\#(I \cup J)$ or $d, d-1, d-3 \in J$ and $d-2, g, g+1 \in I$ and we have at least $\#(I \cup J) - 3$ gaps and we are done. Let the basic $\pm$representation contains only one $\pm$block. Then the following reduction will take place: apply R4 so that $a = k-2$ and then apply R1 to the rest of the $\pm$block. In the worst case $k = 4i+5$ and there are $2i+1$ gaps and we are done. $\square$

Lemma 3. *If in a $\pm$ representation $B_e^+ \neq \emptyset, B_e^- = B_o^+ = B_o^- = \emptyset$ then we can create a new $\pm$representation where the number of gaps is at least $\#(I \cup J) - 3$.*

Proof. Let $B_e^+ \neq \emptyset, B_e^- = B_o^+ = B_o^- = \emptyset$. If $k \in I$, apply R4 so that a is a maximum of any other $\pm$block. We get an odd length $\pm$block $B = k-1, k-2, ..., y$ s.t. $k-1, k-3, k-5, ..., y \in J$ and $k-2, k-4, k-6, ..., y+1 \in I$ and one odd length $\pm$block of form: $u-1, u-2, u-3, u-4, ..., v+1, v$ s.t. $u-1, u-2, u-4, ..., v \in I, u-3, u-5, ..., v+1 \in J$ and $y-1, u+1, u, v-1$ are gap numbers. Now we can use the rule R1 so that $a = k-1, a-1 = k-2$ and as $b, b+1$ can be taken from any other $\pm$block - the only condition is that we take it from the shortest one and from its beginning. If the $\pm$block B is exhausted by the rule R1 or it was empty, we continue as it is in the case where $k \notin I \cup J$ (see below).

If the length of the $\pm$block B is greater than the total length of the other $\pm$blocks (here we do not count the $\pm$blocks of length 1) , then in the place of other $\pm$blocks we get as many gaps as non gap numbers (we also count the original gap numbers). We can apply to the rest of the $\pm$block B the reduction rule R1 so that the $b+1, b$ of the rule is taken from the beginning of the rest and $a, a-1$ from the end of the rest of the $\pm$block B. Either we get a $\pm$block $y, y-1 \in J$ or a $\pm$block $y, y-1, y-2, y-3$ where $y, y-1, y-3 \in J, y-2 \in I$. In the first case, there are as many gap numbers as non gap numbers in place of the original $\pm$block with the maximum k and we are done. In the second case, the number of gap numbers is smaller by two than the number of the non gap numbers in the place of the original $\pm$block with maximum k and we are done.

Let $k \notin I$. Let us use the reduction rule R1 in such a way that $a, a-1$ we take from the shorter $\pm$blocks and $b+1, b$ i we take from the longer $\pm$blocks.

If one $\pm$block remains, then we apply the rule so that $a, a-1$ is from the beginning of the $\pm$block and $b+1, b$ from its end. In the worst case, a $\pm$block remains composed of: $a, a-1, a-3 \in J$ and $a-2 \in I$ such that $a+1, a-4 \notin I \cup J$ or a $\pm$block composed of: $a, a-1, a-3 \in I$ and $a-2 \in J$ such that $a+1, a-4 \notin I \cup J$. As there are at least $\#(I \cup J) - 3$ gaps, we are done.

If two same length $\pm$blocks remain, then we apply the reduction rule R1 at the maximal possible extent (if the length is 2, we do not apply it) so that we take $a, a-1$ from the first of the blocks and $b+1, b$ from the second of the blocks. In this case there are at least $\#(I \cup J) - 3$ gaps in the new $\pm$representation.

If three same length $\pm$blocks remain, then we apply R2, if possible, so that we take c from the end of one of the $\pm$blocks and then we apply R1 similarly as in the previous case. If rule R2 is impossible to apply then there are at least 2 gaps behind each of at least two of these $\pm$blocks (it means that behind these two $\pm$blocks are in the basic $\pm$representation $\pm$blocks which are associated with 2 gaps - not only 1). By application of R1 at maximal possible extent we get (one should consider the above mentioned 2 superficial gaps from the basic $\pm$representation) a $\pm$representation with at least $\#(I \cup J)$ gaps.

If four $\pm$blocks remain, we can apply R2, if possible, and R1 in the similar way as in the case of three $\pm$blocks or we can argue similarly if R2 can not be used so that for these four $\pm$blocks the number of gaps is the same as the numbers in $I \cup J$. It means that in the new $\pm$representation we have at least $\#(I \cup J) - 1$ gaps.

Finally, if $4j + i, 0 \leq i \leq 3$, $\pm$blocks remain, we can create a new $\pm$representation with at least $\#(I \cup J) - 3$ gaps so that we apply R2 and R1 j times in the way as for four blocks and then for the rest i blocks we apply the previous cases accordingly. $\qquad\square$

The cases: $B_o^- \neq \emptyset, B_e^+ = B_e^- = B_o^+ = \emptyset$ and $B_e^- \neq \emptyset, B_e^+ = B_o^+ = B_o^- = \emptyset$ are symmetrical (we have to use R3 rule in place of R2) to the cases discussed in the Lemma 2 and in the Lemma 3, respectively.

In the resulting $\pm$representation, if $k = 2l$ then $\#(I \cup J) \leq l + 1$ and if $k = 2l + 1$ then $\#(I \cup J) \leq l + 2$. In both cases we have $\#(I \cup J) \leq \lceil (k+2)/2 \rceil$.

3.2   Odd Case

Let x be an odd k-bit number. The basic $\pm$representation of x is created as follows:

Let the three least significant bits of the binary representation of x be 001. Then the basic $\pm$ representation of x is defined as follows:

$$x = \sum_{t=1}^{m} (2^{i_t} - 2^{j_t}) + 2^2 - 3,$$

where $k \geq i_1 > j_1 > i_2 > j_2 > \dots > i_m > j_m \geq 3$. We determine a new $\pm$representation so that we use the same procedure as in the case of x even to the $k - 2$ most significant bits of the binary representation of x. Such a $\pm$representation is at most $\lceil k/2 \rceil$ long. This $\pm$representation completed by $2^2 - 3$ is a new $\pm$representation of x s.t. its length is at most $\lceil k/2 \rceil + 1 = \lceil (k+2)/2 \rceil$.

Let the three least significant bits of the binary representation of x be 101. Then the basic $\pm$represention of x is as follows:

$$x = \sum_{t=1}^{m} (2^{i_t} - 2^{j_t}) + 2^3 - 3,$$

where $k \geq i_1 > j_1 > i_2 > j_2 > \dots > i_m > j_m \geq 4$. According to the previous subsection we can find a $\pm$representation of the $k - 3$ most significant bits of the binary representation of x s.t. its length is at most $\lceil (k-1)/2 \rceil$. This $\pm$representation together with $2^3 - 3$ is a new $\pm$representation of x s.t. its length is at most $\lceil (k-1)/2 \rceil + 1 < \lceil (k+2)/2 \rceil$ as k is odd.

If the three least significant bits are 011 or 111 then

$$x = \sum_{t=1}^{m} (2^{i_t} - 2^{j_t}) + 2^1 - 3 (\mathrm{mod}\ n),$$

such that $k \geq j_1 > i_1 > j_2 > i_2 > \dots > j_m > i_m \geq 2$. Observe that $2^1 - 3 (\mathrm{mod}\ n) = 2^k - 1$ sets ones in all bits of the binary representation of x and $2^{i_t} -$

2^{j_t}, for each $1 \leq t \leq m$, represents a block of zeroes in the binary representation of x.

First, assume that the three least significant bits are 111. A $\pm$representation of zeroes in the $k-2$ most significant bits is symmetrical to a $\pm$representation of ones in a $k-2-$bit even number. As such a $\pm$representation of length at most $\lceil k/2 \rceil$ exists, there also exists a $\pm$representation of the zeroes in the $k-2$ most significant bits of the same length. This $\pm$representation together with $2^1 - 3$ taken mod n is at most of length $\lceil k/2 \rceil + 1 = \lceil (k+2)/2 \rceil$.

Now, assume that the three least significant bits are 011. If the most significant ($k-1-$st) bit of x is 1, then our problem is similar to the previous one: a $\pm$representation of the zeroes in a $k-2-$bit odd number is symmetrical to a $\pm$representation of the ones in a $k-2-$bit even number. This $\pm$representation together with $2^1 - 3$ taken mod n is at most of length $\lceil k/2 \rceil + 1 = \lceil (k+2)/2 \rceil$.

If the most significant bit of x is 0, then the problem of representation of the $k-1$ most significant bits is symmetrical to the problem of representation of a $k-1$ bit even number with ones in the most significant bit and in the second least significant bit. In this case we can find a $\pm$representation (see Lemmas 2 and 3, parts of the proofs with application of the rule R4) where the number of gaps is at least equal to $\#(I \cup J) - 2$. The symmetrical $\pm$representation together with $2^1 - 3$ taken mod n is of length at most $\lceil k/2 \rceil + 1 = \lceil (k+2)/2 \rceil$.

4 Lower Bound

In this section we present the lower bound part of the proof giving examples of vertices whose distance from 0 is at least $\lfloor (k+2)/2 \rfloor$.

Let $0 \leq x < 2^k$ be an integer. Recall that the sequence of consecutive ones in the binary representation of x is called a block.

Lemma 4. *The number of blocks in $x \pm 2^i \pmod{2^k}, 0 \leq i \leq k$ is by at most 1 greater than the number of blocks in x.*

Proof. Simple case analysis. □

Assume k is even. Set $x = 2^{k-1} + 2^{k-3} + 2^{k-5} + .. + 2^1$. Then x has $k/2$ blocks. Let

$$x = \sum_{t=1}^{m}(2^{i_t} - 2^{j_t}) \pmod{2^k},$$

where $1 \leq i_t, j_t \leq k, t = 1, 2, 3..., m$, be a shortest $\pm$representation of x. Notice that $I = \{i_1, i_2, ..., i_m\}$, and $J = \{j_1, j_2, ..., j_m\}$ may be multisets now.

1. If $a, a+1 \in I$ or $a, a \in I$ then $2^a + 2^{a+1} \pmod{2^k}$ and $2^a + 2^a \pmod{2^k}$ has only one block. Starting with $a, a+1$ or a, a and adding the remaining terms and noting Lemma 4, we can conclude that the number of blocks in x is at most $1 + 2(m-1) \geq k/2$ which gives $2m \geq (k+2)/2$. Similar analysis holds if I is replaced by J.

2. If $a \in I$ and $b \in J$, $a > b$, exist, then $2^a - 2^b$ has only one block and we continue as in the previous case.
3. If the previous two cases do not occur, then it must hold that $j_m > j_{m-1} > ... > j_1 > i_m > i_{m-1} > ... > i_1$. Then clearly

$$x = 2^k - \sum_{t=1}^{m} 2^{j_t} + \sum_{t=1}^{m} 2^{i_t} = 2^{k-1} + 2^{k-3} + 2^{k-5} + .. + 2^1.$$

This implies that $i_t = 2t - 1$, for $t = 1, 2, ..., m$ and $2^k - \sum_{t=1}^{m} 2^{j_t} = 2^{k-1} + 2^{k-3} + ... + 2^{2m+1}$. Then $\sum_{t=1}^{m} 2^{j_t} = 2^k - (2^{k-1} + 2^{k-3} + ... + 2^{2m+1})$ contains two consecutive ones in the binary representation, a contradiction.

Let k be odd. Set $x = 2^{k-1} + 2^{k-3} + 2^{k-5} + .. + 2^2 + 1$. Then x has $(k+1)/2$ blocks. Let

$$x = (\sum_{t=1}^{m}(2^{i_t} - 2^{j_t}) + 2^{i_{m+1}} - 3) \pmod{2^k},$$

where $0 \leq i_t, j_t \leq k, t = 1, 2, 3..., m$, and $0 \leq i_{m+1} \leq k$ be a shortest representation of x. $I = \{i_1, i_1, ..., i_{m+1}\}$, and $J = \{j_1, j_1, ..., j_m\}$ may be multisets, again. Wlog assume that $i_1 \leq i_2 \leq ... \leq i_{m+1}$, and $j_1 \leq j_2 \leq ... \leq j_m$.

1. If there exists $j_s \in J$ s.t. $i_{m+1} > j_s$, then the number $2^{i_{m+1}} - 2^{j_s} + 2^{i_1} - 3$, has 3 blocks. Therefore, the total number of blocks in x is at most $3 + 2(m-1) \geq (k+1)/2$. This implies that the diameter is at least $2m + 1 \geq (k+1)/2$. Assume that the diameter equals $(k+1)/2$. Consider the number $y = 2^{k-1} + 2^{k-3} + ... + 2^2$. It has $(k-1)/2$ blocks. Let

$$y = \sum_{t=1}^{m}(2^{i_t} - 2^{j_t}) \pmod{2^k},$$

be its shortest representation. Similarly, as in the case (k even) we prove that the number of blocks in y is at most $1 + 2(m - 1) \geq (k - 1)/2$, which implies $2m \geq (k+1)/2$. It means that the distance between the node 0 and y and between 0 and x is $(k+1)/2$. Consider the shortest paths between x and y in the graph. The length of this path is an odd number. Then the 3 shortest paths $0 - x, x - y, y - 0$ produce an odd-cycle, a contradiction, as our graph is bipartite.
2. Finally assume that the previous case is not valid. If $a, a + 1 \in I$ or $a, a \in I$ or $a, a + 1 \in J$, or $a, a \in J$ or $a \in I, a + 1 \in J$, we continue as in case k even, 1.(a) and conclude that the diameter is at least $2m + 1 \geq (k+1)/2$. Suppose that the diameter is $(k + 1)/2$. Consider the number y from the case k odd 1. Similarly, we prove that the distance between 0 and y is $(k + 1)/2$ and force an odd cycle, a contradiction.
   Hence, we may assume that $j_m > j_{m-1} > ... > j_1 > i_{m+1} > i_m > i_{m-1} > ... > i_1$ and $j_1 > i_{m+1} + 1$. Then clearly

$$x = 2^k - \sum_{t=1}^{m} 2^{j_t} + \sum_{t=1}^{m} 2^{i_t} - 3 = 2^{k-1} + 2^{k-3} + 2^{k-5} + .. + 2^2 + 1.$$

This implies that $\sum_{t=1}^{m} 2^{i_t} - 3 = 2^{2m-2} + 2^{2m-4} + ... + 2^2 + 1$ and $2^k - \sum_{t=1}^{m} 2^{j_t} = 2^{k-1} + 2^{k-3} + ... + 2^{2m}$. Then $\sum_{t=1}^{m} 2^{j_t} = 2^k - (2^{k-1} + 2^{k-3} + ... + 2^{2m})$ contains two consecutive ones in the binary representation, a contradiction.

5 Conclusions

In spite of the fact that the Knödel graph is an important interconnection network and has important broadcast and gossip properties, its diameter has not been determined yet. In the present work we have shown that the diameter of the Knödel graph W_{2^k} is equal to $\lceil (k+2)/2 \rceil$. Besides the Knödel graph there are two other examples of the minimum gossip and broadcast garphs on 2^k vertices: the well known hypercube and the recursive circulant graph [13]. In particular, the diameter of the recursive circulant graph is $\lceil (3k-1)/4 \rceil$. The main reason for offering the recursive circulant graph as a new topology was its performance (number of vertices to diameter ratio) which could compete with the k−dimensional hypercube. Since the Knödel graph $W_{k,2^k}$ has the diameter $\lceil (k+2)/2 \rceil$, it could compete with both graphs.

The more general question of determining the diameter of the Knödel graph W_n for any even n is still open. Open question is also how to find the shortest path for any pair of nodes of the Knödel graph.

Finally, we note that our result (or rather proof technique) can be interpreted as a result from the area of computational arithmetic. In certain applications of modular arithmetic (for example in cryptography) it is important to quickly evaluate powers of the form y^x where y is an element of a finite group and x is an integer exponent. One can speed up the computation by expressing x as a short sum of terms of the form $\pm 2^i$. In [2] a notion of optimal representation of this type is defined and a fast (linear time) algorithm to find such representations was constructed. In fact, the authors of [2] reduce the problem of finding shortest paths in an efficiently constructible directed acyclic graph. The latter problem can then be solved fast by Dijkstra's well known algorithm. One of the consequences of our result is that the number of $\pm 2^i$ terms in such a representation of a k−bit x is at most $\lceil k + 2/2 \rceil$.

References

1. de Rumeur, J., Communications dans les Réseaux de Processeurs. Masson, Paris, 1994.
2. Demetrovics, J., Pethő, A., Rónyai, L., On ±1 representations of integers, submitted for publication.
3. Decker, T., Monien, B., Preis, R., Towards optimal load balancing topologies, in: *Proc. EUROPAR'2000*, Lecture Notes in Computer Science, Springer Verlag, Berlin, 2000.

4. Dinneen, M. J., Fellows, M. R., Faber, V., Algebraic constructions of efficient broadcast networks, in: *Proc. of Applied Algebra, Algorithms and Error Correcting Codes,* Lecture Notes in Computer Science 539, Springer Verlag, Berlin, 1991, 152–158.

5. Fertin, G., Raspaud, A., Families of graphs having broadcasting and gossiping properties, in: *Proc. of the 24th Intl. Workshop on Graph-Theoretic Concepts in Computer Science,* Lecture Notes in Computer Science 1517, Springer Verlag, Berlin, 1998, 63–77.

6. Fraigniaud, P., Lazard, E., Methods and problems of communication in usual networks, *Discrete Applied Mathematics* **53** (1994), 79-133.

7. Fraigniaud, P., Peters, J.G., Minimum Linear Gossip Graphs and Maximal Linear (Δ, k)-Gossip Graphs. Technical Report 94-06, School of Computing Science, Simon Fraser Univ., 1994. (ftp://fas.sfu.ca/pub/cs/techreports/1994/CMPT94-06.ps.gz)

8. Hedetniemi, S. M., Hedetniemi, S. T., Liestman, A. L., A survey of gossiping and broadcasting in communication networks. *Networks* 18(1988), 319–349.

9. Hromkovič, J., Klasing, R., Monien, R., Peine, R., Dissemination of information in interconnection networks (Broadcasting and Gossiping), in: *Combinatorial Network Theory* Eds.: Frank Hsu, Ding-Zhu Du, 1995, Kluwer, Academic Publishers, 125–212.

10. Khachatrian, L. H., Haroutunian, H. S., Construction of new classes of minimal broadcast networks, in: *Proc. of the 3rd Intl. Colloquium on Coding Theory,* 1990, 69–77.

11. Knödel, W., New gossips and telephones, *Discrete Mathematics* **13** (1975), 95.

12. Labahn, R., Some minimum gossip graphs, *Networks* **23** (1993), 333–341.

13. Park, J.-H., Chwa, K.-Y., Recursive circulant : a new topology for multicomputers networks (extended abstract), in : *Proc. of the Intl. Symposium on Parallel Architectures, Algorithms and Networks,* 1994, 73–80.

On the Domination Search Number[*]

Fedor Fomin[1], Dieter Kratsch[2], and Haiko Müller[3]

[1] Faculty of Mathematics and Mechanics, St.Petersburg State University,
Bibliotechnaya sq. 2, 198904 St. Petersburg, Russia. fomin@gamma.math.spbu.ru
[2] Université de Metz, Laboratoire d'Informatique Théorique et Appliquée,
Île du Saulcy, 57045 Metz Cedex 01, France. kratsch@lrim.sciences.univ-metz.fr
[3] Institut für Informatik, Friedrich-Schiller-Universität Jena, Ernst-Abbe-Platz 1-4,
07743 Jena, Germany. hm@minet.uni-jena.de

Abstract. We introduce the domination search game which can be seen
as a natural modification of the well-known node search game. Various results concerning the domination search number of a graph are presented.

1 Introduction

Graph searching problems have attracted the attention of researchers from Discrete Mathematics and Computer Science for a variety of nice and unexpected
applications in different and seemingly unrelated fields. There is a strong resem-
blance of graph searching to certain pebble games [13] that model sequential computation. Other applications of graph searching can be found in the VLSI theory:
the game-theoretic approach to some important parameters of graph layouts
such as cutwidth [17], topological bandwidth [16], bandwidth [9], profile [10] and
vertex separation number [8] is very useful for the design of efficient algorithms.
Also let us mention the connection between graph searching, pathwidth and tree-
width. These parameters play a very important role in the theory of graph minors
developed by ROBERTSON and SEYMOUR [2,7,20]. Some search problems also
have applications in motion coordination of multiple robots [21] and in problems
of privacy in distributed environments with mobile eavesdroppers ('bugs') [12].

In this paper, we introduce a domination search game which can be regarded
as a natural modification of the well-known node search game. In node searching
at every step some searchers are placed or are removed from vertices of a graph
G. The purpose of searching is to find an invisible and fast fugitive moving from
vertex to vertex in G. The searchers find the fugitive if some of them succeeded
to occupy the same vertex as the fugitive.

In the domination search game the searchers have more power, they find
the fugitive if one of them can 'see' it, i.e. the fugitive stands on a vertex of
the closed neighborhood of a vertex occupied by the searcher. For a survey on
similar 'see-catch' problems on graphs with searchers having 'radius of capture'

---

[*] Most of this research was done during a visit of F. Fomin at the F.-Schiller-
Universität Jena which was supported by a fellowship of the DAAD (Kennziffer:
A/99/09594).

U. Brandes and D. Wagner (Eds.): WG 2000, LNCS 1928, pp. 161–171, 2000.
© Springer-Verlag Berlin Heidelberg 2000

in different graph metrics as well as variants of 'see-catch' pursuit-evasion games on grids we refer to [11]. Domination searching can also be regarded as a version of the art-gallery problem with mobile intruder.

The paper is organized as follows. In § 2 we give definitions and preliminaries. In § 3 we establish a relation between the domination search game and the well-known node search game. In this section we also observe some complexity results. In § 4 we discuss upper bounds for the domination search number that can be obtained by making use of spanning trees. In § 5 we present our main theorem which settles a very interesting connection between domination graph searching and a relatively new graph parameter called dominating target number. (Some proofs will be omitted due to space restrictions.)

2   Preliminaries

We use standard graph-theoretic terminology compatible with [3] to which we refer the reader for basic definitions. $G = (V, E)$ is an undirected, simple (without loops and multiple edges) and finite graph with the vertex set V and the edge set E. We denote by $G[W]$ the subgraph of $G = (V, E)$ induced by $W \subseteq V$. As customary we consider *connected components* (or short *components*) of a graph as maximal connected subgraphs as well as vertex subsets. A vertex set $S \subseteq V$ of a graph G is said to be *connected* if the subgraph of G induced by S is connected.

We denote by $\overline{G}$ the complement of a graph G. The *(open) neighborhood* of a vertex v is $N(v) = \{u \in V : \{u, v\} \in E\}$ and the *closed neighborhood* of v is $N[v] = N(v) \cup \{v\}$. For a vertex set $S \subseteq V$ we put $N[S] = \bigcup_{v \in S} N[v]$ and $N(S) = N[S] \setminus S$. A vertex set $D \subseteq V$ of a graph $G = (V, E)$ is said to be a *dominating set* of G if for every vertex $u \in V \setminus D$ there is a vertex $v \in D$ such that $\{u, v\} \in E$. Thus D is a dominating set iff $N[D] = V$. The minimum cardinality of a dominating set of a graph G is denoted by $\gamma(G)$. We also say that $A \subseteq V$ *dominates* $B \subseteq V$ in the graph $G = (V, E)$ if $B \subseteq N[A]$. The *distance* $d_G(u, v)$ between two vertices u and v of G is the length of the shortest path between u and v in the graph G. For a graph $G = (V, E)$ let G^k be the graph with vertex set V and two vertices u and v are adjacent in G^k if and only if $d_G(u, v) \leq k$.

For our purpose it is more convenient to describe the graph searching in terms of clearing graph vertices. Initially, all vertices are contaminated (uncleared or are occupied by invisible fugitive). A contaminated vertex is cleared once after placing a searcher on a vertex from its closed neighborhood. A clear vertex v is recontaminated if there is a path avoiding closed neighborhoods of vertices occupied by searchers leading from v to a contaminated vertex.

More precisely, a *domination search program* Π on a graph $G = (V, E)$ is a sequence of pairs (also considered as the *steps* of Π) $(D_0, A_0), (D_1, A_1), \ldots,$ (D_{2m-1}, A_{2m-1}) such that

   1. for all $i \in \{0, 1, \ldots, 2m-1\}$, $D_i \subseteq V$ and $A_i \subseteq V$;
   2. $D_0 = \varnothing$, $A_0 = \varnothing$;

3. for every $i \in \{1, 2, \ldots, m\}$ at the $(2i-1)$-th step we *place new searchers and clear vertices*: $D_{2i-2} \subset D_{2i-1}$ and $A_{2i-1} = A_{2i-2} \cup N[D_{2i-1}]$;

4. for every $i \in \{1, 2, \ldots, m-1\}$ at the $2i$-th step we *remove searchers with possible recontamination*: $D_{2i-1} \supset D_{2i}$ and A_{2i} is the subset of A_{2i-1} satisfying that for every vertex $v \in A_{2i}$ every path containing v and a vertex from $V \setminus A_{2i-1}$ contains a vertex from $N[D_{2i}]$. If $A_{2i-1} \setminus A_{2i} \neq \varnothing$ we say that the vertices of $A_{2i-1} \setminus A_{2i}$ are *recontaminated* at the $2i$-th step and that *recontamination occurs* at the $2i$-th step.

It is useful to consider D_i as the set of vertices occupied by searchers and A_i as the set of *cleared vertices after the ith step*. Hence $V \setminus A_i$ is the set of *contaminated vertices* after the ith step.

Notice that a program $\Pi = ((D_0, A_0), (D_1, A_1), \ldots, (D_{2m-1}, A_{2m-1}))$ is fully determined by the sequence $(D_0, D_1, \ldots, D_{2m-1})$ thus we shall mainly use the shorter description $\Pi = (D_0, \ldots, D_{2m-1})$. All proofs in our paper could be given in terms of the formal definition of a search program $\Pi = (D_0, D_1, \ldots, D_{2m-1})$ but often we prefer a more 'informal' (but equivalent) way of proving.

We call a domination search program Π *winning* if $A_{2m-1} = V$. A search program $\Pi = (D_0, D_1, \ldots, D_{2m-1})$ is *monotone* if $A_i \subseteq A_{i+1}$ for all $i \in \{0, \ldots, 2m-2\}$, i.e. no recontamination occurs. We say that a domination search program Π uses k searchers if $\max_{i \in \{0, \ldots, 2m-1\}} |D_i| = k$.

We define the *domination search number* by $ds(G) = \min_\Pi \max_{i \in \{0, \ldots, 2m-1\}} |D_i|$ where the minimum is taken over all winning programs Π.

Problem 1. Is there a graph G such that every monotone winning domination search program on G uses more than $ds(G)$ searchers? In other words, does recontamination help to search a graph?[1]

Notice that the domination search number of a disconnected graph G is equal to the maximum domination search number taken over all components of G. Thus we may assume our considered graphs to be connected.

Obviously $ds(G) \leq \gamma(G)$. To establish a more interesting relation, let $G = (V, E)$ be a graph and let $S_2(G) = (V(S_2(G)), E(S_2(G)))$ be the graph obtained from G by replacing every edge $\{u, v\}$ of G by an (u, v)-path of length three. We call the vertices of $V \subseteq V(S_2(G))$ *original* vertices and the vertices of $V(S_2(G)) \setminus V$ *middle* vertices.

Theorem 1. *For any graph $G = (V, E)$, let H be the graph with vertex set $V(S_2(\overline{G}))$ and edge set $E \cup E(S_2(\overline{G}))$, i.e. H is obtained from G by connecting every two nonadjacent vertices by a path of length three. Then*

$$\gamma(G) \leq ds(H) \leq \gamma(G) + 1.$$

Proof. Let D be a dominating set in G. We first put $|D|$ searchers on the vertices of D in H and then using one additional searcher we clear all middle vertices of $S_2(\overline{G})$. Hence $ds(H) \leq \gamma(G) + 1$.

---

[1] During discussions at WG'2000 Stefan Dobrev (Bratislava, Slovakia) provided an example showing that recontamination may indeed help.

Let us prove now that $\gamma(G)-1$ searchers cannot clear the graph H by showing that at every step of searching there exists a contaminated original vertex. In fact, let $\Pi = (D_0, D_1, \ldots, D_{2m-1})$ be a domination search program on the graph H using at most $\gamma(G)-1$ searchers. $|D_1| \leq \gamma(G)-1$ implies that there is at least one contaminated original vertex after the first step of Π. Suppose that for some $i \in \{1, 2, \ldots, m\}$ there is contaminated original vertex v after the $(2i-1)$-th step. Hence $v \notin N[D_{2i-1}]$. If at any later step $2j-1$, $j > i$, for the first time since the $(2i-1)$-th step a searcher is placed on v or one of its neighbors (in H), then there is an original vertex $w \notin N[D_{2j-1}]$. Notice that either w is adjacent to v or there is a (v,w)-path of length 3 in H with two interior middle vertices such that no vertex of this path belongs to $N[D_{2j-2}]$. Therefore w is contaminated after step $2j-1$.   □

3   Node Search Number

The domination search game can be regarded as a generalization of the well known *node search game*, see [2] for a survey. In the node search game at every step some searchers are placed or are removed from vertices. Initially, all vertices are contaminated. The difference between node and domination searching is that a contaminated vertex v in node searching is cleared once after placing a searcher on v. A clear vertex v is recontaminated if there is a path avoiding vertices occupied by searchers leading from v to a contaminated vertex.

The following theorem establishes a relation between node searching and domination searching.

Theorem 2. *For any graph* $G = (V, E)$, $\mathrm{ns}(G) = \mathrm{ds}(S_2(G))$.

Since the problem NODE SEARCH NUMBER is NP-complete, Theorem 2 implies

Corollary 1. *The problem* DOMINATION SEARCHING: *'Given a graph G and an integer k, decide whether $\mathrm{ds}(G) \leq k$ or not' is NP-hard.*

Finally by combining a result on the non-approximibility of the domination number by a factor of $c \log n$ for some $c > 0$ [19] with Theorem 1 we obtain

Corollary 2. *There is a constant $c > 0$ such that there is no polynomial time algorithm to approximate the domination search number of a graph within a factor of $c \log n$, unless $P = NP$.*

4   Spanning Trees

Let $\delta = (v_1, v_2, \ldots, v_n)$ be a vertex ordering of a graph $G = (V, E)$. The *width* of the ordering δ of G is $\mathrm{b}(G, \delta) := \max\{|i - j| : \{v_i, v_j\} \in E\}$, and the *bandwidth* of G is $\mathrm{bw}(G) := \min\{\mathrm{b}(G, \delta) : \delta$ is a vertex ordering of $G\}$.

In the proof of the next theorem we use a theorem of ANDO et al. [1] stating that for any tree T with l leaves $\mathrm{bw}(G) \leq \lceil \frac{1}{2} l \rceil$. We denote the set of leaves, i.e. vertices of degree 1, of a tree T by $V_1(T)$.

Theorem 3. *Let $T = (V, E(T))$ be a spanning tree of a graph $G = (V, E)$ such that $G \subseteq T^{k+1}$ and let l be the number of leaves of the tree $T_1 = T - V_1(T)$. Then*

$$\mathrm{ds}(G) \leq \lceil \tfrac{1}{2} l \rceil (k+1) + 1.$$

Proof. Let $\delta = (v_1, v_2, \ldots, v_r)$ be an ordering of the vertices of T_1 such that $\mathrm{b}(T_1, \delta) \leq b = \lceil \tfrac{1}{2} l \rceil$. Note that such an ordering exists by [1].

Our domination search program works as follows. We put searchers on the first $b(k+1) + 1$ vertices of δ. Then we remove the searcher from v_1 and place it on $v_{b(k+1)+2}$. Suppose that after removing the searcher from v_1 recontamination occurs. Hence there is a vertex x such that $\{x, v_1\} \in E$ and there is a contaminated vertex y such that $\{x, y\} \in E$. Since searchers occupy all vertices with indices at most $b(k+1) + 1$ unless v_1, there is a vertex v_i, $i > b(k+1) + 1$, such that $y \in N_T[v_i]$. Now $G \subseteq T^{k+1}$ implies that the (x, y)-path in T has length at most $k+1$.

We distinguish three cases. In Case 1 we assume that x is a leaf of T and $\{x, v_1\} \in E(T)$. Then the length of the (v_1, v_i)-path in T is at most k since the (x, y)-path in T passes through v_1. In Case 2 we assume that x is a leaf of T and $\{x, v_1\} \notin E(T)$. Hence the neighbor of x in T is a vertex v_j for some $j > b(k+1) + 1$. $G \subseteq T^{k+1}$ implies $d_T(v_1, x) \leq k+1$. This implies that the length of the (v_1, v_j)-path in T is at most k since the (v_1, x)-path in T passes through v_j. Finally in Case 3 we assume that x is not a leaf of T. Therefore $x = v_j$ for some $j > b(k+1) + 1$. Consequently $\{v_1, v_j\} \in E$ and we obtain $d_T(v_1, v_j) \leq k+1$.

Therefore in all three cases there is a $j > b(k+1) + 1$ such that $d_T(v_1, v_j) \leq k+1$. By 'pigeonhole principle' there is an edge $\{v_p, v_q\}$ in the (v_1, v_j)-path of T such that $|p - q| > b$ which contradicts the choice of the ordering δ.

By the same arguments we can remove a searcher (without recontamination) from v_2 and put it on $v_{b(k+1)+3}$, and so on. Finally, every vertex of G is cleared once by a searcher because T is a spanning tree of G. Since no recontamination occurs when a searcher is placed on v_n all vertices of G are cleared. $\square$

If the spanning tree T is a caterpillar then the results of Theorem 3 can be slightly improved. A *caterpillar* is a tree which consists of a path, called the *backbone*, and leaves adjacent to vertices of the backbone.

Theorem 4. *Let T be a spanning caterpillar of a graph G and let k be an integer such that G is a subgraph of T^{k+1}. Then $\mathrm{ds}(G) \leq \max\{2, k\}$.*

Proof. The backbone $P = (v_1, v_2, \ldots, v_m)$ of T is a dominating path of G, i.e., every vertex of G is adjacent to a vertex of P. First we suppose $k \geq 2$. Consider the following domination search program Π using k searchers. Initially we put searchers on the first k vertices of P. Then we remove the searcher from v_1 and put it on v_{k+1}, then remove searcher from v_2 and put it on v_{k+2} and so on.

Let us show that Π is a monotone program. Assume the converse. Suppose that after removing a searcher from a vertex, say v_j, the first recontamination occurs. Then there is a vertex x such that $\{x, v_j\} \in E$ and $\{x, y\} \in E$ for some contaminated vertex y. Because y is not cleared yet, we conclude that $y \in N_T[v_i]$ for some $i > j + k - 1$. Moreover, if $y = v_i$ then $i > j + k$. Therefore the length of the (x, y)-path in T is at least $k + 2$. But this contradicts the definition of T; hence Π is monotone. Since every vertex of G was once cleared by a searcher we obtain that Π is winning.

If $k \leq 1$ then $G \subseteq T^{k+1} \subseteq T^2 \subseteq T^3$ and two searchers are sufficient to clear G as shown above.   □

An independent set of three vertices is called an *asteroidal triple* if every two of them are connected by a path avoiding the neighborhood of the third. A graph is *AT-free* if it does not contain an asteroidal triple. Asteroidal triples where introduced to characterize interval graphs and comparability graphs, see [4] for references. In their fundamental paper [6] CORNEIL et al. investigate AT-free graphs. Among others properties they prove that every connected AT-free graph contains a *dominating pair*, i.e. a pair of vertices u and v such that every (u, v)-path is dominating. A dominating pair of a connected AT-free graph can be detected by a simple linear time algorithm [5]. In [14] it was shown how to use a dominating path found by that algorithm as backbone of a caterpillar T such that $T \subseteq G \subseteq T^4$. We obtain the following corollary of Theorem 4.

Corollary 3. *Let G be an AT-free graph. Then* $ds(G) \leq 3$.

Conjecture 1. Let G be an AT-free graph. Then $ds(G) \leq 2$.

For some classes of AT-free graphs we are able to prove that the domination search number of every graph in this class is at most two.

A graph G is a comparability graph if and only if G has a transitive orientation of its edges. Cocomparability graph are the complements of comparability graphs. Every interval graph is a cocomparability graph, and every cocomparability graph is AT-free, see [4] for references and a survey on different graph classes.

Theorem 5. *Let G be a cocomparability graph. Then* $ds(G) \leq 2$.

A graph isomorphic to $K_{1,3}$ is referred to as a *claw*, and a graph that does not contain an induced claw is said to be *claw-free*.

Corollary 4. *The domination search number of AT-free claw-free graphs is at most two.*

PARRA and SCHEFFLER [18] proved that for every AT-free claw-free graph G, $ns(G) - 1 = bw(G)$. Combining this and Corollary 4 one can obtain the following interesting result.

Theorem 6. *Let G be an AT-free claw-free graph. Then*

$$\tfrac{1}{2}\Delta(G) \leq bw(G) \leq 2\Delta(G) - 1.$$

5   Dominating Targets

We start with a result on graphs with dominating pair, a class of graphs containing as we have mentioned all connected AT-free graphs. Let us recall the definition of a dominating pair. A *dominating pair* is a pair of two (not necessarily different) vertices u and v of a connected graph G such that the vertex set of every (u,v)-path in G is a dominating set of G.

Lemma 1. *The domination search number of a connected graph with dominating pair is at most* 4.

Proof. Let d_1, d_2 be a dominating pair in G and let P be a shortest (d_1,d_2)-path. Then P is the backbone of a spanning caterpillar T in G. Since P is a shortest path, we have that $G \subseteq T^5$. Then by Theorem 4, $\mathrm{ds}(G) \le 4$.   □

Problem 2. Determine the maximum domination search number of a graph with dominating pair.

Dominating targets have been introduced in [15] as a generalization of dominating pairs. A *dominating target* is a vertex set $T \subseteq V$ of a connected graph $G = (V, E)$ such that every connected superset of T is a dominating set. The *dominating target number* of a graph G, denoted by $\mathrm{dt}(G)$, is the smallest size of a dominating target of G. Hence graphs with dominating pair have dominating target number at most 2. The following theorem that extends Lemma 1 is one of the main results of our paper. To prove it we shall need some technical results about dominating targets as well as a notion of a dominating target for a vertex set of a graph.

A vertex set $T \subseteq V$ of a graph $G = (V, E)$ is a *dominating target for a vertex set* $B \subseteq V$ *in* G if $B \subseteq N[S]$ for every connected superset S of T. Clearly T is a dominating target of a graph $G = (V, E)$ if and only if T is a dominating target for V in G.

Lemma 2. *Let T be a dominating target for a vertex set $B \subseteq V$ of a graph $G = (V, E)$ and let D be a vertex set. Then for any vertex $v \in B \setminus N[T \cup D]$ the following two statements hold.*

1. *The number of connected components of $G - N[v]$ containing vertices of T is at least two.*
2. *Let $C_1, C_2, \ldots, C_k$ be the connected components of $G - N[v]$ with $T \cap C_i \ne \varnothing$. For every $i \in \{1, 2, \ldots, k\}$ let $y_i \in N[v]$ be a vertex with $N[y_i] \cap C_i \ne \varnothing$. Then for every $i \in \{1, 2, \ldots, k\}$ the set $T_i = (T \cap C_i) \cup \{v\}$ is a dominating target for the vertex set $B_i = (B \cap C_i) \setminus N[D_i]$ in the graph $G[C_i \cup N[v]]$, where $D_i = (D \cap (C_i \cup N[v])) \cup \{v\} \cup \{y_j : j \ne i\}$.*

Proof. Let $v \in B \setminus N[T \cup D]$. Suppose there is a connected component C of $G - N[v]$ with $T \subseteq C$. Then C is a connected superset of T that does not dominate the vertex v — a contradiction. Hence at least two components of $G - N[v]$ contain vertices of T.

For the proof of the second statement suppose that T_i is not a dominating target for B_i in the graph $G[C_i \cup N[v]]$. Then there is a connected superset $S_i \subseteq C_i \cup N[v]$ of T_i and a vertex $w \in B_i$ that has no neighbor in S_i. We extend S_i to a connected superset S of T in G by adding for every $j \neq i$ the vertex y_j and (for simplicity) C_j. Hence $w \in B$ is not adjacent to a vertex of S which implies that T is not a dominating target for B in G — a contradiction.   □

Theorem 7. *For every connected graph G, $\mathrm{ds}(G) \leq 2\,\mathrm{dt}(G) + 3$.*

Proof. By induction on $k = |T|$, we prove the following stronger statement that implies the theorem when taking a dominating target T of G such that $|T| = \mathrm{dt}(G)$, i.e. $B = V$ and $D = \varnothing$.

> Let T be a dominating target for $B \subseteq V$ in the graph $G = (V, E)$ and let $D \subseteq V$ be a vertex set such that $N[D] \supseteq V \setminus B$. Then there is a winning domination search program Π on G using at most $|D| + 2|T| + 3$ searchers such that at the first step of Π searchers are placed on all vertices of D and after that on each vertex of D there is a searcher throughout all steps of the search program Π.

If T is a dominating set of $G[B]$ then $T \cup D$ is a dominating set of G. Hence the statement is true since we simply place searchers on all vertices of $T \cup D$. Notice that $|T| = 1$ implies that T is a dominating set of $G[B]$.

If $|T| = 2$ then we place a searcher on each vertex of D and by Lemma 1, we can clear all vertices of B using 4 additional searchers to be moved along a shortest path P between the two vertices of T in G.

Suppose inductively that for all $D \subseteq V$ and all dominating targets T for B in G with $N[D] \supseteq V \setminus B$ the statement is true if $|T| \leq k - 1$. Let T be a dominating target for $B \subseteq V$ in the graph $G = (V, E)$ with $|T| = k$ and let $D \subseteq V$ be a vertex set such that $N[D] \supseteq V \setminus B$.

We may assume that $T \cup D$ is a not a dominating set of $G[B]$, thus $S = B \setminus N[T \cup D] \neq \varnothing$. Then by the first part of Lemma 2, for every vertex $v \in S$ at least two components of $G - N[v]$ contain vertices of T.

Consider first the easy case in which there is a vertex $v \in S$ such that every component of $G - N[v]$ contains at least 2 vertices of T. Let $C_1, C_2, \ldots, C_m$, $m \geq 2$, be the components of $G - N[v]$ with $T \cap C_i \neq \varnothing$ such that $k - 2 \geq |C_1 \cap T| \geq |C_2 \cap T| \geq \ldots \geq |C_m \cap T| \geq 2$. For every $j \in \{1, 2, \ldots, m\}$ choose a vertex $y_j \in N[v]$ with $N[y_j] \cap C_j \neq \varnothing$.

The domination search program works as follows. We place searchers on all vertices of D and a searcher on v that will never be removed. Then the components $C_1, C_2, \ldots, C_m$ are cleared individually one by one, where each component C_i is cleared as follows:

Place searchers on all vertices y_j with $j \neq i$. By the second part of Lemma 2, we may conclude that $T_i = (T \cap C_i) \cup \{v\}$ is a dominating target for the vertex set $B_i = ((B \cap C_i) \setminus \bigcup_{j \neq i} N[y_j]) \setminus N[D_i]$ in the graph $G[C_i \cup N[v]]$ where $D_i = (D \cap (C_i \cup N[v_i])) \cup \{v\} \cup \{y_j : j \neq i\}$ satisfies $N[D_i] \supseteq C_i \setminus B_i$. Since $|C_i \cap T| + 1 \leq$

$k - 1$ the graph $G[C_i \cup N[v]]$ can be searched (resp. C_i can be cleared) using $2(|C_i \cap T| + 1) + 3 + |D_i|$ searchers by our inductive assumption.

Since the components are cleared individually and $|D_i| \leq |D| + 1 + (m - 1)$ we have that the number of searchers needed is at most $|D| + m + 2(|C_1 \cap T| + 1) + 3$. Each component C_i contains at least two vertices of T and we obtain $|C_1 \cap T| + m \leq k$. Combined with $|C_1 \cap T| \leq k - 2$ this implies that the number of searchers is at most $|D| + 2k - 2 + 5 = |D| + 2k + 3$.

Now we can concentrate our efforts on the hard case in which for every vertex $v \in S = B \setminus N[T \cup D]$ at least one component of $G - N[v]$ contains exactly one vertex of T.

We say that a vertex $v \in S$ is t-separating in G for a vertex $t \in T$ if there is a component C of $G - N[v]$ such that $C \cap T = \{t\}$.

Suppose that the dominating target T for B in G contains a vertex t for which no vertex $v \in B$ is t-separating. Then $T \setminus \{t\}$ is a dominating target for $B \setminus N[t]$ in the graph G. Thus placing a searcher on vertex t and then inductively using the claim for a dominating target of cardinality $k - 1$, we easily obtain that $1 + |D| + 2k + 1 = |D| + 2k + 2$ searchers are sufficient to search G.

From now on we may assume that for every vertex $t \in T$ there is a vertex $v \in S$ which is t-separating. Notice that every vertex $v \in S$ is t-separating for some $t \in T$ (otherwise we are in the easy case) and that a vertex v may be t-separating for various vertices of T. It will be convenient to assume that $T = \{t_1, t_2, \ldots, t_k\}$.

For every $i \in \{1, 2, \ldots, k\}$, let $R(t_i)$ be the set of all vertices $v \in S$ that are t_i-separating. Notice that $R(t_i) \neq \varnothing$ for all $i \in \{1, 2, \ldots, k\}$ and that $\bigcup_{1 \leq i \leq k} R(t_i) = V \setminus N[D \cup T]$.

For vertices $a, b \in R(t_i)$ we say that $a \prec_i b$ if a and t_i are in one component of $G - N[b]$ but b and t_i are not in one component of $G - N[a]$. For every $i \in \{1, 2, \ldots, k\}$, we choose a $\prec_i$-maximal element v_i, i.e. a maximal element of the partially ordered set $(R(t_i), \prec_i)$. (Notice that $v_i = v_j$ for $i \neq j$ is possible.) Let C_i be the component of $G - N[v_i]$ containing t_i, i.e. $T \cap C_i = \{t_i\}$.

The domination search program works as follows. We place searchers on all vertices of D and on all vertices of $\{v_1, v_2, \ldots v_k\}$ that will never be removed throughout the search.

Then the components $C_1, C_2, \ldots, C_k$ will be cleared individually one by one, where each component C_i is cleared as follows. Let $C_1^i, C_2^i, \ldots, C_{m_i}^i$ be all components of $G - N[v_i]$ containing a vertex of T except the component C_i, thus $m_i < k - 1$. For each component C_j^i, $j \in \{1, 2, \ldots, m_i\}$, we choose a vertex $y_j^i \in N[v_i] \cap N(C_j^i)$. By the second part of Lemma 2 and similar to the easy case we may conclude that for every $i \in \{1, 2, \ldots, k\}$, $\{v_i, t_i\}$ is a dominating target for the vertex set $B_i = ((B \cap C_i) \setminus \bigcup_{j \neq i} N[y_j^i]) \setminus N[D_i]$ in the graph $G[C_i \cup N[v_i]]$ where $D_i = ((D \cap (C_i \cup N[v_i])) \cup \{v_i\} \cup \{y_j^i : j \neq i\}$ satisfies $N[D_i] \supseteq C_i \setminus B_i$. We place searchers on all the vertices $y_1^i, y_2^i, \ldots, y_{m_i}^i$. Finally as consequence of Lemma 1 (already obtained at the beginning of this proof), four additional searchers are sufficient to clear C_i. Finally we remove all searchers from $y_1^i, y_2^i, \ldots, y_{m_i}^i$. Altogether this clears C_i.

Hence all components $C_1, C_2, \ldots, C_k$ can be cleared by using at most $|D|+k+ (k-1)+4 = |D|+2k+3$ searchers. After clearing the components $C_1, C_2, \ldots, C_k$ there remain $|D| + k$ searchers on G. The current set of cleared vertices is $N[D] \cup \bigcup_{1 \leq i \leq k} C_i \cup \bigcup_{1 \leq i \leq k} N[v_i]$. Since $N[t_i] \subseteq C_i$ and $\bigcup_{1 \leq i \leq k} R(t_i) = V \setminus N[T \cup D]$ we may conclude that all contaminated vertices belong to the set $U = (\bigcup_{1 \leq i \leq k} R(t_i) \setminus (N[v_i] \cup C_i)) \setminus N[D]$.

Let us show how the vertices from the set U can be cleared by placing k additional searchers on vertices of G. For every $i \in \{1, 2, \ldots, k\}$, we choose a vertex $u_i \in N[v_i]$ having a neighbor in C_i. We claim that every vertex of U is adjacent to a vertex of the set $\{u_1, u_2, \ldots, u_k\}$ which would imply that additional k searchers placed on all vertices of $\{u_1, u_2, \ldots, u_k\}$ together with the k searchers already placed on $\{v_1, v_2, \ldots, v_k\}$ will clear all vertices of U. In fact, choose a vertex $w \in U$.

Then $w \in R(l)$ for some $l \in \{1, 2, \ldots, k\}$, thus w is t_l-separating. Choose a (t_l, v_l)-path P having all vertices in C_l except u_l and v_l. Since v_l is a $\prec_l$-maximal element, v_l does not belong to the component of $G - N[w]$ containing t_l and therefore w is adjacent to a vertex p of the path P. On the other hand, since $w \notin C_l \cup N[v_l]$ we may conclude that $p \notin C_l$. Therefore $p = u_l$ and the claim follows.

Finally we place searchers on all vertices of $\{u_1, u_2, \ldots u_k\}$ thus clearing vertices from U by our above claim. This needs at most $2k + |D|$ searchers.

Summarizing, the number of searchers needed is at most $|D| + 2k + 3$. This completes the proof.   □

Corollary 5. *For every disconnected graph G, $ds(G) \leq 3 + 2 \cdot \max_C dt(C)$, where the maximum is taken over all components C of G.*

To illustrate the strength of Theorem 7 let us consider the complete graph G on n vertices and the graph $S_2(G)$ obtained from G by replacing each edge by a path of length 3. Clearly $dt(S_2(G)) = n$ and by Theorem 2 we have $ds(S_2(G)) = n$. This leads to the following conjecture which is up to our knowledge the strongest possible strengthening of our theorem.

Conjecture 2. $ds(G) \leq dt(G)$ for all graphs G.

References

1. K. ANDO, A. KANEKO and S. GERVACIO, *The bandwidth of a tree with k leaves is at most $\lceil \frac{k}{2} \rceil$*, Discrete Mathematics, 150 (1996), pp. 403–406.
2. D. BIENSTOCK, *Graph searching, path-width, tree-width and related problems (a survey)*, DIMACS Series in Discrete Mathematics and Theoretical Computer Science, 5 (1991), pp. 33–49.
3. J. A. BONDY, *Basic graph theory: Paths and circuits*, in Handbook of Combinatorics, Vol. 1, R. L. Graham, M. Grötschel, and L. Lovász, eds., Elsevier Science B.V., 1995, pp. 3–110.

4. A. BRANDSTÄDT, V. B. LE and J. P. SPINRAD, *Graph classes: a survey*, SIAM Monographs on Discrete Mathematics and Applications, Society for Industrial and Applied Mathematics, Philadelphia, 1999.

5. D. G. CORNEIL, S. OLARIU and L. K. STEWART, *Linear time algorithms for dominating pairs in asteroidal triple-free graphs*, SIAM Journal on Computing, 28 (1999), pp. 1284–1297.

6. D. G. CORNEIL, S. OLARIU and L. K. STEWART, *Asteroidal triple-free graphs*, SIAM Journal on Discrete Mathematics, 10 (1997), pp. 399–430.

7. N. D. DENDRIS, L. M. KIROUSIS and D. M. THILIKOS, *Fugitive-search games on graphs and related parameters*, Theoretical Computer Science, 172 (1997), pp. 233–254.

8. J. A. ELLIS, I. H. SUDBOROUGH and J. TURNER, *The vertex separation and search number of a graph*, Information and Computation, 113 (1994), pp. 50–79.

9. F. FOMIN, *Helicopter search problems, bandwidth and pathwidth*, Discrete Applied Mathematics, 85 (1998), pp. 59–71.

10. F. FOMIN, *Search demands and interval graphs* (in russian), Discrete Analysis and Operations Research, Series 1, 3 (1998), pp. 70–79.

11. F. V. FOMIN and N. N. PETROV, *Pursuit-evasion and search problems on graphs*, Congressus Numerantium, 122 (1996), pp. 47–58.

12. M. FRANKLIN, Z. GALIL and M. YUNG, *Eavesdropping games: A graph-theoretic approach to privacy in distributed systems*, in Proceedings of the 34th Annual Symposium on Foundations of Computer Science, Palo Alto, California, 1993, IEEE, pp. 670–679.

13. L. M. KIROUSIS and C. H. PAPADIMITRIOU, *Searching and pebbling*, Theoretical Computer Science, 47 (1986), pp. 205–218.

14. T. KLOKS, D. KRATSCH and H. MÜLLER, *Approximating the bandwidth of asteroidal-triple free graphs*, Journal of Algorithms, 32 (1999), pp. 41–57.

15. T. KLOKS, D. KRATSCH and H. MÜLLER, *On the structure of graphs with bounded asteroidal number*, Technical Report Math/Inf/97/22, Friedrich-Schiller-Universität, Jena, Germany, 1997.

16. F. S. MAKEDON, C. H. PAPADIMITRIOU and I. H. SUDBOROUGH, *Topological bandwidth*, SIAM Journal on Algebraic and Discrete Methods, 6 (1985), pp. 418–444.

17. F. S. MAKEDON and I. H. SUDBOROUGH, *On minimizing width in linear layouts*, Discrete Applied Mathematics, 23 (1989), pp. 243–265.

18. A. PARRA and P. SCHEFFLER, *Characterizations and algorithmic applications of chordal graph embeddings*, Discrete Applied Mathematics, 79 (1997), pp. 171–188.

19. R. RAZ and S. SAFRA, *A sub-constant error-probability low-degree test, and sub-constant error-probability PCP characterization of NP*, in Proceedings of the Annual ACM Symposium on Theory of Computing, vol. 29, The Association for Computing Machinery, 1997, pp. 475–484.

20. N. ROBERTSON and P. D. SEYMOUR, *Graph minors — a survey*, in Surveys in Combinatorics, I. Anderson, ed., Cambridge Univ. Press, 1985, pp. 153–171.

21. K. SUGIHARA and I. SUZUKI, *Optimal algorithms for a pursuit-evasion problem in grids*, SIAM Journal on Discrete Mathematics, 2 (1989), pp. 126–143.

Efficient Communication in Unknown Networks

Luisa Gargano[1,4], Andrzej Pelc[2], Stephane Perennes[3], and Ugo Vaccaro[1,4]

[1] Dipartimento di Informatica ed Applicazioni, Università di Salerno, 84081
Baronissi(SA), Italy. {lg,uv}@dia.unisa.it. Supported in part by CNR.
[2] Département d'Informatique, Université du Québec à Hull, Hull, Québec J8X 3X7,
Canada. pelc@uqah.uquebec.ca. Supported in part by NSERC grant OGP 0008136.
[3] SLOOP I3S-CNRS/INRIA, Université de Nice-Sophia Antipolis, F-06902 Sophia
Antipolis Cedex, France. Stephane.Perennes@sophia.inria.fr
[4] Research done during this author's visit at INRIA Sophia Antipolis

Abstract. We consider the problem of disseminating messages in networks. We are interested in information dissemination algorithms in which machines operate independently without any knowledge of the network topology or size. Three communication tasks of increasing difficulty are studied. In *blind broadcasting* (BB) the goal is to communicate the source message to all nodes. In *acknowledged blind broadcasting* (ABB) the goal is to achieve BB and inform the source about it. Finally, in *full synchronization* (FS) all nodes must simultaneously enter the state *terminated* after receiving the source message. The algorithms should be efficient both in terms of the time required and the communication overhead they put on the network. We limit the latter by allowing every node to send a message to at most one neighbor in each round. We show that BB is achieved in time at most $2n$ in any n-node network and show networks in which time $2n - o(n)$ is needed. For ABB we show algorithms working in time $(2+\epsilon)n$, for any fixed positive constant ϵ and sufficiently large n. Thus for both BB and ABB our algorithms are close to optimal. Finally, we show a simple algorithm for FS working in time $3n$.The optimal time of full synchronization remains an open problem.

1 Introduction

Broadcasting is the task of disseminating a message from a source node to all other nodes of a communication network. The goal is to complete the task as fast as possible using a moderate amount of communication in the network.

Broadcasting is an important and basic communication primitive in many multiprocessor systems. Application domains of broadcasting include scientific computation, network management protocols, database transactions, and multimedia applications. Due to their widespread use, broadcasting algorithms need to be efficient in terms of time and network communication. The dissemination of the source message should be accomplished quickly, without posing too heavy communication load on the network.

We are interested in broadcasting (and more general information diffusion) algorithms which operate correctly and efficiently when there is no centralized

U. Brandes and D. Wagner (Eds.): WG 2000, LNCS 1928, pp. 172–183, 2000.

control in the network, and machines operate independently without *any a priori* knowledge of the network topology or even size.

We assume that communication proceeds in synchronous parallel *rounds*. A round is the time necessary for a machine to transfer a message to a neighbor. A message sent by a node in a given round becomes available at the neighbor in the next round. The number of rounds necessary to carry out a given information dissemination task is called the *time* needed for this task.

Under our hypothesis, even the size of the network is unknown. Therefore, a first problem is how the machines know that the dissemination process is completed. It could be very well finished but no node may be aware of this fact. This is different from the scenario when all necessary information about the network is available to nodes. In that case the time of broadcasting can be known in advance and thus all nodes can be aware of the termination of broadcasting as soon as it is completed. Since this is not the case in our setting, we study the following three information dissemination tasks of increasing difficulty.

- **Blind Broadcasting** (BB): the goal is simply to communicate the source message to all nodes.
- **Acknowledged Blind Broadcasting** (ABB): the goal is to achieve BB and inform the source that the process is terminated.
- **Full Synchronization** (FS): all nodes must (simultaneously) enter the state *terminated* after receiving the source message.

ABB may be crucial, e.g., when the source has several messages to disseminate and nobody should learn the next message until everybody learns the previous one. FS implies common knowledge about completion of broadcasting (all nodes know that everybody got the message, they know that everybody knows that everybody got the message, etc.). It is the first time when a concerted action of all nodes can be undertaken after learning the source message.

The BB, ABB, and FS communication tasks are closely related to some information diffusion problems which have been extensively studied in the literature. Blind broadcasting is related to Propagation of Information introduced in [17]. Acknowledged blind broadcasting is related to Propagation of Information with Feedback (PIF) [17], also called broadcasting with echo [18]. Full synchronization is related to the Firing Squad Problem (cf., e.g., [18]).

The main difference between our approach and that in [1,17,18] is that while the latter focus on formal description and validation of protocols and on the message complexity for these tasks in the asynchronous setting, in the present paper we study the worst case number of rounds needed to complete these tasks.

Moreover, we stress that the tasks described above are classically accomplished by flooding the network, that is, by protocols in which nodes send at each step all the information they know to *all* of their neighbors (e.g. [1]). This implies that the communication complexity of such algorithms is very large, thus resulting in a significant communication overhead placed on the network. We study algorithms that place less overhead on the network while keeping low the number of rounds.

1.1   Related Work

Coping with incomplete information in the design of network algorithms is an important problem whose many variations have been extensively studied.

In particular, Topology Discovery and Network Measurement constitute a class of problems, recently studied in the literature, related to the need of gathering information about an unknown network structure; well studied tasks in this area are the discovery of the physical topology of the network or the router-level map of the Internet [4,10]. The above tasks are considered of fundamental importance for an efficient solution of many critical network management problems. One general requirement for discovery algorithms is the need of being executed without imposing a significant overhead on the network. In particular, an important problem in this area, the Resource Discovery problem in which machines in a network initially unaware of each other need to learn both the existence of each other and the topology of the entire network, has been recently studied under communication assumptions similar to the ones we use in this paper [11]. Some of the techniques we develop apply also to the general problem of learning the topology of the network.

Recently, a few papers [6,7,8] have been devoted to the study of broadcasting in networks in which neither edges nor nodes have *a priori* assigned labels, and thus in networks both anonymous and devoid of sense of direction. However, in these papers it was assumed that nodes know the topology of the network, although they lack some information necessary to orient it, and thus to carry out broadcasting in the most efficient way.

While randomized broadcasting protocols have been studied under assumption that nodes are ignorant of the network (cf., e.g., [3,9]), deterministic broadcasting in point-to-point networks has been always considered assuming at least partial knowledge of the network (see the survey papers [13,14,15]). For example in [2] the number of messages needed for broadcasting was studied assuming that nodes know at least their immediate neighborhood. To the best of our knowledge, ours is the first paper devoted to the study of time of deterministic information dissemination algorithms in arbitrary point-to point networks, under network ignorance scenario (see [12] for similar investigations in specific topologies and [5] for similar investigations in arbitrary radio networks).

1.2   Our Results

Flooding based algorithms present the disadvantage of putting a significant communication overhead on the network. In order to reduce the number of messages sent in each round we will consider algorithms in which each machine is allowed to send a message to at most one of its neighbors in each round. We refer to the above reduced communication power of the algorithms as the *restricted communication* model (cf. a similar assumption in [11]). We study the minimum time (i.e., number of rounds) needed for the BB, ABB, and FS tasks under this model.

Since nodes are ignorant of network topology, they cannot distinguish between yet unused adjacent links and consequently the decision on which of them the message should be sent belongs to the adversary, similarly as in [6,7,8]. We consider the worst case time, i.e., the time resulting from the most detrimental behavior of the adversary.

For BB we show an asymptotically optimal algorithm. More precisely, we show that BB is achieved in time at most $2n$ in any n-node network and show networks in which time $2n - o(n)$ is needed for BB (and hence also for the other two tasks). For ABB we show algorithms achieving this goal in any n-node network in time $(2 + \epsilon)n$, for any fixed positive constant ϵ and sufficiently large n. These algorithms use messages of logarithmic size. Using large messages, the source can additionally learn the entire topology of the network and time can be reduced to $2n$. Finally, we show a simple algorithm for FS working in time $3n$ for any n-node network. This algorithm also uses messages of logarithmic size. We also mention that we can reduce this time to slightly less than $2.9n$, using large messages, and simultaneously inform all nodes about the topology of the network. The modification details will be given in the full version of the paper.

2   Terminology

A network is modelled by a simple connected undirected graph with a distinguished node called the *source*. The *distance* between two nodes is the length of the shortest path between them. The *depth* of the network is the largest distance between the source and any other node. The ith *level* of a network is the set of all nodes at distance i from the source. Clearly a node in level i can be adjacent only to nodes in levels $i-1$, i and $i+1$. Every node in level i is adjacent to a node in level $i - 1$. Call an edge *saturated* if the source message has been transmitted on it in at least one way.

3   Blind Broadcasting

The following observation will be essential in further considerations.

Proposition 1. *Accomplishing BB in a communication network is equivalent to transmitting the source message on each link in at least one direction.*

A proof can be found in [16]. A link on which the source message has been transmitted will be called *saturated*. Establishing an upper bound for BB is equivalent to showing that all edges of any network will be saturated in the given number of rounds, regardless of the behavior of the adversary, while establishing a lower bound is equivalent to constructing a network and showing an adversary strategy for it which prevents saturation of all edges before the given time.

The following natural algorithm for BB is asymptotically optimal.

Protocol BB

Phase 1. The source sends its message on all adjacent edges in consecutive rounds starting in round 1.

Phase 2. Every node v that got the source message sends the source message in consecutive rounds on all yet unsaturated edges (in any order) and then stops.

Theorem 1. *For any n-node network of depth d, Protocol BB accomplishes BB in worst case time $2n - d - 2$.*

Proof. Let s be the source and L_i, for $i = 1, ..., d$, be the ith level of the network. Denote by a_i the size of L_i. After a_1 rounds since the beginning of broadcasting, all edges between the source and L_1 are saturated. After at most $a_1 - 1 + a_2$ further rounds, all internal edges in L_1 and all edges between L_1 and L_2 are saturated. After at most $a_2 - 1 + a_3$ further rounds, all internal edges in L_2 and all edges between L_2 and L_3 are saturated. Proceeding in this way we conclude that after at most $a_1 + (a_1 - 1 + a_2) + (a_2 - 1 + a_3) + \cdots + (a_{d-1} - 1 + a_d)$ rounds, all edges of the network, except possibly the internal edges in L_d, are saturated. Further $a_d - 1$ rounds are sufficient to saturate all these edges. Thus total time is at most $2(a_1 + \cdots + a_d) - d = 2(n - 1) - d = 2n - d - 2$.   □

Theorem 2. *For any positive integer n and any algorithm for BB, there exists an n-node network and an adversary strategy for which this algorithm requires time $2n - o(n)$.*

Proof. We assume without loss of generality that no message is sent on a saturated edge. Whenever a node sends a message, the adversary chooses among unsaturated edges adjacent to v.

We show the argument for n of the form $4k^2$. Modifications in the general case are easy. First we define the network. The set of all n nodes is divided into $m = 2k$ disjoint sets $A_1, ..., A_m$ of size m. Edges join every pair of nodes in A_i, for all $i = 1, ..., m$, and every pair u, v of nodes, such that $u \in A_i$ and $v \in A_{i+1}$, for all $i = 1, ..., m - 1$. Let $A_1 = \{x_0, ..., x_{m-1}\}$. The source is x_0.

Next we define the adversary strategy. In the ith round, $i = 1, ..., m - 1$, all nodes $x_0, ..., x_{i-1}$ send a message to x_i. In the mth round every node $x_i \in A_1$ sends a message to a different node $y_i \in A_2$. In the next $m - 1$ rounds the adversary strategy is as follows. In round $m + j$, $j = 1, ..., m - 1$, nodes x_i and $y_{i+j(\bmod m)}$, $i = 0, ..., m - 1$, send messages to each other. Hence, after $2m - 1$ rounds since the start of broadcasting, saturated edges are exactly those between pairs of nodes in A_1 and between A_1 and A_2. Moreover, informed nodes with adjacent unsaturated edges are exactly those in A_2.

It is well known that the set of edges of a complete graph of even size m can be partitioned into $m - 1$ pairwise disjoint perfect matchings. Let $M_1, ..., M_{m-1}$ be these matchings for the clique on A_2. In the further $m - 1$ rounds, nodes in A_2 send messages to each other along edges from M_i in round $(2m - 1) + i$. The next m rounds are spent sending messages between A_2 and A_3, etc. In general, saturating all internal edges of A_i takes $m - 1$ rounds, and saturating all edges between A_i and A_{i+1} takes m rounds. Hence the total number of rounds is $m(m - 1) + (m - 1)m = 2m^2 - 2m = 2n - 2\sqrt{n}$.   □

4   Acknowledged Blind Broadcasting

In this section we describe a family of protocols achieving acknowledged blind broadcasting. Their aim is to broadcast the source message to all other nodes of the network, and inform the source that all nodes got the message. For any *a priori* given positive constant ϵ, we show a protocol working in worst case time $(2 + \epsilon)n$, for n-node networks, where n is sufficiently large. The protocol uses messages of size at most $\log n$ and additionally informs the source of the number of nodes in the network.

Let ϵ be a positive constant fixed throughout this section. The protocol consists of two phases, the first of which is modified blind broadcasting and the second is the acknowledgement phase. The aim of the first phase is to broadcast the source message to all nodes, and simultaneously create a spanning tree of the network, of height equal to the depth of the network. Then information about completion of broadcasting and the number of nodes is efficiently sent to the source bottom-up using this tree.

The tree is constructed dynamically, and the depth of each node in it decreases during the execution of the protocol, to finally reach the level number of the node in the network. This guarantees efficient execution of the acknowledgement phase but requires updating labels which indicate the current depth, and updating current parents. The above actions must be carefully scheduled to avoid long delays in spreading the source message. We show how this can be done so that the additional delay caused by these actions be at most ϵn.

Another problem is avoiding confusion among various versions of the constructed tree. Nodes do not know when the tree is final and hence send information about completion of broadcasting and about size of parts of the network already using earlier versions of the tree. Thus messages concerning various versions of the tree circulate simultaneously and there is an *a priori* danger of forgetting a node or counting it more than once. This could happen if information from different rounds were mixed: a node earlier counted in one branch could subsequently change parents and be also counted in a different branch. For the same reason a node might not be counted at all. To avoid this confusion we use "time stamps" indicating which version of the tree the message refers to. At the end, the source makes sure to consider a single version of the tree.

Protocol ABB(ϵ)

Phase 1. Let c be an integer larger than $6/\epsilon$. c is appended to the original source message and forms the message M which is disseminated. The source assigns itself label 0 and sends it together with M on consecutive yet unsaturated edges. When a node gets M for the first time from a node with label i, it assigns itself label $i + 1$ and sends it together with M on consecutive incident edges.

The crucial idea is label updating. Whenever a node v gets a new label, it sends it (together with M), to all neighbors u from which it didn't hear, or from which it heard that u's label is higher than v's current label. Such neighbors u will be called *compulsory* with respect to the current label of v. If a node has label i and hears from a node with label $j < i - 1$, it updates its own label to $j + 1$. Thus labels of each node are decreased every time they change, and the smallest label a node can get is the number of its level.

Label updating is combined with dynamic construction of a spanning tree. At every round every informed node has a single *parent* which can change several times in the course of the protocol. The current parent of node v is its neighbor whose message caused v to modify its label most recently (or whose message woke up v, in case when v still has its first label).

A node v acknowledges its current parent by sending to it a message "you became my parent" and cancels the old parent by sending to it a message "you stopped being my parent". This is combined with label updating as follows.

After getting a new label l, a node v starts process Spread(l) described as follows. In the first $c-1$ rounds after getting label l, v sends this label (together with M) to consecutive compulsory neighbors. In the next two rounds, it first cancels the old parent, then acknowledges the new parent. In further rounds it continues sending its label to other compulsory neighbors. If during process Spread(l), v gets a new label l', Spread(l) is halted and v starts process Spread(l') from the beginning.

A node involved in some process Spread(l) is called *busy*. Otherwise it is *idle*. Clearly, an idle informed node has all its incident edges saturated.

A node considers as its *current children* all nodes from which it got an acknowledgement "you became my parent" that has not been cancelled afterwards.

A node v that becomes idle at some round t', starts a *waiting period* of $c+1$ rounds. If it does not change its label before the end of the waiting period, v becomes *confirmed* in round $t = t' + c + 1$. If at this round v does not have current children, v considers itself a *leaf* of the tree corresponding to round t, and enters Phase 2. (Notice that a confirmed node can subsequently loose its status by getting a new label. In this case it becomes busy, then idle, and then confirmed again.)

Phase 2. Messages in phase 2 consist of a *time stamp* and of a *content* of the form "my subtree has size x". Every confirmed node keeps a variable *con* whose value is the round number in which it became (most recently) confirmed. Such a node also keeps lists of all messages obtained from all current children.

Phase 2 is started by confirmed nodes which consider themselves leaves of the tree corresponding to round t. In every round $i > t$ such a node v sends a message "my subtree has size 1", with time stamp $i-1$, to its current parent. This is done as long as v remains confirmed.

In every round every confirmed node v that has current children looks at all lists of messages obtained from all of them. If there exists a common time stamp t' in all these lists, larger than the current value of v's variable *con*, call the corresponding messages t'-*healthy*. Let t be the maximum of such t'. Node v takes all contents "my subtree has size x_i" corresponding to all t-healthy messages from all its current children c_i, $i = 1, ... r$, and then sends the message "my subtree has size $x_1 + \cdots + x_r + 1$ " with time stamp t to its parent.

As soon as the source detects t-healthy messages from all its children u_i, $i = 1, ..., k$, for some time stamp t, it knows that all nodes of the network got the source message, and that the total number of nodes is $n = y_1 + \cdots + y_k + 1$, where "my subtree has size y_i" are the corresponding contents.

Theorem 3. *Protocol ABB(ϵ) correctly accomplishes acknowledged blind broadcasting and works in worst case time at most $(2+\epsilon)n$, in any n-node network, for sufficiently large n.*

Proof. In order to prove the correctness of Protocol ABB(ϵ), it is enough to show that detecting t-healthy messages by the source, for some time stamp t, implies that in round t broadcasting has been completed and this information reached the source bottom-up using the spanning tree frozen at time t.

Indeed, according to the description of process Spread(l), a node may be acknowledged as a parent at most $c+1$ rounds after becoming it. Hence, the waiting period imposed on a node v before it becomes confirmed, guarantees that every node w whose parent is v, is already counted by v among its current children. In particular, nodes considering themselves leaves of the tree corresponding to round t are indeed leaves of this tree.

When a confirmed non-leaf v gets t-healthy messages "my subtree has size x_i", it knows that all of its children corresponding to round t reported the sizes of their subtrees corresponding to this round, and consequently v's subtree in the tree corresponding to round t had size $x_1 + \cdots + x_r + 1$. Node v reports this information (still with time stamp t) to its parent, that did not change since round t. Consequently, the rule of conveying information by non-leaves guarantees that only snapshots corresponding to the same round are relayed up the tree. Hence the total number of nodes computed by the source is indeed the number of nodes in the spanning tree corresponding to some round t, i.e., the number of nodes in the network.

Note that between round t and the round when t-healthy messages are detected by the source, the tree could change many times. Nevertheless the source gets a correct and complete snapshot of the situation occurring in round t.

It remains to show that the source will detect t-healthy messages for some $t \leq (2+\epsilon)n$.

Call a node *large* if its degree is at least c, and *small* otherwise. According to the description of process Spread(l), the actions of cancelling and acknowledging parents do not delay blind broadcasting in small nodes, and each such action delays it by at most one round in large nodes. Let a_i be the size of the i-th level of the network. Define the round

$$T_i = a_1 + (a_1 - 1 + a_2) + ... + (a_{i-1} - 1 + a_i) + 2j_{i-1},$$

where j_{i-1} is the number of levels before the i-th that contain large nodes.

By round T_i all nodes in the i-th level may have changed their labels several times but they will certainly get the message from a neighbor in level $i-1$, containing its final label $i-1$. Hence such a node, in case when it is large, may only loose two further rounds to cancel the old parent and acknowledge the new one (which from now on will never change), and then it will pass its final (freshly updated) label i further.

A computation similar to that in the proof of Theorem 1 shows that all edges will become saturated in round at most $2n-d+2x$, where x is the number of levels containing large nodes. Clearly, $x \leq 3n/c$. The waiting period is $c+1$. Hence after the round $2n-d+2x+c+1$, phase 2 must start, and at this point all labels are final, the spanning tree has height d (equal to the depth of the network) and never changes again. This implies that by round $2n-d+2x+c+1+d$ the source will detect t-healthy messages, for $t = 2n - d + 2x + c + 1$, if it has not detected such messages for smaller t at some earlier round. We conclude that Protocol ABB(ϵ) works in worst case time at most $2n-d+2x+c+1+d \leq 2n+6n/c+c+1$, which does not exceed $(2 + \epsilon)n$, for sufficiently large n. $\square$

It should be noticed that at the time when the source learns that all nodes got the message in Protocol ABB(ϵ), other nodes still send messages with new time stamps at each round. As mentioned in the introduction, ABB should be considered as one cycle in an ongoing process of broadcasting several messages, in which a new message should be disseminated only after everybody learned the previous one. When the source sends the next message, it simultaneously orders to stop all messages concerning the previous cycle.

We conclude this section by describing a protocol working even faster than Protocol ABB(ϵ) (in time at most $2n$ for an n-node network) and accomplishing much more. Upon its completion, all nodes get the source message, and the source learns not only that and the number of nodes, as in the previous case, but in fact learns the topology of the entire network. The drawback of this protocol is that the circulating messages are large, as opposed to logarithmic size in Protocol ABB(ϵ). However, large messages are hard to avoid if the goal is to learn fast the entire topology.

We present the protocol assuming that, when a node is informed, it gets a name, all names being distinct. In the full paper it will be shown how this naming procedure can be efficiently added to the protocol.

The protocol works in two phases. Similarly as before, the aim of the first phase is to broadcast the source message to all nodes and simultaneously create a spanning tree of the network, of height equal to the depth of the network. This is done by a mechanism of label updating, as in Protocol ABB(ϵ), with the exception that current parents are not acknowledged (this permits to save ϵn time). In phase 2, nodes send all topological information they learned to date, up the tree, toward the source.

Protocol Learn-All

Phase 1. The source assigns itself label 0 and sends it together with the original message on consecutive yet unsaturated edges. Nodes get and update labels in the same way as in Protocol ABB. When a node v gets a new label, it sends it to all neighbors compulsory with respect to this label (i.e., to all neighbors u from which it didn't hear or from which it heard that u's label is higher than v's current label).

The current parent of a node is defined as before. Any node that got all of its edges saturated and sent its current label to all compulsory neighbors, enters phase 2. (As opposed to Protocol ABB, phase 2 is started by every node independently, not only by the leaves of the current tree.)

Phase 2. Every node v keeps a record $CLK(v)$ called *current local knowledge*. It consists of the name v of the node, of its degree $\deg(v)$, and of a list of neighbors of v, initialized as empty. In the beginning of phase 2, every node v sends $CLK(v)$ to its current parent. Whenever v gets a phase 2 message from a neighbor u, it updates $CLK(v)$ by adding u to its list of neighbors and then sends its own $CLK(v)$ together with all $CLK(w)$ that it previously learned, to its current parent. In fact, nodes keep track of what they have sent to the current parent. If the parent did not change and v obtained no new information, it does not send anything in the given round.

It should be noted that messages from phase 1 can still be sent after a node started phase 2. If v changes its label, it first sends it to all compulsory neighbors and then sends all available $CLK(w)$ (together with its own) to the new parent.

The source continuously updates all incoming records $CLK(v)$ by adding new nodes to the respective neighbors lists, and applies the following *termination rule*. At each round, for any v for which it got a record $CLK(v)$, the source counts the number $N(v)$ of nodes u such that either u is in the neighbors list in $CLK(v)$ or v is in the neighbors list in $CLK(u)$. As soon as $N(v) = \deg(v)$ for all these v, the source concludes that all nodes in the network got the message and reconstructs the entire network as follows. The nodes of the network are precisely those nodes v for which it got records $CLK(v)$, and nodes u and v are adjacent, iff either u is in the neighbors list in $CLK(v)$ or v is in the neighbors list in $CLK(u)$.

Theorem 4. *Upon completion of Protocol Learn-All, the source knows that all nodes got the original message and has a correct knowledge of the topology of the network. The protocol works in worst case time at most $2n$, for any n-node network.*

5   Full Synchronization

We present a simple protocol for FS working in worst case time at most $3n$, for any n-node network. The advantage of this protocol is that it uses only messages of size logarithmic in n.

The idea of the protocol is the following. In the first phase, blind broadcasting is accomplished and simultaneously a spanning tree (of uncontrolled depth and shape) is constructed. In the second phase, gossiping (i.e., all-to-all broadcasting) is done in this tree, thus informing all nodes that everybody got the source message, and permitting them to simultaneously enter the state *terminated*.

Protocol Simple-FS

Phase 1. [It is a simplified version of phase 1 of protocol ABB]. Nodes of the network perform blind broadcasting. When a node v gets the source message for the first time, and the sender of the message is neighbor u, v sends the message "you are my parent" to u in the next round, and then relays the source message on consecutive yet unsaturated edges, as usual. (As opposed to protocol ABB, parents never change.) If node v gets the source message for the first time simultaneously from many neighbors, it chooses as its parent one of these nodes, arbitrarily. After getting the acknowledgement, u considers v as its child. The parent and all children of a node are called its *tree neighbors*. They are indeed its neighbors in the spanning tree which is being constructed.

Phase 2. Phase 2 is started by nodes that got all their adjacent edges saturated and did not get the message "you are my parent" in the next round. In the special case when the source has degree 1, phase 2 is also started by the source, after it has send the message on its unique edge. The above nodes are *leaves* in the (unoriented) spanning tree constructed in phase 1. They start phase 2 by sending the message "my subtree has size 1" to their unique tree neighbor. All other nodes v whose all adjacent edges are saturated, proceed as follows. Suppose that the tree neighbors of v are w, $v_1,...,v_k$, and v got messages "my subtree has size x_i" from all nodes v_i, $i = 1,...,k$. Then v sends the message "my subtree has size $x_1 + \cdots + x_k + 1$" to node w.

Any node that got messages "my subtree has size y_i" from all its tree neighbors u_i, $i = 1,...,k$, knows that all nodes have already got the source message and that the total number of nodes in the network is $n = y_1 + \cdots + y_k + 1$. It then waits until round $3n$ and enters the state *terminated*.

Notice that the size of messages should only permit sending numbers smaller than n, thus size logarithmic in n suffices.

Theorem 5. *Protocol Simple-FS correctly accomplishes full synchronization and works in worst case time at most $3n$, for any n-node network.*

Proof. Acknowledging the parent in phase 1 delays blind broadcasting by one round in every node, except the source. Hence, using the computation of the upper bound for BB from the proof of Theorem 1, we conclude that phase 1 of Protocol Simple-FS will be completed at the latest in round $(2n-d-2)+d < 2n$. Leaves need to wait at most 1 more round to learn that they are indeed leaves. Hence phase 2 is initiated by all leaves before round $2n$.

Phase 2 is completed in at most n further rounds because its duration is limited by the longest travel time between any pair of nodes. Hence before round $3n$ every node knows that all nodes have already got the source message and knows n. Hence all nodes compute $3n$ before round $3n$, and consequently are able to enter the state *terminated* simultaneously in round $3n$. □

Our second protocol for full synchronization, Protocol Faster-FS, is slightly faster than Simple-FS - it works in worst-case time at most $2.9n$, for sufficiently large n - but it uses large messages, similarly as Protocol Learn-All. Its additional advantage is that it informs all nodes about the entire topology of the network. Since the description and analysis of the protocol are fairly involved, we do not include them in this extended abstract.

6   Conclusion and Open Problems

We considered the number of rounds required for three communication tasks of increasing difficulty, performed by nodes entirely ignorant of the network.

The optimal time of full synchronization remains an open problem. However, it should be noted that if the source knows *a priori* the size n of the network, full synchronization can be easily achieved in time at most $2n$. Indeed, it is sufficient to perform blind broadcasting, adding the starting time and the value n to the original message. In view of Theorem 1 all nodes get the message before round $2n$. Then, knowing n, they wait until round $2n$ and enter state *terminated*.

It seems interesting to study also other communication problems, such as gossiping (i.e., all-to-all broadcasting) or multicasting, under the network ignorance scenario. Other important questions concern the impact of partial knowledge of the network (e.g., nodes knowing only such parameters as the size, the diameter, or the maximum degree of the network) on the efficiency of our communication protocols.

References

1. H. Attiya and J. Welch, *Distributed Computing*, McGraw-Hill.
2. B. Awerbuch, O. Goldreich, D. Peleg and R. Vainish, A Tradeoff Between Information and Communication in Broadcast Protocols, J. of ACM 37, (1990), 238–256.
3. R. Bar-Yehuda, O. Goldreich, and A. Itai, On the time complexity of broadcast in radio networks: An exponential gap between determinism and randomization, Proc. 6th ACM PODC (1987), 98–108.
4. Y. Beritbart, M. Garofalakis, R. Rastogi, S. Seshadri, A. Silbershatz, Topology Discovery in Heterogeneous IP networks, Proceedings of INFOCOM'99.
5. B.S. Chlebus, L. Gasieniec, A. Gibbons, A. Pelc, W. Rytter, Deterministic broadcasting in unknown radio networks, Proc. SODA'2000, 861–870.
6. K. Diks, E. Kranakis and A. Pelc, Broadcasting in unlabeled tori, Par. Proc. Lett. 8 (1998), 177–188.
7. K. Diks, E. Kranakis and A. Pelc, Perfect broadcasting in unlabeled networks, Discrete Applied Mathematics 87 (1999), 33–47.
8. K. Diks, S. Dobrev, E. Kranakis, A. Pelc and P. Ruzicka, Broadcasting in unlabeled hypercubes with linear number of messages, IPL 66 (1998), 181–186.
9. U. Feige, D. Peleg, P. Raghavan and E. Upfal, Randomized broadcast in networks, Random Structures and Algorithms 1 (1990), 447–460.
10. R. Gavindan, H. Tangmunarunkit, Heuristics for Internet map discovery, Proceedings of INFOCOM'99.
11. M. Harchol-Balter, T. Leighton, D. Lewin, Resource Discovery in Distributed Networks, Proceedings of *PODC'99*, 229–237.
12. H. Harutyunyan and A. Liestman, Messy broadcasting, PPL 8 (1998), 149–160.
13. S.M. Hedetniemi, S.T. Hedetniemi and A.L. Liestman, A survey of Gossiping and Broadcasting in Communication Networks, Networks 18 (1988), 319–349.
14. J. Hromkovič, R. Klasing, B. Monien, and R. Peine, Dissemination of Information in Interconnection Networks, in: Ding-Zhu Du and D. Frank Hsu (Eds.) *Combinatorial Network Theory*, Kluwer, 1995, 125–212.
15. A. Pelc, Fault Tolerant Broadcasting and Gossiping in Communication Networks, *NETWORKS* 28 (1996), 143–156.
16. R. Reischuk and M. Koshors, Lower bound for Synchronous Systems and the Advantage of Local Information, Proc. of 2nd WDAG, (1987), pp. 374–387.
17. A. Segall, Distributed Network Protocols, IEEE Trans. Inf. Th. 29 (1983), 23–35.
18. G. Tel, Introduction to distributed algorithms, Cambridge University Press, 1994.

Graph Coloring on a Coarse Grained Multiprocessor

(Extended Abstract)

Assefaw Hadish Gebremedhin[1], Isabelle Guérin Lassous[2],
Jens Gustedt[3], and Jan Arne Telle[1]

[1] University of Bergen, Norway. {assefaw,telle}@ii.uib.no
[2] INRIA Rocquencourt, France. Isabelle.Guerin-Lassous@inria.fr
[3] LORIA & INRIA Lorraine, France. Jens.Gustedt@loria.fr

Abstract. We present the first efficient algorithm for a coarse grained
multiprocessor that colors a graph G with a guarantee of at most $\Delta_G + 1$
colors.

1   Introduction and Overview

The problem of graph coloring is crucial both for the applications of graph algorithms to real world problems *and* for the domain of parallel graph algorithms
itself. For the latter, graph colorings using a bounded number of colors are often
used in a theoretical setting to ensure the independence of tasks that are to be
accomplished on the vertices of a graph, i.e. since the color classes form independent sets that don't interact each one of them can be treated in parallel. For
a long time, no efficient parallel implementation of a graph coloring heuristic
with good speedups was known [1]. However, in a recent result, Gebremedhin
and Manne [4,5] present an algorithm and an implementation for a *shared memory computer* that proves to be theoretically and practically efficient with good
speedups.

In this paper we make this successful approach feasible for a larger variety of
architectures by extending it to the more general setting of coarse grained multiprocessors (CGM) [3]. This model of parallel computation makes an abstraction
of the interconnection network between the processors of a parallel machine (or
network) and tries to capture the efficiency of a parallel algorithm using only
a few parameters. Several experiments show that the CGM model is of practical relevance: implementations of algorithms formulated in the CGM model in
general turn out to be feasible, portable, predictable and efficient [7].

This paper is organized as follows. In the next section we review the coarse
grained models of parallel computation and the basics of graph coloring heuristics. Then, we present our algorithm together with an average case analysis
of its time and work complexity. Finally, we show how to handle high degree
vertices and how to alter the algorithm to achieve the same good time and work
complexity also in the worst-case.

U. Brandes and D. Wagner (Eds.): WG 2000, LNCS 1928, pp. 184–195, 2000.

1.1   Coarse Grained Models of Parallel Computation

In recent years several efforts have been made to define models of parallel (or distributed) computation that are more realistic than the classical PRAM models. In contrast to the PRAM, these new models are *coarse grained*, i.e. they assume that the number of processors p and the size of the input N of an algorithm are orders of magnitudes apart, $p \ll N$. By that assumption these models map much better on existing architectures where in general the number of processors is at most some thousands and the size of the data that are to be handled goes into millions and billions.

This branch of research got its kick-off with Valiant [8] introducing the so-called *bulk synchronous parallel machine*, BSP, and was refined in different directions for example by Culler et al. [2] to LogP, and by Dehne et al. [3] to CGM.

We place ourselves in the context of CGM which seems to be well suited for a design of algorithms that are not too dependent on an individual architecture. We summarize the assumptions of this model as follows:

- All algorithms perform in so-called supersteps. A superstep consists of one phase of interprocessor communication and one phase of local computation.
- All processors have the same size $M = O(N/p)$ of memory.
- The communication network between the processors can be arbitrary.

The goal when designing an algorithm in this model is to keep the individual workload, time for communication and idle time of each processor within $T/s(p)$ where T is the runtime of the best sequential algorithm on the same data and $s(p)$, the *speedup*, is a function that should be as close to p as possible. To be able to do so, it is considered as good idea to keep the number of supersteps of such an algorithm as low as possible, preferably $o(M)$.

The rationale for that is that for the communication time there are at least two invariants of the architecture that come into play: the *latency*, i.e. the minimal time a communication needs to *startup* before any data reach the other end, and the *bandwidth*, i.e. the overall throughput per time unit of the communication network for large chunks of data. In any superstep there are at most $O(p)$ communications for each processor and so a number of supersteps of $o(M)$ ensures that the latency can be neglected for the performance analysis of such an algorithm. The bandwidth restriction of a specific platform must still be observed, and here the best strategy is simply to reduce the communication volume as much as possible. See Guérin Lassous et al. [7] for an introduction and overview on algorithms, code and experiments within the coarse grained setting.

As a legacy from the PRAM model it is usually assumed that the number of supersteps should be polylogarithmic in p, but there seems to be no real world rationale for that. In fact, no relationship of the coarseness models to the complexity classes NC^k have been found, and algorithms that simply ensure a number of supersteps that are a function of p (and not of N) perform quite well in practice [6].

To be able to organize the supersteps well it is natural to assume that each processor can keep a vector of p-sized data for each other processor and so the coarseness requirement translates into

$$p^2 < M \approx N/p \tag{1}$$

1.2   Graph Coloring

A graph coloring is a labeling of the vertices of a graph $G = (V, E)$ with positive integers, the *colors*, such that no two neighbors obtain the same color. An important parameter of such a coloring C is the number of colors $\chi(C)$ that the coloring uses. This can be viewed equivalently as searching for a partition of the vertex set of the graph into *independent sets*. Even though coloring a graph with the fewest number of colors is an NP-hard problem, in many practical and theoretical settings a coloring using a bounded number of colors, possibly far from the minimum, may suffice. Particularly in many PRAM graph algorithms, a bounded coloring (resp. partition into independent sets) is needed as a subroutine. However, up to recently no reasonable coloring had shown to perform well in practical parallel settings.

One of the simplest but quite efficient sequential heuristics for coloring is the so-called *list coloring*. It iterates over $v \in V$ and colors v with the least color that has not yet been assigned to one of its neighbors. It is easy to see that this heuristic always uses at most $\Delta_G + 1$ colors, where $\Delta_G = \max_{v \in V}\{\text{degree of } v\}$. A parallelization of list coloring has been shown by Gebremedhin and Manne [5] to perform well both in theory and by experiment on shared memory machines. They show that distributing the vertices evenly among the processors and running list coloring concurrently on the processors, while checking for color compatibility with already colored neighbors, creates very few conflicts: the probability that two neighbors are colored at exactly the same instance of the computation is quite small. Their algorithm was proven to behave well on expectation and the conflicts could easily be resolved thereafter with very low cost.

1.3   The Distribution of Data on the Processors

Even more than for sequential algorithms, a good organization of the data is crucial for the efficiency of parallel code. We will organize the graphs that we handle as follows:

- Every processor P_i is responsible for a specific subset U_i of the vertices. For a given vertex v this processor is denoted by P_v.
- Every edge $\{v, w\}$ is represented as arcs (v, w), stored at P_v, and (w, v), stored at P_w.
- For every arc (v, w) processor P_v stores the identity of P_w and thus the location of the arc (w, v). This is to avoid a logarithmic blow-up due to searching for P_w.
- The arcs are sorted lexicographically and stored as a linked list per vertex.

For convenience, we will also assume that the input size N for any graph algorithm is in fact equal to the amount of edges $|E|$. Up to a constant which corresponds to the encoding size of an edge this will always be achieved in graphs that don't have isolated vertices. If the input that we receive is not of the desired form it can be efficiently tranformed into one by the following steps:

- Generate two arcs for each edge as described above,
- Radix sort (see Guérin Lassous et al. [7] for a CGM radix sort) the list of arcs such that each processor receives the arcs (v, w) if it is responsible for vertex w,
- Let every processor note its identity on these sibling arcs,
- Radix sort the list of arcs such that every processor receives its proper arcs (arcs (v, w) if it is responsible for vertex v).

On a somewhat simplified level, the parallelization of list coloring for shared memory machines as given by Gebremedhin and Manne [5] worked by tackling the list of vertices numbered from 1 to n in a 'round robin' manner: at a given time $1 \leq t \leq \frac{n}{p}$ processor P_i colors vertex $(i - 1) \cdot \frac{n}{p} + t$. The shared memory assumptions ensure that P_i may access the color information of any vertex at unit cost of time. The only problems that can occur are with the neighbors that are in fact handled at exactly the same time. Gebremedhin and Manne [5] show that the number of such conflicts is small on expectation, and that they can easily be handled a posteriori. In a coarse grained setting we have to be more careful about the access to data that are situated on other processors.

2   The Algorithm

As the best sequential algorithm for an approximate graph coloring is linear in the size of the graph $|G|$, our aim is to design a parallel algorithm in CGM with a work (total number of local computations) per processor in $O(\frac{|G|}{p})$ and overall communication costs in $O(|G|)$.

Algorithm 1 colors any graph G such that $\Delta_G \leq M$ recursively. Section 3 shows how to get rid of high degree vertices. The first type of high degree vertices are the vertices v such that $deg(v) > \frac{N}{p}$. The second type of high degree vertices are the vertices v such that $\frac{N}{pk} < deg(v) \leq \frac{N}{p}$. Note that the first call to Algorithm 1 is applied on the whole graph G ($G' = G$).

In Algorithm 1, we group the vertices that are to be handled on each processor into k different *timeslots*, see the figure and **group vertices** in the algorithm. The number $1 < k \leq p$ of timeslots is a parameter of our algorithm. For each such timeslot we group the messages to the other processors together, see **send to neighbors** and **receive from neighbors**, and thus in general we produce few, even if possibly large, messages. Figure 1 shows a graph on 72 vertices distributed onto 6 processors and 4 timeslots.

A naive parallel list coloring will respect adjacencies between timeslots, but may violate adjacencies listed in bold, inside a timeslot. Our algorithm avoids

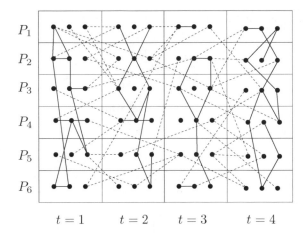

Fig. 1. Graph on 72 vertices distributed onto 6 processors and 4 timeslots.

this by coloring the bold subgraphs recursively until they are small enough to fit onto a single processor. In the recursive call, a vertex will belong to the same processor but probably to a different timeslot. Note that in general some processors may receive more vertices than others.

By proceeding like this, we may generate more conflicts than in the shared memory algorithm; we may accidentally color two neighbors in the same timeslot with the same color. We find it convenient to resolve these conflicts *beforehand*, see **identify conflicts** and **resolve conflicts**, by a recursive call to the same algorithm. We must ensure that these recursive calls do not produce a blow-up in computation and communication, and in fact the main part of the technical details in the paper are dedicated to ensure this.

In the recursive calls we must handle the restrictions that are imposed by previously colored vertices. We extend the problem specification and assume that a vertex v also has a list F_v of forbidden colors that is initially empty. An important issue for the complexity bounds will be that we will only add individual forbidden colors to F_v as the knowledge about them arrives on P_v. The list F_v as a whole will only be touched once, namely when v is finally colored.

Observe also that the recursive calls in line **resolve conflicts** are not issued synchronized between the different processors: it is not necessary (nor desired) that the processors start recursion at exactly the same moment in time. When the calls reach the communication parts of the algorithm during recursion, they will be synchronized automatically when waiting for the data of each other.[1]

---

[1] In fact, when writing this the authors got more and more convinced of the importance of asynchronicity between those of us working in Europe and the one visiting Australia.

Algorithm 1: List coloring on a CGM with p processors

Input: Subgraph $G' = (V', E')$ of a base graph $G = (V, E)$ with M' edges per processor such that $\Delta_{G'} \leq M'$, M the initial input size per processor and lists F_v of forbidden colors for the vertices.

Output: A valid coloring of G' with at most $\Delta_G + 1$ colors.

bottom:

if *the size of G' is less than* $\frac{N}{kp^2}$ **then** solve the problem sequentially on processor P_1;

else

> high degree:
>
> Get rid of high degree vertices of second type (see Section 3);
>
> group vertices:
>
> Each processor P_i groups its vertices U_i into k timeslots $U_{i,t}$, $t = 1, \ldots, k$, such that the degree sum in all timeslots is about the same; For each vertex v denote the index of its timeslot by t_v;
>
> **for** $t = 1$ **to** k **do**
>
>> **foreach** *processor P_i* **do**
>>
>>> identify conflicts:
>>>
>>> Consider all arcs $e = (v, w)$ with $v \in U_{i,t}$ and $t_v = t_w = t$; Name this set S_i and consider the vertices V_{S_i} that have such an arc;
>>>
>>> resolve conflicts:
>>>
>>> Recursively color the graph $(\bigcup V_{S_i}, \bigcup S_i)$;
>>>
>>> color timeslot:
>>>
>>> **foreach** *uncolored vertex v with $t_v = t$* **do** color v with least possible color;
>>>
>>> send to neighbors:
>>>
>>> **foreach** *arc (v, w) with $v \in U_i$, $t_v = t$ and $t_w > t$* **do** collect the color of v for P_w in a send buffer; send out the tuples $(w, color)$;
>>>
>>> receive from neighbors:
>>>
>>> receive the colors from the other processors;
>>> **foreach** *arc (v, w) with $w \in U_i$, $t_v = t$ and $t_w > t$* **do** add the color of v to F_w;

Another issue that we have to face is the possible degree variation among the vertices. Whereas for the shared memory algorithm, different degrees of the vertices that are handled in parallel just causes a slight asynchronicity of the execution of the algorithm, in a CGM setting it might result in a severe imbalance of charge and even in a memory overflow of individual processors. We will see in Section 3 how to handle this.

2.1   An Average Case Analysis

To get a high level understanding of the algorithm, let us first assume that we deal with the best of all worlds, that is:

- we never have high degree vertices of first and second type (d is a high degree if $d > \frac{M'}{k}$),
- the events of having an edge between a pair of vertices are all random and totally independent of each other.

In Section 3, we show how to handle high degree vertices and in Section 4 we show what processing we have to add if we don't assume randomness for the edges.

Lemma 1. *Consider Algorithm 1. For any edge $\{v, w\}$, the probability that $t_v = t_w$ is $\frac{1}{k}$.*

Proof. The expected size of the degree sums of the timeslots $U_{i,t}$ is the same. Since there are k timeslots, when fixing the timeslot of v the probability that w is in the same timeslot on its processor is $\frac{1}{k}$. $\square$

Lemma 2. *The expected size of all subgraphs over all $t = 1, \ldots, k$ in* **resolve conflicts** *is N/k.*

Proof. Every processor has M edges so each timeslot is expected to contain M/k edges, M/k^2 of which may create conflicts. So on each processor, considering all the timeslots, we expect M/k conflict edges and in total N/k. $\square$

Lemma 3. *For any value $1 < k \leq p$ the expected number of supersteps is linear in p.*

Proof. Each call can initiate k recursive calls. The maximal recursion depth of our algorithm is the minimum value d such that $N/k^d \leq M = N/p$, i.e. $k^d \geq p$, i.e. $d = \lceil \log_k p \rceil$. The total number of supersteps in each call is $c \cdot k$, for some constant c, one for each timeslot plus some to get rid of high degree vertices. Then, the total number of supersteps on recursion level i is given by $c \cdot k^i$ and so the total number of supersteps is

$$\sum_{i=1}^{\lceil \log_k p \rceil} c \cdot k^i \approx c \cdot k^{\log_k p} = c \cdot p \tag{2}$$

$\square$

Lemma 4. *Besides the cost for* **bottom** *and* **high degree**, *for any value of $1 < k \leq p$ the expected work and communication per processor is $O(M)$.*

Proof. First we show that the work and communication that any processor has to perform is a function in the number of its edges, i.e. M.

Algorithm 2: Solve the problem sequentially on processor P_1

Input: M the initial input size per processor, subgraph $G' = (V', E')$ of a base
graph $G = (V, E)$ with $|E'| \leq M$ and lists F_v of forbidden colors for
the vertices.

collect colors:

  foreach *processor P_i* **do**

  Let $U'_i = U_i \cap V'$ be the vertices that are stored on P_i;

  For each $v \in U'_i$ let $d(v)$ be the degree of v in G';

  Compute a sorted list A_v of the least $d(v) + 1$ allowed colors for v;

  Communicate E' and all lists A_v to P_1;

solve sequentially:

  for *processor P_1* **do**

  Collect the graph G' together with the lists A_v;

  Color G' sequentially;

  Send the resulting colors back to the corresponding processors;

retransmit colors:

  foreach *processor P_i* **do**

  Inform all neighbors of U_i of the colors that have been assigned;

  Receive the colors from the other processors and update the lists F_v accordingly;

Inserting new forbidden colors into an unsorted list F_v can be done in constant time per color. Any edge adds an item to the list of forbidden colors of one of its end-vertices at most once, so the size of such a list is bounded by the degree of the vertex. Thus, the total size of these lists on any of the processors will never exceed the input size M.

To find the least available color in **color timeslot** we then have to process the list as a whole. This happens only once for each vertex v, and so one might hope to get away with just sorting the list F_v. But sorting here can be too expensive, comparison sort would impose a time of $M \log M$ whereas counting sort would lead to $|F_v| + M$ per vertex.

But nevertheless the work for this can be bound as follows. Each processor maintains a Boolean vector *colors* that is indexed with the colors and that will help to decide for a vertex v on the least color to be taken. Since we got rid of high degree vertices we know that no list F_v will be longer than M/k and so a length of $M/k + 1$ suffices for *colors*.

Later, when relaxing this condition in Section 3 we will need at most p colors for vertices of degree greater than N/p and ensure to add no more than $\Delta' + 1$ colors, where Δ' is the maximum degree among the remaining vertices ($\Delta' \leq M$).

In total this means that we have at most $p + M + 1$ colors and so our vector *colors* still fits on a processor. This vector is initialized once with all values "true". Then when processing a vertex v we run through its list of forbidden colors and set the corresponding items of *colors* to "false". After that, we look for the first item in *colors* that still is true and choose that color for v. Then, to revert the changes we run through the list again and set all values to "true".

Algorithm 3: Compute the allowed colors A_v of a vertex v.

Input: v together with its actual degree $d(v)$ and its (unordered) list F_v of forbidden colors; A Boolean vector *colors* with all values set to *true*.

foreach $c \in F_v$ **do** Set $colors[c] = false$;
for $(c = 1; |A_v| < d(v); + + c)$ **do** **if** $colors[c]$ **then** $A_v = c + A_v$;
foreach $c \in F_v$ **do** Set $colors[c] = true$;

This then clearly needs at most a time of $p + M + 1$ plus the sizes of the list, so $O(M)$ time in total.

As seen above, on any processor the total fraction of edges going into recursion is expected to be M/k, so the Main Theorem for divide and conquer algorithms shows that the total costs are $O(M)$. □

2.2 The Bottom of Recursion

At first one might be tempted to think that the bottom of the recursion should easily stay within the required bounds and only communicates as much data as there are edges in the corresponding subgraph. But such an approach doesn't count for the lists F_v of forbidden colors that the vertices might already have collected during higher levels of recursion. The size of these lists may actually be too large and their union might not fit on a single processor. To take care of that situation we proceed in three steps, see Algorithm 2.

In the first step **collect colors**, for each vertex $v \in V'$ we produce a short list of *allowed* colors. In fact, the idea is that when we color the vertex later on we will not use more than its degree $+1$ colors so a list of $d(v) + 1$ allowed colors suffices to take all restrictions of forbidden colors into account. With the same trick as in the previous section, we can get away with a computation time of $|F_v| + d(v)$ to compute the list A_v, see Algorithm 3. It is also easy to see that we can use this trick again when we sequentially color the graph on P_1. We summarize:

Lemma 5. *The bottom of the recursion for graph $G' = (V', E')$ with lists of forbidden colors F_v can be done in a constant number of communication steps with overall work that is proportional to $|G'|$ and the lists F_v and with a communication that is proportional to $|G'|$.*

Note that there will be $k^{\lceil \log_k p \rceil}$ calls to **bottom**, therefore P_1 handles at most $k^{\lceil \log_k p \rceil} \frac{N}{kp^2}$ edges and $k^{\lceil \log_k p \rceil} \frac{N}{kp^2} \leq k^{1+\log_k p} \frac{N}{kp^2} \leq kp \frac{N}{kp^2} = M$. That implies a total time for **bottom** of $O(M)$.

3 Getting Rid of High Degree Vertices

Line **group vertices** of Algorithm 1 groups the vertices into $k \leq p$ timeslots of about equal degree sum. Such a grouping would not be possible if the variation

---

Algorithm 4: Get rid of high degree vertices of second type.

> **foreach** *processor P_i* **do**
>> find all $v \in U_i$ with degree higher than M'/k (Note: all degrees less than N/p);
>> send the names and the degrees of these vertices to P_1;
>
> **for** *processor P_1* **do**
>> Receive lists of high degree vertices;
>> Group these vertices into $k' \leq k$ timeslots $W_1, ..., W_{k'}$ of at most p vertices each and of a degree sum of at most $2N/p$ for each timeslot;
>> Communicate the timeslots to the other processors;
>
> **foreach** *processor P_i* **do**
>> Receive the timeslots for the high degree vertices in U_i;
>> Communicate these values to all the neighbors of these vertices;
>> Receive the corresponding information from the other processors;
>> Compute $E_{t,i}$ for $t = 1, \ldots, k'$ where one endpoint is in U_i;
>
> **for** $t = 1$ *to* k' **do**
>> **foreach** *processor P_i* **do** Communicate $E_{t,i}$ to processor P_1;
>> **for** *processor P_1* **do**
>>> Receive $E_t = \bigcup_{1 \leq i \leq p} E_{t,i}$ and denote by $G_t = (W_t, E_t)$ the induced subgraph of high degree vertices of timeslot t;
>>> Solve the problem for G_t sequentially, see Algorithm 2;

---

in the degrees of the vertices is too large. For example, if we have one vertex of very large degree, it would always dominate the degree sum of its timeslot and we cannot achieve a balance. So we will ensure that the degrees of all vertices is fairly small, namely smaller than M/k. Observe that this notion of 'small' depends on the input size M and thus the property of being of small degree may change during the course of the algorithm. This is why we have to have the line **high degree** in every recursive call and not only for the top level call. On the other hand this choice of M/k will leave us enough freedom to choose k in the range of $2, \ldots, p$ as convenient.

We distinguish two different kinds of high degree vertices. The **first type** we have to handle in a preprocessing step that is only done once on the top level of recursion. These are vertices for which the degree is greater than $M = N/p$. In fact, these vertices can't have all their arcs stored at one processor alone. Clearly, overall we can have at most p such vertices, otherwise we would have more than N edges total. Thus the subgraph induced by these vertices has at most p^2 edges. Because of (1) this induced subgraph fits on processor P_1 and a call to Algorithm 2 in a preprocessing step will color it.

The **second type** of high degree vertices, that we indeed have to treat in each recursive call, are those vertices v with $N/(pk) = M'/k < deg(v) \leq M' = N/p$, see Algorithm 4. Every processor holds at most k such vertices, otherwise it would hold more than $(M'/k) \cdot k = M'$ edges. So in total there are at most $p \cdot k$ such vertices.

Algorithm 5: Determine an ordering on the $k = 2$ timeslots on each processor.

---

foreach *processor P_i* **do**

    foreach *edge (v, w)* **do** inform the processor of w about the timeslot of v;

    for $s = 1 \ldots p$ **do**

        for $r, r' = 0, 1$ **do** set $m_{is}^{rr'} = 0$;

    foreach *edge (v, w)* **do** add 1 to $m_{is}^{rr'}$, where P_s is the processor of w and r and r' are the timeslots of v and w;

    Broadcast all values $m_{is}^{rr'}$ for $s = 1, \ldots, p$ to all other processors;

    $inv[1] = false$;

    for $s = 2$ **to** p **do**

        $A^{||} = 0; A^{\times} = 0$;

        for $s' < s$ **do**

            if $\neg inv[s']$ **then**

                $A^{||} = A^{||} + m_{ss'}^{00} + m_{ss'}^{11}$;

                $A^{\times} = A^{\times} + m_{ss'}^{01} + m_{ss'}^{10}$

            else

                $A^{||} = A^{||} + m_{ss'}^{01} + m_{ss'}^{10}$;

                $A^{\times} = A^{\times} + m_{ss'}^{00} + m_{ss'}^{11}$

        if $A^{\times} < A^{||}$ **then** $inv[s] = true$;

        else $inv[s] = false$;

It is again easy to see that the probability for a vertex v to become a high degree vertex of second type is small: even if it actually has M'/k edges, with high probability only M'/k^2 have their other endpoint in the same timeslot, and so v will not become of high degree on the next recursion level. So on expectation Algorithm 4 will only contribute little to the total number of supersteps, workload and communication.

4   An Add-On to Achieve a Good Worst-Case Behavior

So far for a possible implementation of our algorithm we have a degree of freedom in the number of timeslots k. If we are heading for just a guarantee on expectation as shown above we certainly would not like to bother with recursion and can choose $k = p$. This gives an algorithm that has $3p$ supersteps.

To give a deterministic algorithm with a worst case bound we choose the other extreme, namely $k = 2$. This enables us to bound the number of edges that go into the recursion. We have to distinguish two different types of edges: edges that have both endpoints on the same processor, *internal* edges, and those that have them on different processors, *external* edges.

Here we only describe how to handle external edges. In fact, the ideas for handling internal edges during the partition of the vertices into the two different timeslots are quite similar, but we omitted them for this short abstract.

To handle the external edges we add a call to Algorithm 5 after **group vertices** in Algorithm 1. This algorithm counts the number $m_{is}^{rr'}$ of edges between all possible pairs of timeslots on different processors, and broadcasts these values to all processors. Then a quick iterative algorithm is executed in parallel on all processors that decides on the processors for which the roles of the two timeslots are interchanged.

After having decided whether or not to interchange the roles of the timeslots on processors $P_1, \ldots, P_{i-1}$ we compute two values for processor P_i: $A^{\parallel}$ the number of edges that would go into recursion if we would keep the role of the two timeslots, and $A^{\times}$ the same number if we would interchange the role of the two timeslots.

References

1. J. R. Allwright, R. Bordawekar, P. D. Coddington, K. Dincer, and C. L. Martin. A comparison of parallel graph coloring algorithms. Technical Report SCCS-666, Northeast Parallel Architecture Center, Syracuse University, 1995.
2. D. Culler, R. Karp, D. Patterson, A. Sahay, K.E. Schauser, E. Santos, R. Subramonian, and T. von Eicken. LogP: Towards a Realistic Model of Parallel Computation. In *Proceeding of 4-th ACM SIGPLAN Symp. on Principles and Practises of Parallel Programming*, pages 1–12, 1993.
3. F. Dehne, A. Fabri, and A. Rau-Chaplin. Scalable parallel computational geometry for coarse grained multicomputers. *International Journal on Computational Geometry*, 6(3):379–400, 1996.
4. Assefaw Hadish Gebremedhin and Fredrik Manne. Parallel graph coloring algorithms using OpenMP (extended abstract). In *First European Workshop on OpenMP*, pages 10–18, Lund, Sweden, September 30 – October 1, 1999.
5. Assefaw Hadish Gebremedhin and Fredrik Manne. Scalable, shared memory parallel graph coloring heuristics. Technical Report 181, Department of Informatics, University of Bergen, 5020 Bergen, Norway, December 1999.
6. M. Goudreau, K. Lang, S. Rao, T. Suel, and T. Tsantilas. Towards efficiency and portability: Programming with the BSP model. In *8th Annual ACM symposium on Parallel Algorithms and Architectures (SPAA'96)*, pages 1–12, 1996.
7. Isabelle Guérin Lassous, Jens Gustedt, and Michel Morvan. Feasability, portability, predictability and efficiency: Four ambitious goals for the design and implementation of parallel coarse grained graph algorithms. Technical report, INRIA, 2000.
8. Leslie G. Valiant. A bridging model for parallel computation. *Communications of the ACM*, 33(8):103–111, 1990.

The Tree-Width of Clique-Width Bounded Graphs without $K_{n,n}$

Frank Gurski[*] and Egon Wanke

Department of Computer Science, Mathematical Institute, Heinrich-Heine-University
Düsseldorf, 40225 Düsseldorf, Germany. {gurski,wanke}@cs.uni-duesseldorf.de

Abstract. We proof that every graph of clique-width k which does not
contain the complete bipartite graph $K_{n,n}$ for some $n > 1$ as a subgraph
has tree-width at most $3k(n - 1) - 1$. This immediately implies that
a set of graphs of bounded clique-width has bounded tree-width if it is
uniformly l-sparse, closed under subgraphs, of bounded degree, or planar.

1 Introduction

The clique-width of a graph is defined by composition mechanisms for vertex-
labeled graphs, see [5]. The operations are the vertex disjoint union of labeled
graphs, the addition of edges between vertices controlled by some label pair,
and a relabeling of the vertices. The used number of labels corresponds to the
clique-width of the defined graph. Clique-width bounded graphs are especially
interesting from an algorithmic point of view. A lot of NP-complete graph problems can be solved in polynomial time for graphs of bounded clique-width if
the composition tree of the graphs is explicitly given. For example, the set of all
graph properties which are expressible in monadic second order logic with quantifications over vertices and vertex sets (MSO_1-logic) can be solved in linear
time on clique-width bounded graphs [4]. The MSO_1-logic has been extended by
counting mechanisms which allow the expressibility of optimization problems,
see [4]. All these problems expressible in the extended MSO_1-logic can be solved in polynomial time on clique-width bounded graphs. Furthermore, a lot of
NP-complete graph problems which are not expressible in MSO_1-logic or extended MSO_1-logic like Hamiltonicity and the simple max cut problem can also be
solved in polynomial time on clique-width bounded graphs, see [14].

In [14] the notion of NLC-width is defined by a composition mechanism for
vertex-labeled graphs which is similar to that for clique-width. Every graph of
clique-width at most k has NLC-width at most k and every graph of NLC-
width at most k has clique-width at most $2k$, see [11]. The only essential difference between the composition mechanisms of clique-width bounded graphs and
NLC-width bounded graphs is the addition of edges. In an NLC-composition
the addition of edges is combined with the union operation. The union operation applied to two graphs G and J is controlled by a set S of label pairs such

---

[*] The work of the first author was supported by the German Research Association
(DFG) grant WA 674/9-1.

that for each pair (a, b) all vertices of G labeled by a will be connected with all vertices of J labeled by b. The tree structure of the NLC-expression is a very suitable representation for the efficient processing of the graph with respect to a graph property or an optimization problem. This shows especially the algorithmic framework introduced in [14], and the proof of the main theorem of this paper.

The most famous class of graphs for which a lot of NP-complete graph problems can be solved in polynomial time is the class of tree-width bounded graphs, see Bodlaender [2] for a survey. All graph properties expressible in monadic second order logic with quantifications over vertex sets and edge sets (MSO_2-logic) can be solved in linear time for tree-width bounded graphs, see [6]. The MSO_2-logic has also been extended by counting mechanisms to express optimization problems which can then be solved in polynomial time for tree-width bounded graphs, see [1].

It is already known that each graph of tree-width at most k has clique-width at most $2^{k+1} + 1$, see [5], and NLC-width at most $2^{k+1} - 1$, see [14]. However, the set of all graphs of clique-width at most 2 is equivalent to the set of all graphs of NLC-width 1, and equivalent to the set of all cographs. Since the set of all cographs contains all complete graphs, the set of all graphs of clique-width at most 2 and the set of all graphs of NLC-width 1 do not have bounded tree-width.

In this paper, we proof that each graph of NLC-width k which does not contain the complete bipartite graph $K_{n,n}$ for some $n > 1$ as a subgraph has tree-width at most $3k(n-1) - 1$. This result immediately implies a lot of further characterizations of tree-width bounded graphs. For example a set L of clique-width bounded graphs has bounded tree-width if (1.) the graphs of L do not contain arbitrary large complete bipartite subgraphs (2.) the graphs of L are uniformly l-sparse, (3.) the set of all minors of the graphs of L is not the set of all graphs, (4.) the graphs of L are planar or have bounded degree, (5.) the set of all bipartite planar subgraphs, subgraphs, or minors of the graphs of L has bounded clique-width, (6.) L is closed under bipartite planar subgraphs, subgraphs, or minors. For case (1.), (2.), (3.), and (5.) the reverse direction holds also true.

In [5] it is already shown that there is some function f such that a set of graphs of clique-width at most k which does not contain some $K_{n,n}$ as a subgraph has tree-width at most $f(n, k)$. However, Courcelle and Olariu proofed only the existence of such a function f. In this paper we explicity give a bound on the tree-width and additionally the corresponding tree-decomposition.

2 Basic Definitions

We work with finite undirected *graphs* $G = (V_G, E_G)$, where V_G is a finite set of *vertices* and $E_G \subseteq \{\{u, v\} \mid u, v \in V_G, u \neq v\}$ is a set of *edges*. Graph $J = (V_J, E_J)$ is a *subgraph* of G if V_J is a subset of V_G and E_J is a subset of E_G. J is an *induced subgraph* of G if additionally $E_J = \{\{u, v\} \in E_G \mid u, v \in V_J\}$. We say a set of graphs L is *closed* under taking subgraphs or induced subgraphs if for every graph $G \in L$ all subgraphs or induced subgraphs of G, respectively, are in L.

Next we recall the definitions of tree-width, clique-width, and NLC-width of a graph. To distinguish between the vertices of general graphs and trees, we call the vertices of the decomposition-trees or expression-trees *nodes*.

Definition 1 (Tree-width, [13]). *A* tree decomposition *of a graph* $G = (V_G, E_G)$ *is a pair* $(\mathcal{X}, T)$ *where* $T = (V_T, E_T)$ *is a tree and* $\mathcal{X} = \{X_u \mid u \in V_T\}$ *is a family of subsets* $X_u \subseteq V_G$ *one for each node* u *of* T *such that*

1. $\bigcup_{u \in V_T} X_u = V_G$.
2. *For every edge* $\{w_1, w_2\} \in E_G$, *there is some node* $u \in V_T$ *such that* $w_1 \in X_u$ *and* $w_2 \in X_u$.
3. *For every vertex* $w \in V_G$ *the subgraph of* T *induced by the nodes* $u \in V_T$ *with* $w \in X_u$ *is connected.*

The width *of a tree decomposition* $(\mathcal{X} = \{X_u \mid u \in V_T\}, T = (V_T, E_T))$ *is* $\max_{u \in V_T} |X_u| - 1$. *The* tree-width *of a graph* G *is the minimum tree-width of all tree decompositions of* G.

Courcelle and Olariu define in [5] the notion of clique-width for labeled graphs. Let $[k] := \{1, \ldots, k\}$ be the set of all integers between 1 and k. A k-*labeled graph* $G := (V_G, E_G, \mathrm{lab}_G)$ is a graph (V_G, E_G) whose vertices are labeled by some mapping $\mathrm{lab}_G : V \to [k]$. A labeled graph $J = (V_J, E_J, \mathrm{lab}_J)$ is a subgraph of G if $V_J \subseteq V_G$, $E_J \subseteq E_G$ and $\mathrm{lab}_J(u) = \mathrm{lab}_G(u)$ for all $u \in V_J$. A labeled graph which consists of a single vertex labeled by $t \in [k]$ is denoted by $\bullet_t$.

Definition 2 (Clique-width, [5]). *Let* k *be some positive integer. The class* CW_k *of labeled graphs is recursively defined as follows.*

1. *The single vertex graph* $\bullet_t$ *for some* $t \in [k]$ *is in* CW_k.
2. *Let* $G = (V_G, E_G, \mathrm{lab}_G) \in CW_k$ *and* $J = (V_J, E_J, \mathrm{lab}_J) \in CW_k$ *be two vertex disjoint labeled graphs. Then* $G \oplus J := (V', E', \mathrm{lab}') \in CW_k$ *defined by* $V' := V_G \cup V_J$, $E' := E_G \cup E_J$, *and*

$$\mathrm{lab}'(u) := \begin{cases} \mathrm{lab}_G(u) \text{ if } u \in V_G \\ \mathrm{lab}_J(u) \text{ if } u \in V_J \end{cases}, \quad \forall u \in V'.$$

3. *Let* $i, j \in [k]$ *be two distinct integers and* $G = (V_G, E_G, \mathrm{lab}_G) \in CW_k$ *be a labeled graph then*
   a) $\rho_{i \to j}(G) := (V_G, E_G, \mathrm{lab}') \in CW_k$ *defined by*

$$\mathrm{lab}'(u) := \begin{cases} \mathrm{lab}_G(u) \text{ if } \mathrm{lab}_G(u) \neq i \\ j \qquad\quad\; \text{ if } \mathrm{lab}_G(u) = i \end{cases}, \quad \forall u \in V_G$$

   and
   b) $\eta_{i,j}(G) := (V_G, E', \mathrm{lab}_G) \in CW_k$ *defined by*

$$E' := E \cup \{\{u, v\} \mid \mathrm{lab}(u) = i, \ \mathrm{lab}(v) = j\}.$$

The clique-width *of a labeled graph* G *is the smallest integer* k *such that* $G \in CW_k$.

Wanke defines in [14] the notion of NLC-width[1] of labeled graphs.

Definition 3 (NLC-width, [14]). *Let* k *be some positive integer. The class* NLC_k *of labeled graphs is recursively defined as follows.*

1. *The single vertex graph* $\bullet_t$ *for some* $t \in [k]$ *is in* NLC_k.
2. *Let* $G = (V_G, E_G, lab_G) \in NLC_k$ *and* $J = (V_J, E_J, lab_J) \in NLC_k$ *be two vertex disjoint labeled graphs and* $S \subseteq [k]^2$, *then* $G \times_S J := (V', E', lab') \in NLC_k$ *defined by* $V' := V_G \cup V_J$,

$$E' := E_G \cup E_J \cup \{\{u, v\} \mid u \in V_G, \ v \in V_J, \ (lab_G(u), lab_J(v)) \in S\},$$

and

$$lab'(u) \ := \ \begin{cases} lab_G(u) \ if \ u \in V_G \\ lab_J(u) \ if \ u \in V_J \end{cases}, \ \forall u \in V'.$$

3. *Let* $G = (V_G, E_G, lab_G) \in NLC_k$ *and* $R : [k] \to [k]$, *then* $\circ_R(G) := (V_G, E_G, lab') \in NLC_k$ *defined by* $lab'(u) := R(lab(u)), \ \forall u \in V_G$.

The NLC-width *of a labeled graph* G *is the smallest integer* k *such that* $G \in NLC_k$.

An unlabeled graph $G = (V_G, E_G)$ has clique-width k (NLC-width k) if there is some labeling $lab_G : V_G \to [k]$ such that (V_G, E_G, lab_G) has clique-width k (NLC-width k, respectively). Since a relabeling of vertices does not change the clique-width or NLC-width of a graph, we can assume that the vertices in unlabeled graphs are all equally labeled. This allows us to use the notation graph without any confusion for labeled and unlabeled graphs.

There is a close relation between the clique-width and the NLC-width of a graph as the next theorem states.

Theorem 1 ([10], [11]). *Every graph of clique-width* k *has NLC-width at most* k *and every graph of NLC-width* k *has clique-width at most* $2k$.

The set of graphs of NLC-width at most k is closed under induced subgraphs and edge complement, i.e., if $G = (V_G, E_G, lab_G) \in NLC_k$ then $\overline{G} = (V_G, \overline{E_G}, lab_G) \in NLC_k$ for $\overline{E_G} = \{\{u, v\} \mid u, v \in V_G, \ u \neq v, \ \{u, v\} \notin E_G\}$, see [14]. If G has clique-width at most k then $\overline{G}$ has clique-width at most $2k$, see [5]. The set of graphs of clique-width at most 2 and NLC-width 1 is exactly the set of all cographs, see [5,14]. Thus the tree-width of a graph is not bounded by its clique-width or NLC-width. Distance hereditary graphs have clique-width and NLC-width at most 3, see [9]. The clique-width and NLC-width of permutation

---
[1] The abbreviation NLC results from the *node label controlled* embedding mechanism originally defined for graph grammars.

graphs, interval graphs, grids and planar graphs is not bounded by some fixed integer k, see [9]. If a graph has tree-width k then it has clique-width at most $2^{k+1} + 1$, see [5], and NLC-width at most $2^{k+1} - 1$, see [14]. A graph with n vertices has clique-width at most $n - r$, if $2^r < n - r$, and NLC-width at most $\lceil \frac{n}{2} \rceil$, see [11]. The recognition problem for graphs of clique-width at most k and graphs of NLC-width at most k is still open for $k \geq 4$ and $k \geq 3$, respectively. Clique-width of at most 2 and NLC-width of at most 1 are decidable linear time, see [8]. Clique-width of at most 3 and NLC-width of at most 2 are decidable in polynomial time, see [3,12].

3   The Main Theorem

The expression-tree of an NLC-expression is an ordered rooted tree whose nodes are labeled by the operations of the expression and whose arcs are pointing to the roots of the expression-trees of the involved sub-expressions.

Definition 4 (Expression-tree). *The expression-tree T of $\bullet_t$ consists of a single node r (the root of T) labeled by $\bullet_t$. The expression-tree T of $\circ_R(G)$ consists of the expression-tree T' of G with an additional node r (the root of T) labeled by $\circ_R$ and an additional arc from r to the root of T'. The expression-tree T of $G \times_S J$ consists of the disjoint union of the expression-trees T_G and T_J of G and J, respectively, with an additional node r (the root of T) labeled by $\times_S$ and two additional arcs from node r to the roots of T_G and T_J. The left son of r is the root of T_G, the right son of r is the root of T_J.*

Note that there is a one-to-one correspondence between the vertices of G and the leafs of the expression-tree T_G of an NLC-expression for G. Let $T_G(u)$ for some node u of T_G be the subtree of T_G induced by node u and all nodes v for which there is directed path from u to v in T_G. The tree $T_G(u)$ is an ordered rooted tree with root u. The expression of $T_G(u)$ defines a (possibly) relabeled induced subgraph $G(u)$ of G. The vertices of $G(u)$ are the vertices of G corresponding to the leafs of the subtree $T_G(u)$. The edges of $G(u)$ are those edges of G for which both end vertices are in $G(u)$. The labels of the vertices in $G(u)$ are defined by the expression of $T_G(u)$ which is a sub-expression of the expression of T_G. These labels are not necessarily the final labels of the vertices as in graph G, because the vertices of $G(u)$ can be relabeled by the operations of the nodes on the path from the root of T_G to the father of u.

Theorem 2. *Let G be a graph of NLC-width k such that the complete bipartite graph $K_{n,n}$ for some $n > 1$ is not a subgraph of G, then G has tree-width at most $3k(n - 1) - 1$.*

Proof. We define a tree-decomposition $(\mathcal{X} = \{X_u \mid u \in V_T\}, T = (V_T, E_T))$ of width at most $3k(n - 1) - 1$ from a given expression-tree $T_G = (V_{T_G}, E_{T_G})$ for some graph $G = (V_G, E_G)$ of NLC-width k that does not contain the complete bipartite graph $K_{n,n}$ as a subgraph.

The tree $T = (V_T, E_T)$ of the tree-decomposition is defined from the expression-tree T_G for G as follows.

1. The node set of T is the node set of T_G.
2. There is an edge $\{u, v\}$ in T if and only if there is an arc from u to v in T_G.

To define the sets X_u for the nodes u of T, we need some further notations. For a node u of T_G and a label $t \in [k]$, let $A(u, t)$ be the set of all vertices that appear in $G(u)$ with label t. Let $B(u, t)$ be the set of all vertices of G that are not in $G(u)$ but which are adjacent in G to some vertex of $A(u, t)$. Note that all vertices of $A(u, t)$ are adjacent to all vertices of $B(u, t)$, because equal labeled vertices of $G(u)$ will be treated in the same way by all the operations of the nodes on the path from the root of T_G to the father of u. That is, either set $A(u, t)$ or set $B(u, t)$ or both sets have less than n vertices, otherwise G contains some $K_{n,n}$ as a subgraph which contradicts our assumption.

Now we define the sets $X_u \subseteq V_G$ for the nodes u of T depending on the label of u in T_G.

1. If node u is labeled by $\bullet_t$, i.e, if u is a leaf in T_G:
   Then X_u is defined to be the set which consists of the single vertex of G corresponding to the leaf u of T_G.
2. If node u is labeled by $\circ_R$:
   Let v be the son of u in T_G. Then X_u is defined to be the union of all sets $A(v, t)$ and $B(u, t)$ for $t \in [k]$ which contain less than n vertices.
3. If node u is labeled by $\times_S$:
   Let v_1, v_2 be the two sons of u in T_G. Then X_u is defined to be the union of all sets $A(v_1, t)$, $A(v_2, t)$, and $B(u, t)$ for $t \in [k]$ which contain less than n vertices.

It remains to show that $(\mathcal{X}, T)$ is a tree-decomposition of width at most $3k(n-1) - 1$.

1. $\bigcup_{u \in V_T} X_u = V_G$.
   Since each vertex of G corresponds to some leaf of T_G, all vertices of G are already contained in the union of all sets X_u for the leafs u of T.
2. $|X_u| < 3k(n-1)$ for each $u \in V_T$.
   This follows from the fact that each X_u either consists of a single vertex or is the union of at most $3k$ sets of size at most $n - 1$.
3. For each edge $\{w_1, w_2\}$ of G there is some node $u \in V_T$ such that $w_1, w_2 \in X_u$.
   Let $\{w_1, w_2\}$ be an edge of G and u be the node of T_G such that u has two sons v_1 and v_2 such that w_1 is a vertex of $G(v_1)$ and w_2 is a vertex of $G(v_2)$. The connection between w_1 and w_2, i.e., the edge $\{w_1, w_2\} \in E_G$, is due to the $\times_S$ operation of node u. Let t_1 and t_2 be the labels of w_1 and w_2 in graph $G(v_1)$ and graph $G(v_2)$, respectively. If v_1 is the left son and v_2 the right son of u then $(t_1, t_2) \in S$. We also know that $w_1 \in A(v_1, t_1)$, $w_2 \in A(v_2, t_2)$, $w_2 \in B(v_1, t_1)$, and $w_1 \in B(v_2, t_2)$ by the definitions of these sets.

a) If $|A(v_1, t_1)| < n$ and $|A(v_2, t_2)| < n$ then w_1 and w_2 are both in X_u, because then $A(v_1, t_1) \subseteq X_u$ and $A(v_2, t_2) \subseteq X_u$.

b) If $|A(v_1, t_1)| \geq n$ and $|A(v_2, t_2)| \geq n$ then the $K_{n,n}$ is a subgraph of $G(u)$, which contradicts our assumption.

c) Suppose $|A(v_1, t_1)| < n$ and $|A(v_2, t_2)| \geq n$. The case for $|A(v_1, t_1)| \geq n$ and $|A(v_2, t_2)| < n$ runs analogously. If $|A(v_2, t_2)| \geq n$ then $|B(v_2, t_2)| < n$ and $B(v_2, t_2) \subseteq X_{v_2}$, thus $w_1 \in X_{v_2}$. Consider now the nodes in the expression-tree T_G on the path from node v_2 to the leaf which corresponds to vertex w_2 of G. Let r_1 be this leaf, r_{i+1} be the father of r_i for $i \geq 1$, and r_s be the node v_2. Let l_i be the label of vertex w_2 in graph $G(r_i)$ for $i = 1, \dots, s$. Now we have $A(r_i, l_i) \subseteq A(r_{i+1}, l_{i+1})$ and $B(r_{i+1}, l_{i+1}) \subseteq B(r_i, l_i)$ for $i = 1, \dots, s-1$. Let j, $1 \leq j \leq s$, be the least index such that $|A(r_j, l_j)| \geq n$. Then X_{r_j} is the set which contains both vertices w_1 and w_2. Since $|A(r_{j-1}, l_{j-1})| < n$, we know that $A(r_{j-1}, l_{j-1}) \subseteq X_{r_j}$, and thus $w_2 \in X_{r_j}$. Since $|A(r_j, l_j)| \geq n$, we know that $|B(r_j, l_j)| < n$, $B(r_j, l_j) \subseteq X_{r_j}$, and thus $w_1 \in X_{r_j}$.

4. For each vertex $w \in V_G$, the subgraph of T induced by the nodes u with $w \in X_u$ is connected.

   Let w be any vertex of G and let r_1 be the leaf of the expression-tree T_G corresponding to w. Let r_{i+1} be the father of r_i for $i = 1, \dots, h-1$ such that r_h is the root of T_G. For all these nodes r_i, $1 \leq i \leq h$, we know that w is a vertex of $G(r_i)$ and thus $w \notin B(r_i, t)$ for any $t \in [k]$. Let l_i be the label of w in graph $G(r_i)$ for $i = 1, \dots, h$. If $w \in X_{r_i}$ for some $i > 1$ then $w \in A(r_{i-1}, l_{i-1})$. That is, the nodes r_i with $w \in X_{r_i}$ induce a path the in decomposition-tree T. We call the nodes $r_1, \dots, r_h$ the *back bone* of vertex w in T_G. Note that vertex w is in subgraph $G(u)$ for some node u of T_G if and only if u is from the back bone of w in T_G.

   Let v be any node of $V_{T_G} - \{r_1, \dots, r_h\}$ such that $w \in X_v$. Since v is not from the back bone, there has to be some label $t \in [k]$ such that $w \in B(v, t)$ and $|B(v, t)| < n$. That is, subgraph $G(v)$ has at least one vertex w' labeled by t which is connected to w in G. Now we show that there is always a path in T_G from some node r_i with $|A(r_{i-1}, t_{i-1})| < n$ of the back bone of vertex w in T_G to node v. This shows that the subgraph of T induced by all nodes u with $w \in X_u$ is connected.

   a) If the father of v is one of the nodes r_i of the back bone with $|A(r_{i-1}, t_{i-1})| < n$ then nothing more is to show.

   b) If the father of v is one of the nodes r_i of the back bone with $|A(r_{i-1}, t_{i-1})| \geq n$, then the subgraph $G(r_{i-1})$ has at least n vertices labeled by t_{i-1} (the label of vertex w in $G(r_{i-1})$). Since equal labeled vertices are treated in the same way, all these vertices have to be connected in G with vertices w' of $G(v)$. So $A(r_{i-1}, t_{i-1}) \subseteq B(v, t)$ and thus $|B(v, t)| \geq |A(r_{i-1}, t_{i-1})| \geq n$, which contradicts our assumption that $|B(v, t)| < n$.

   c) If the father v' of v is not from the back bone $\{r_1, \dots, r_h\}$ then let t' be the label of the vertices of $G(v')$ which are labeled by t in $G(v)$. Then

$B(v', t') \subseteq B(v, t)$, $|B(v', t')| \leq |B(v, t)| < n$, and thus $B(v', t') \subseteq X_{v'}$. Since w has to be in $B(v', t')$, we have $w \in X_{v'}$. This argumentation can be repeated up to the nodes of the back bone of T_G.

$\square$

Note that the bound of theorem 2 above is tight for $k = 1$ and $n = 2$. The complete graph with 3 vertices has NLC-width 1, does not contain the $K_{2,2}$ as a subgraph, and has tree-width $3k(n-1) - 1 = 2$.

A graph $J = (V_J, E_J)$ is a *minor* of some graph $G = (V_G, E_G)$ if J can be obtained from G by a sequence of edge deletions, edge contractions and deletions of vertices without incident edges. We say a set of graphs L is *closed* under taking minors if for every graph $G \in L$ all minors of G are in L.

A graph $G = (V_G, E_G)$ is *l-sparse* if $|E_G| \leq l \cdot |V_G|$. It is *uniformly l-sparse* if every subgraph of G is l-sparse. A set of graphs is *uniformly l-sparse* if all its graphs are uniformly l-sparse, see [7].

We say a set of graphs L has bounded tree-width (bounded clique-width, bounded NLC-width) if there is some k such that every graph $G \in L$ has tree-width at most k, (clique-width at most k, NLC-width at most k, respectively).

Our main theorem immediately implies the following corollaries.

Corollary 1. *Let G be a graph of clique-width (NLC-width) at most k.*

1. *If the complete bipartite graph $K_{n,n}$ is not a subgraph of G, then G has tree-width at most $3k(n-1) - 1$.*
2. *If G is uniformly l-sparse, then G has tree-width at most $6kl - 1$.*
3. *If there is a graph with n vertices which is not a minor of G, then G has tree-width at most $3k(n-1) - 1$.*
4. *If G is planar, then G has tree-width at most $6k - 1$.*
5. *If G has degree at most d, then G has tree-width at most $3kd - 1$.*

Proof.

1. By theorem 1 and theorem 2.
2. If a graph G is uniformly l-sparse then the complete bipartite graph $K_{2l+1,2l+1}$ is not a subgraph of G.
3. If there is a graph with n vertices which is not a minor of G then the complete graph K_n is not a minor of G, and thus the $K_{n,n}$ is not a minor of G, and thus the $K_{n,n}$ is not a subgraph of G.
4. Planar graphs do not contain the $K_{3,3}$ as a subgraph.
5. Graphs with vertex degree at most d do not contain the $K_{d+1,d+1}$ as a subgraph.

$\square$

Corollary 2. *Let L be a set of graphs of clique-width (NLC-width) at most k.*

1. *If the set L' of all bipartite planar subgraphs, all subgraphs, or all minors of the graphs of L has clique-width at most l, then L has tree-width at most $3k(\lceil \frac{l^2}{2} \rceil - 1) - 1$.*
2. *If L is closed under taking bipartite planar subgraphs, subgraphs, or minors, then L has tree-width at most $3k(\lceil \frac{k^2}{2} \rceil - 1) - 1$.*

Proof.

1. The $l \times l$-grid is bipartite, planar, has clique-width $l+1$, see [9] for the clique-width bound, and is a subgraph of $K_{\lceil \frac{l^2}{2} \rceil, \lfloor \frac{l^2}{2} \rfloor}$. That is, if L' has clique-width l then the $l \times l$ grid is not a subgraph of any graph of L and the $K_{\lceil \frac{l^2}{2} \rceil, \lfloor \frac{l^2}{2} \rfloor}$ is not a subgraph of any graph of L.
2. By (1.).

$\square$

The corollaries above can also be used to characterize sets of graphs of bounded tree-width as follows. Let L be a set of graphs of bounded clique-width. The set L has bounded tree-width if and only if

1. every graph of L does not contain the complete bipartite graph $K_{n,n}$ as a subgraph for some $n > 1$,
2. L is uniformly l-sparse for some l,
3. there is some graph that is not a minor of every graph of L, or
4. the set of all bipartite planar subgraphs, subgraphs or minors of the graphs of L has bounded clique-width.

The if cases follow from corollary 1 (1.) (2.) (3.) and corollary 2 case (1.). The only if cases follow from the following observation. If a graph G has tree-width at most k then (1.) G does not contain the complete bipartite graph $K_{k+1,k+1}$ as a subgraph, (2.) G is uniformly k-sparse, (3.) the complete graph K_{k+2} is not a minor of G, and (4.) each bipartite planar subgraph, subgraph, and minor of G has tree-width at most k and thus clique-width at most $2^{k+1} + 1$.

References

1. S. Arnborg, J. Lagergren, and D. Seese. Easy problems for tree-decomposable graphs. *Journal of Algorithms*, 12:308–340, 1991.
2. H.L. Bodlaender. A partial k-arboretum of graphs with bounded treewidth. *Theoretical Computer Science*, 209:1–45, 1998.
3. B.D.G. Corneil, M. Habib, J.M. Lanlignel, B. Reed, and U. Rotics. Polynomial time recognition of clique-width at most three graphs. In *Proceedings of Latin American Symposium on Theoretical Informatics (LATIN '2000)*, volume 1776 of *LNCS*. Springer-Verlag, 2000.
4. B. Courcelle, J.A. Makowsky, and U. Rotics. Linear time solvable optimization problems on graphs of bounded clique width, extended abstract. In *Proceedings of Graph-Theoretical Concepts in Computer Science*, volume 1517 of *LNCS*, pages 1–16. Springer-Verlag, 1998.
5. B. Courcelle and S. Olariu. Upper bounds to the clique width of graphs. *Discrete Applied Mathematics*, 101:77–114, 2000.
6. B. Courcelle. The monadic second-order logic of graphs I: Recognizable sets of finite graphs. *Information and Computation*, 85:12–75, 1990.
7. B. Courcelle. The monadic second-order logic of graphs XIV: Uniformly sparse graphs and edge set quantifications. submitted for publication, 2000.
8. D.G. Corneil, Y. Perl, and L.K. Stewart. A linear recognition algorithm for cographs. *SIAM Journal on Computing*, 14(4):926–934, 1985.

9. M.C. Golumbic and U. Rotics. On the clique-width of perfect graph classes. In *Proceedings of Graph-Theoretical Concepts in Computer Science*, volume 1665 of *LNCS*, pages 135–147. Springer-Verlag, 1999.

10. F. Gurski. Algorithmische Charakterisierungen spezieller Graphklassen. Diplomarbeit, Heinrich-Heine-Universität, Düsseldorf, Germany, 1998.

11. Ö. Johansson. Clique-decomposition, NLC-decomposition, and modular decomposition - relationships and results for random graphs. *Congressus Numerantium*, 132:39–60, 1998.

12. Ö. Johansson. NLC2 decomposition in polynomial time. In *Proceedings of Graph-Theoretical Concepts in Computer Science*, volume 1665 of *LNCS*, pages 110–121. Springer-Verlag, 1999.

13. N. Robertson and P.D. Seymour. Graph minors II. Algorithmic aspects of tree width. *Journal of Algorithms*, 7:309–322, 1986.

14. E. Wanke. k-NLC graphs and polynomial algorithms. *Discrete Applied Mathematics*, 54:251–266, 1994.

Tree Spanners for Subgraphs and Related Tree Covering Problems

Dagmar Handke[1] and Guy Kortsarz[2]

[1] University of Konstanz, Germany. Dagmar.Handke@uni-konstanz.de
[2] The Open University, Tel Aviv, Israel. guyk@oumail.openu.ac.il

Abstract. For any fixed parameter $k \geq 1$, a *tree k–spanner* of a graph G is a spanning tree T in G such that the distance between every pair of vertices in T is at most k times their distance in G. In this paper, we generalize on this very restrictive concept, and introduce *Steiner tree k–spanners*: We are given an input graph consisting of *terminals* and *Steiner vertices*, and we are now looking for a tree k–spanner that spans all terminals.

The complexity status of deciding the existence of a Steiner tree k–spanner is easy for some k: it is $\mathcal{NP}$-hard for $k \geq 4$, and it is in $\mathcal{P}$ for $k = 1$. For the case $k = 2$, we develop a model in terms of an equivalent tree covering problem, and use this to show $\mathcal{NP}$-hardness. By showing the $\mathcal{NP}$-hardness also for the case $k = 3$, the complexity results for all k are complete.

We also consider the problem of finding a smallest Steiner tree k–spanner (if one exists at all). For any arbitrary $k \geq 2$, we prove that we cannot hope to find efficiently a Steiner tree k–spanner that is closer to the smallest one than within a logarithmic factor. We conclude by discussing some problems related to the model for the case $k = 2$.

1 Introduction

Given a graph G, a *tree k–spanner* T of G is a spanning tree T, such that the distance between any two vertices in T is at most k times longer than the distance in G. The concept of k–spanners has been introduced by [13], and has been studied widely since (see, e.g., [2,6,11,12] and the references therein).

The concept of tree k–spanners turns out to be very restrictive in the sense that the class of graphs that admit a tree k–spanner for a fixed k is quite small. In some settings in the design of subnetworks, however, it is not strictly necessary to span all vertices or edges. In this paper, we introduce a generalized version of tree k–spanners, called *Steiner tree k–spanners*, which models this situation: Similar as in the MINIMUM STEINER TREE Problem in unweighted graphs (cf. [10]), we are given an input graph consisting of *terminals* and *Steiner vertices*. We are now looking for a *tree k–spanner* that includes at least all terminals and spans the edges induced by them.

Steiner spanners have appeared in the literature for example in [1], where the authors deal with absolute lower bounds on the number of edges that an

U. Brandes and D. Wagner (Eds.): WG 2000, LNCS 1928, pp. 206–217, 2000.

arbitrary Steiner k–spanner (not necessarily being a tree) can have. Some authors have considered *delay-bounded minimum Steiner trees*, where one is interested in finding a Steiner tree T such that T fulfills some given distance constraints especially for the distances from a specified root (see for example [14,9]). But there an absolute, fixed delay bound is imposed on the distance of the vertices to the root, i.e. only to one particular vertex. To our knowledge, this is the first time where the generalization of tree k–spanners in terms of Steiner tree k–spanners is examined (see also [6]).

1.1   Formal Definitions

Basic notation. In what follows, $G = (V, E)$ denotes an *unweighted* and *undirected* graph with finite vertex set V and finite edge set E. $V(G)$ (resp. $E(G)$), denotes the vertex set (resp. edge set) of G. We do not allow loops or multiple edges. If R is a subset of V, then $G[R]$ represents the subgraph of G that is induced by R. The *distance* between two vertices u and v in G, i.e., the number of edges in a shortest path, is denoted by $d_G(u, v)$. For a vertex v, denote by $N(v)$ the set of all neighbors of v. A vertex v of a graph G is called *universal* w.r.t. $V(G)$ if $N(v) \cup \{v\} = V(G)$. Let v be universal w.r.t. $V(G)$, then the *star centered at v* is the graph consisting of all vertices of G and all edges incident to v. For a connected graph, an *articulation vertex* is a vertex whose deletion disconnects the graph. A graph is *biconnected* if it has no articulation vertex. A *block* of a graph is a maximal biconnected subgraph.

Let $\mathcal{A}$ be a polynomial-time algorithm that produces a feasible (though not necessarily optimal) solution for an optimization problem Π, and denote by $\mathcal{A}(I)$ the value of the solution that is achieved by $\mathcal{A}$ for the instance I. $\mathcal{A}$ is called a *δ–approximation algorithm* (where $\delta > 1$) for Π if for every instance I of Π, $\max\left\{\frac{\mathcal{A}(I)}{\mathrm{OPT}(I)}, \frac{\mathrm{OPT}(I)}{\mathcal{A}(I)}\right\} \leq \delta$, where $\mathrm{OPT}(I)$ is the value of an optimal solution for I in Π. δ is called *approximation ratio* of $\mathcal{A}$. A problem is *inapproximable* (or *hard to approximate*) within some ratio δ if it is $\mathcal{NP}$-hard to find a δ–approximation algorithm for this problem. See for example [3] for a survey on inapproximability results.

Tree spanners. For any rational $k \geq 1$, a spanning tree T is a *tree k–spanner* of a graph $G = (V, E)$, if $d_T(u, v) \leq k$ for all edges $\{u, v\} \in E \backslash E(T)$. The parameter k is called *stretch factor*.

Observe that we only consider *constant* stretch factors, i.e., stretch factors that are independent of $|V|$ and $|E|$. We say that an edge e that does not belong to T is *spanned* (by T) if there exists a path of length at most k (in T) connecting the end-vertices of e. Such a path is called a *spanning path*.

Since distances in unweighted graphs are integral, it follows that T is a tree k–spanner of G if and only if T is a $\lfloor k \rfloor$–spanner of G. Thus, it suffices to consider *integer* stretch factors k. If we look for a tree k–spanner in a given graph G, it is clear that we only have to consider the *biconnected* components of G. Observe that there are many graphs that, for a fixed stretch factor k, do not contain a tree k–spanner as a subgraph. We say that such a graph does not *admit* a tree k–spanner. For example, $K_{3,3}$ does not admit a tree 2–spanner. The corresponding decision problem is called TREE k–SPANNER.

Note that G may only admit a tree 1–spanner if G is a tree itself. Thus, TREE 1–SPANNER is in $\mathcal{P}$. Furthermore, in [2] a linear algorithm for TREE 2–SPANNER is given, whereas TREE k–SPANNER is shown to be $\mathcal{NP}$-complete for all $k \geq 4$.

Steiner tree spanners. We now define the generalized version of tree k–spanners formally. In the following, the vertex set of the input graph G is partitioned into two disjoint subsets, the set of *terminals* R and the set of *Steiner vertices* S. The edges of $G[R]$ are called *terminal edges*, all other edges are called *Steiner edges*. Denote by $S(T)$, where T is a subgraph, the set of Steiner vertices in G used in T.

Definition 1. *Given a graph* $G = (R \,\dot\cup\, S, E)$ *where* $G[R]$ *is connected, and an arbitrary* $k \geq 1$*, a subtree* T *of* G *is a* Steiner tree k–spanner *of* G *if* $d_T(u, v) \leq k$ *for every edge* $e = \{u, v\}$ *of* $G[R]$*.*

In particular, a Steiner tree k–spanner T must contain all vertices of R, and may include some of the vertices of S too. The edges of T may be a combination of both terminal and Steiner edges. Note that Steiner edges of $G[V(T)]$ do not necessarily have to be spanned in T while terminal edges of $G[V(T)]$ do. Observe that, by this definition, it is possible that the distance between two vertices in T may be shorter than their distance in $G[R]$.

In contrast to tree k–spanners, it is interesting to consider a decision problem as well as an optimization problem defined as follows:

Problem 1. STEINER TREE k–SPANNER

Given: $G = (R \,\dot\cup\, S, E)$ where $G[R]$ is connected.
Problem: Does G admit a Steiner tree k–spanner?

Problem 2. MINIMUM STEINER TREE k–SPANNER

Given: $G = (R \,\dot\cup\, S, E)$ where $G[R]$ is connected, $K \geq 0$.
Problem: Does G admit a Steiner tree k–spanner T such that $|S(T)| \leq K$?

In the latter, we are looking for a Steiner tree k–spanner of the input graph that uses the smallest number of Steiner vertices, if one exists at all. Observe that the number $|E(T)|$ of edges of a Steiner tree k–spanner T is related to the number $|S(T)|$ of Steiner vertices: $|E(T)| = |R| + |S(T)| - 1$. If we now measure the quality of T in terms of $|E(T)|$ instead of $|S(T)|$, then a solution to Problem 2 is also a minimum Steiner tree k–spanner in this sense.

1.2 Results

Since STEINER TREE k–SPANNER is a generalized version of TREE k–SPANNER, the $\mathcal{NP}$-completeness for $k \geq 4$ is immediate. It remains to consider the cases where k is 1, 2, or 3. The case $k = 1$ coincides with TREE 1–SPANNER because no terminal edge can be spanned using some Steiner edges. Thus, it suffices to check whether $G[R]$ is a tree.

It remains to discuss the cases $k = 2, 3$. In Section 2, for $k = 2$, we develop a model in terms of a tree covering problem. Using this, in Section 3 we show that STEINER TREE 2–SPANNER is $\mathcal{NP}$-complete, in contrast to TREE 2–SPANNER. Observe that our model only copes with the case $k = 2$; it is inappropriate for other stretch factors. But we can use the $\mathcal{NP}$-completeness for $k = 2$ to show the $\mathcal{NP}$-completeness for $k = 3$.

If we now assume that we know in advance that a graph G admits a Steiner tree k–spanner for some fixed k, it makes sense to also study the optimization problem. Clearly, the $\mathcal{NP}$-hardness for this follows directly from the results above. In Section 4, we examine the approximability status: MINIMUM STEINER TREE k–SPANNER is hard to approximate within anything better than logarithmic ratio.

Finally, in Section 5, we give an outlook on further tree covering problems that are similar to the one that arises in the light of STEINER TREE 2–SPANNER. These problems do not directly carry over to the context of Steiner tree k–spanners, but we consider them of independent interest.

Due to space limitations, we do not give details of some proofs within this extended abstract but refer to the full version [8].

2 A Model for Steiner Tree 2–Spanners

Instead of trying to solve STEINER TREE 2–SPANNER directly, we first examine the underlying structure of Steiner tree 2–spanners in order to find a suitable model. When dealing with Steiner tree 2–spanners, we have the following situation: a terminal edge can either be spanned by itself or by a path of length 2 that consists either of two terminal edges or two Steiner edges. That means that exactly one Steiner vertex may be used for spanning a terminal edge unless it is spanned by itself or by two terminal edges.

2.1 The Role of the Blocks

Steiner tree 2–spanners in a block. As shown in [2], there is an efficient algorithm to decide whether or not a graph admits a tree 2–spanner. Among others, this algorithm uses the fact that the blocks may be treated separately. Also in the case of Steiner tree 2–spanners, the blocks of $G[R]$ are important:

Lemma 1. *Let $G = (R \;\dot\cup\; S, E)$, and let $G[R]$ be a biconnected graph that does not admit a tree 2–spanner. Then a subgraph T of G is a Steiner tree 2–spanner of G if and only if there is a Steiner vertex $s \in S$ such that $\{s, r\} \in E(T)$ for all terminals $r \in R$ (i.e., s is a universal vertex w.r.t. R).*

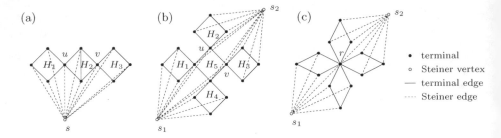

Fig. 1. Importance of the block structure.

Note that this lemma heavily uses the stretch factor of 2; it does not hold for larger k. By the previous lemma, it is shown that STEINER TREE 2–SPANNER can be solved polynomially when $G[R]$ is biconnected.

Now consider a graph G where $G[R]$ is connected and consists of several blocks. Let T be a Steiner tree 2–spanner of G. Then, for every block there is *at most one* universal Steiner vertex in a T:

Lemma 2. *Let G be a graph where $G[R]$ is connected. If T is a Steiner tree 2–spanner of G then for every block H of $G[R]$, $T[V(H)]$ is either a tree 2–spanner of H or there is* exactly one *Steiner vertex $s \in S$ such that $\{s,r\} \in E(T)$ for all $r \in V(H)$.*

In the following, we say that a Steiner vertex s *spans a block H* of $G[R]$ whenever s is universal for H and the star centered at s is a subgraph of the Steiner tree 2–spanner.

Unfortunately, the existence of a tree 2–spanner or universal Steiner vertex for every block is not sufficient. It does *not* suffice to examine every block separately and then simply combine the Steiner tree 2–spanners of every block (as it is the case for tree 2–spanners). For example, even if a block admits a tree 2–spanner, it may be inevitable to span this block by a Steiner vertex. See Figure 1(a) for an example. Figure 1(b) shows an example of a graph where every block has some universal Steiner vertex, but there is no Steiner tree 2–spanner for the whole graph. The situation in Figure 1(c) is different: r is an articulation vertex that is shared by four blocks. Here, the two stars centered at s_1 and s_2, respectively, form a Steiner tree 2–spanner. Thus, we have to take into account the whole block structure of $G[R]$.

The block-graph. Observe that the blocks of a connected graph have a tree-like structure. We model this by a so-called ①–②–*tree* (see also [7]) consisting of two different types of nodes: a ①–node for every block, and a ②–node for every articulation vertex. We connect the nodes in a way such that a ②–node is connected to a ①–node by an edge, whenever the corresponding articulation vertex is contained in the respective block. For reasons of simplicity, we also create an additional ②–node for every block that contains only one articulation vertex. We refer to this graph as *block-graph*, denoted by $B(G)$.

Altogether, we have the following straightforward properties: $B(G)$ is a tree, and the leaves of $B(G)$ are ②–nodes. Moreover, every ①–node (or ②–node, respectively) is adjacent only to ②–nodes (or ①–nodes, respectively).

Certainly, there are also other straightforward methods to consider the block structure. But we will see later that we really have to distinguish between articulation vertices and real blocks. In the sequel, we abbreviate the notation $B(G[R])$ as $B(G)$. Concerning Steiner tree 2–spanners, we get the following observation:

Lemma 3. *If two blocks H_1 and H_3 of $G[R]$ are spanned in a Steiner tree 2–spanner T by the same universal Steiner vertex s then also all other blocks that lie on the path from H_1 to H_3 in $B(G)$ are spanned by s in T.*

2.2   An Equivalent Tree Covering Problem

Our goal is to model STEINER TREE 2–SPANNER in terms of an equivalent problem that is easier to handle. For this, we need some more notation for the block-graph: A subtree of a ①–②–tree B is called ①–②–*subtree*. A ①–②–subtree T of a ①–②–tree B is called *proper* if the following holds: whenever a ①–node b belongs to T then also all ②–nodes that are adjacent to b in B belong to T. The intuition here is as follows: whenever a block (corresponding to a ①–node) is selected, we also select all incident articulation vertices (corresponding to a ②–node). Therefore, in a proper ①–②–subtree, the neighborhood of every ①–node is preserved. A collection of proper ①–②–subtrees that are node disjoint is called a *proper ①–②–forest*.

Before starting with the model, we modify the given instance as follows: For every block that admits a tree 2–spanner, we create a new Steiner vertex (called a *fake* Steiner vertex, in contrast to *real* Steiner vertices). Note that every block of $G[R]$ that consists of only a single edge admits a tree 2–spanner and is thus treated here. It is easy to see that the new instance is equivalent to the original instance. In the sequel, we always use the extended set of Steiner vertices.

We use the block-graph as a basis for our model. For every Steiner vertex s of G, we denote by $F^{(s)}$ the subgraph of $B(G)$ that is induced by the ①–nodes that correspond to blocks that may be spanned by s, together with their adjacent ②–nodes. Then, $F^{(s)}$ is a proper ①–②–forest of $B(G)$, consisting of several maximal proper ①–②–subtrees of $B(G)$. Denote these proper ①–②–subtrees by $\{T_1^{(s)}, \dots, T_\ell^{(s)}\}$, and let $\mathcal{F}$ be the collection of the $F^{(s)}$ for all (real or fake) Steiner vertices s.

Our goal is to define a tree covering problem in the following sense: Given a tree and a collection of subtrees thereof, find a cover of the tree, i.e., a collection of the subtrees such that each vertex of the tree is included in a subtree, subject to some further constraints. Here, we take $B(G)$ as the underlying tree and $\mathcal{F}$ as the collection of subgraphs. In the sequel, we develop the constraints such that a covering selection of subtrees from $\mathcal{F}$ induces a Steiner tree 2–spanner for G and vice versa. For this, we translate the facts about Steiner tree 2–spanners that we have compiled above into the context of the $(B(G), \mathcal{F})$ tree cover problem:

1. Lemma 3 indicates that all 'intermediate' blocks must be spanned whenever two blocks are to be spanned by one distinguished Steiner vertex. Stated the other way round: whenever a block H_1 is spanned by a Steiner vertex s_1, and another block H_2 is spanned by a different Steiner vertex s_2, then no other block H_3 such that H_2 lies on the path from H_1 to H_3 in the block-graph is spanned by s_1. In the context of the tree cover problem, this means that we may choose at most one $T_i^{(s)}$ of $F^{(s)}$ for every s.

2. If we select one particular Steiner vertex s, it is not necessary that we use s for all blocks that are potentially spanned by s. In other words, we may pick only some of these blocks, as long as these blocks induce a connected component within $B(G)$. In the context of the tree cover problem, this corresponds to choosing subtrees of the trees contained in each $F^{(s)}$.

3. By Lemma 2, each block has to be spanned by exactly one Steiner vertex. That means that each ①–node of $B(G)$ has to be covered exclusively. Since articulation vertices may be touched repeatedly, ②–nodes may be covered more than once. We can achieve this by considering edge disjoint proper ①–②–subtrees: Two proper ①–②–subtrees of the same ①–②–tree that are edge disjoint never share a ①–node. They may share at most one ②–node.

Using this, we can model STEINER TREE 2–SPANNER as a problem of an edge disjoint cover of the ①–②–tree by proper ①–②–subtrees:

Problem 3. EDGE DISJOINT ①–②–SUBTREE COVER

Given: A ①–②–tree B and a set of ①–②–forests $\mathcal{F} = \{F^{(1)}, \ldots, F^{(n)}\}$ such that each $F^{(j)}$ consists of a set of node disjoint proper ①–②–subtrees $T_i^{(j)}$ of B.

Problem: Find a cover of the nodes of B consisting of edge disjoint proper ①–②–subtrees $\widehat{T}^{(j)}$ of a $T_i^{(j)}$ where every j is chosen at most once.

Using the remarks above, we can prove that both problems are equivalent:

Lemma 4. *G admits a Steiner tree 2–spanner if and only if there is a solution to* EDGE DISJOINT ①–②–SUBTREE COVER *with input* $(B(G), \mathcal{F})$ *as above.*

Furthermore, any instance of EDGE DISJOINT ①–②–SUBTREE COVER can be re-interpreted and re-constructed in terms of STEINER TREE 2–SPANNER.

3   Complexity of Finding Steiner Tree k–Spanners

Theorem 1 implies the $\mathcal{NP}$-completeness of STEINER TREE 2–SPANNER:

Theorem 1. EDGE DISJOINT ①–②–SUBTREE COVER *is $\mathcal{NP}$-complete.*

Proof. It is clear that EDGE DISJOINT ①–②–SUBTREE COVER is in $\mathcal{NP}$. We show the $\mathcal{NP}$-completeness of the problem by a reduction from the DOMATIC NUMBER Problem: We are given a graph $G = (V = \{v_1, \ldots, v_{|V|}\}, E)$ and a positive integer $K \leq |V|$. The problem is to decide whether V can be partitioned

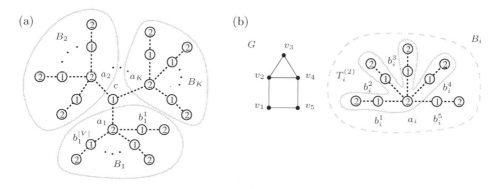

Fig. 2. (a) The construction of B, and (b) of the ①-②-subtree $T_i^{(2)}$.

into $\ell \geq K$ disjoint sets $V_1, V_2, \ldots, V_\ell$ such that every V_i is a dominating set for G (i.e., such that for every $1 \leq i \leq \ell$, every vertex in V is adjacent to at least one vertex in V_i)? DOMATIC NUMBER is $\mathcal{NP}$-complete, and remains so for any fixed $K \geq 3$ (see [5]). W.l.o.g., we can restrict ourselves to the case $\ell = K$.

Let us first review the problem: In a feasible solution $V_1, V_2, \ldots, V_K$ for the instance G of DOMATIC NUMBER, every vertex v_j appears in exactly one of the V_i. The V_i can be viewed as bins, and every v_j is put into exactly one of them. Moreover, the cumulated, closed neighborhood of all of the vertices in each of the bins is V; i.e., $V_i \cup \bigcup_{v_j \in V_i} N(v_j) = V$ for all $i \in \{1, \ldots, K\}$.

Consequently, each of the bins 'induces' an isomorphic copy of V.

In the reduction from DOMATIC NUMBER to EDGE DISJOINT ①-②-SUBTREE COVER, the idea is as follows: Each isomorphic copy of V that is induced by one of the bins, say V_i, is modeled by an individual component B_i of the block-graph, which each contains one ①-node for every vertex of V. The adjacencies in G are reflected in $\mathcal{F}$. Formally, given an instance of DOMATIC NUMBER, we construct the instance of EDGE DISJOINT ①-②-SUBTREE COVER as follows:

- The ①-②-tree B consists of one central ①-node c and K isomorphic block-graph components B_i for $1 \leq i \leq K$: Create a ②-node a_i and connect it to $|V|$ new ①-nodes b_i^j for $1 \leq j \leq |V|$; add a new ②-node to every ①-node as a leaf. Finally, connect each a_i to the central ①-node c. See Figure 2(a).
- $\mathcal{F}$ consists of $1 + |V|$ ①-②-forests $F^{(0)}, \ldots, F^{(|V|)}$ as follows:
  - $F^{(0)}$ consists of just one proper ①-②-subtree $T^{(0)}$ induced by the nodes a_i and by c.
  - For $1 \leq j \leq |V|$, $F^{(j)}$ consists of K disjoint (but isomorphic) proper ①-②-subtrees $T_i^{(j)}$ for $1 \leq i \leq K$, one in each block-graph component B_i. $T_i^{(j)}$ contains the ②-node a_i and the ①-node b_i^ℓ if v_j is adjacent to v_ℓ in G, or if $\ell = j$. Add the corresponding ②-nodes together with all induced edges. See Figure 2(b) for an example.

The adjacencies of a vertex v_j in G are modeled by $F^{(j)}$: Each $T_i^{(j)}$ contains all ①-nodes b_i^j that correspond to the closed neighborhood of v_j in G (i.e., $\{v_j\} \cup N(v_j)$). Observe that all $T_i^{(j)}$ are isomorphic and node disjoint for a fixed j, and that none contains the central ①-node c. Moreover, every $T_i^{(j)}$ contains only nodes from one block-graph component B_i.

Thus, if $T_i^{(j)}$ or a proper ①-②-subtree thereof is selected in a solution of $(B, \mathcal{F})$, then this exactly mirrors the case where v_j is put into bin V_i in a solution of DOMATIC NUMBER. In particular, ②-node a_i mirrors the bin V_i: V_i contains a vertex v_j if a_i is covered by $T_i^{(j)}$ (or a subtree thereof). Since we are only allowed to choose at most one ①-②-tree from every ①-②-forest $F^{(j)}$, this corresponds to the exclusive selection of vertices to bins.

This construction gives a feasible instance of EDGE DISJOINT ①-②-SUBTREE COVER, and it can be constructed in polynomial time. Furthermore, all ①-nodes b_i^j are covered by at least one of the proper ①-②-subtrees. Lemma 5 shows the equivalence of the instances.

Lemma 5. *Let $(B, \mathcal{F})$ be as constructed above from the instance G of* DOMATIC NUMBER. *G has a solution of* DOMATIC NUMBER *if and only if $(B, \mathcal{F})$ has a feasible edge cover.*

A solution for the instance of DOMATIC NUMBER can be computed efficiently from the solution of an instance of EDGE DISJOINT ①-②-SUBTREE COVER and vice versa.   □

Discussion of special cases. We now discuss how some special cases of STEINER TREE 2–SPANNER can be solved polynomially. Observe that we have already shown that the problem is polynomially solvable if $G[R]$ is biconnected. Furthermore, if the number of Steiner vertices $|S|$ or the number of blocks of $G[R]$ is bounded by a constant, we can use dynamic programming or even exhaustive search to get polynomial algorithms.

Now consider the case that the block-graph $B(G)$ is a ①-②-star centered at a ②-node. In this case, it is sufficient to check if every block can be spanned by at least one Steiner vertex. But, as we will see in the next subsection, even in this simple case, we cannot hope for finding a solution to the optimization problem (MINIMUM STEINER TREE 2–SPANNER) that is anything better than a logarithmic factor of the optimal solution.

Hardness for $k=3$. We can use the hardness of STEINER TREE 2–SPANNER to show the $\mathcal{NP}$-hardness of STEINER TREE 3–SPANNER.

Theorem 2. STEINER TREE 3–SPANNER *is $\mathcal{NP}$-complete.*

4   Finding Minimum Steiner Tree k–Spanners

We now turn to optimization problem MINIMUM STEINER TREE k–SPANNER. As a consequence of Theorem 1, MINIMUM STEINER TREE 2–SPANNER is also $\mathcal{NP}$-complete. Hence, even if we know in advance that the given instance admits some Steiner tree 2–spanner, we cannot hope for finding efficiently one that uses the minimum number of Steiner vertices. We now strengthen this result by proving an inapproximability result. We do this for the general case of arbitrary $k \geq 2$. In particular, we show that it is even $\mathcal{NP}$-hard to find a Steiner tree k–spanner in which the number of Steiner vertices is guaranteed to be within anything better than a logarithmic factor of the number of Steiner vertices in an optimal solution.

Since we now consider arbitrary k, in this section, we cannot use the equivalent formulation as an edge disjoint subtree covering problem. Instead, we show the result directly by proving that finding an (approximate) minimum Steiner tree k–spanner is as hard as solving MINIMUM HITTING SET: Given a collection C of subsets of a finite set F, and a positive integer $K \leq |F|$, the problem is to decide whether there is a subset $F' \subseteq F$ with $|F'| \leq K$ such that F' contains at least one element from each subset in C.

Using the inapproximability results of [4] for the equivalent MINIMUM SET COVER Problem, there is no $((1 - \epsilon) \log |F|)$–approximation algorithm for MINIMUM HITTING SET for any $\epsilon > 0$, unless $\mathcal{NP} \subset \mathcal{DTIME}(n^{\log \log n})$. Note that $\mathcal{NP} \not\subset \mathcal{DTIME}(n^{\log \log n})$ means that $\mathcal{NP}$ does not have quasi-polynomial deterministic algorithms, a slightly weaker assumption than $\mathcal{P} \neq \mathcal{NP}$.

Theorem 3. *For any fixed integer $k \geq 2$, MINIMUM STEINER TREE k–SPANNER cannot be approximated within ratio $(1 - \epsilon) \log |S|$ for any $\epsilon > 0$, unless $\mathcal{NP} \subset \mathcal{DTIME}(n^{\log \log n})$.*

Proof. We prove the hardness of MINIMUM STEINER TREE k–SPANNER by showing that finding an (approximate) minimum Steiner tree k–spanner is as hard as solving MINIMUM HITTING SET. Starting from an instance (C, F) of MINIMUM HITTING SET where $C = \{C_1, \ldots, C_m\}$ and $F = \{f_1, \ldots, f_n\}$, the idea is as follows: A subset C_i corresponds to a block in G that does not contain a tree k–spanner (but a Steiner tree k–spanner), whereas the elements of F are modeled by Steiner vertices. We construct the graph G for MINIMUM STEINER TREE k–SPANNER as follows:

- Create terminals and terminal edges: For every $C_i \in C$ for $1 \leq i \leq m$, create a simple cycle R_i of length $k + 2$ consisting of vertices $V(R_i) = \{r_i^c, r_i^1, \ldots, r_i^{k+1}\}$. Melt together all vertices r_i^c to form one central vertex r.
- Create Steiner vertices and edges: For every $f_\ell \in F$ for $1 \leq \ell \leq n$, create a new Steiner vertex s_ℓ. For every $f_\ell \in C_i$, connect s_ℓ to r and to r_i^j for $1 \leq j \leq k + 1$.

We get the graph $G = (R \,\dot\cup\, S, E)$ with Steiner vertices $S = \{s_1, \ldots, s_n\}$ and terminals $R = \{r\} \cup \{r_i^j \mid 1 \leq i \leq m, \, 1 \leq j \leq k + 1\}$. Observe that $|S| = |F|$.

For every $C_i \in C$, there is a block R_i in G, and the members of C_i represent Steiner vertices that may span R_i. Observe that $G[R]$ is connected, that the blocks of $G[R]$ are formed by the R_i for $1 \leq i \leq m$, and that the blocks R_i do not admit a tree k–spanner. Moreover, every block R_i admits a Steiner tree k–spanner. Lemma 6 completes the proof.

Lemma 6. *There is a hitting set of cardinality K for the instance (C, F) of* MINIMUM HITTING SET*, if and only if G as constructed above admits a Steiner tree k–spanner containing K Steiner vertices.*

$\square$

Observe that the instance G of MINIMUM STEINER TREE k–SPANNER as constructed above has a block-graph $B(G)$ that is a ①–②–star centered at a ②–node. As stated in Section 3 above, in this special case at least STEINER TREE 2–SPANNER is solvable efficiently.

In the definition of MINIMUM STEINER TREE k–SPANNER, we have measured the quality of a Steiner tree k–spanner T in terms of the *number of Steiner vertices*. Let us now reconsider the optimization problem with respect to the *total number of edges*. As mentioned above, for any Steiner tree k–spanner T, $|E(T)| = |R| + |S(T)| - 1$, and hence a minimum Steiner tree k–spanner with respect to the number of Steiner vertices is also a minimum Steiner tree k–spanner with respect to the total number of edges. By this, the modified optimization problem clearly remains $\mathcal{NP}$-complete. But note that we cannot deduce similar results concerning the (in)approximability of this problem from our reduction from MINIMUM HITTING SET, because the approximation ratio may tend to 1.

5   Related Tree Covering Problems

Motivated by the subtree covering problem that has appeared in the context of STEINER TREE 2–SPANNER, we now examine different variations of this kind of covering problems. Observe however, that these new problems cannot be translated to the context of Steiner tree k–spanners. We deal with them here since we consider them interesting in themselves. In particular, we consider two aspects:

- **Edge disjoint ①–②–tree Cover by ①–②–trees:**
  Instead of considering the possibility of choosing a proper subtree of a selected ①–②–subtree, we are forced to take the whole selected ①–②–subtree. Observe that Problem 3 is not a direct generalization of this problem. The choice of whole ①–②–subtrees of the $F^{(j)}$ instead of picking parts thereof significantly changes the objective. Hence, the approaches of Theorem 1 do not apply here. However, this variant also turns out to be hard.
- **Trees instead of ①–②–trees:**
  In contrast to the problems considered so far, we now generalize to arbitrary trees, not ①–②–trees. This yields the corresponding problems EDGE DISJOINT SUBTREE COVER and EDGE DISJOINT TREE COVER. For both problems, the complexity status remains $\mathcal{NP}$-complete.

References

1. I. Althöfer, G. Das, D. Dobkin, D. Joseph, and J. Soares. On sparse spanners of weighted graphs. *Discrete Computational Geometry*, 9:81–100, 1993.
2. L. Cai and D.G. Corneil. Tree spanners. *SIAM J. Discrete Math.*, 8(3):359–387, 1995.
3. P. Crescenzi and V. Kann. A compendium of NP optimization problems. Technical Report SI/RR-95/02, Univ. di Roma La Sapienza, 1995. http://www.nada.kth.se/theory/problemlist.html.
4. U. Feige. A threshold of $\ln n$ for approximating set cover. In *Proc. 28th Annual ACM Symp. Theory of Computing, STOC'96*, pages 314–318. ACM Press, 1996.
5. M.R. Garey and D.S. Johnson. *Computers and Intractability: A Guide to the Theory of $\mathcal{NP}$-Completeness*. W H Freeman & Co Ltd, 1979.
6. D. Handke. *Graphs with Distance Guarantees*. PhD thesis, Universität Konstanz, Germany, 1999. http://www.ub.uni-konstanz.de/kops/volltexte/2000/377.
7. F. Harary. *Graph Theory*. Addison–Wesley, 1969.
8. D. Handke and G. Kortsarz. Tree spanners for subgraphs and related tree covering problems. Konstanzer Schriften in Mathematik und Informatik 44, Universität Konstanz, 1997, revised version of Feb. 2000. http://www.fmi.uni-konstanz.de/Schriften/.
9. B.K. Haberman and G.N. Rouskas. Cost, delay, and delay variation conscious multicast routing. Technical Report TR-97-03, Dept. of Computer Science, North Carolina State University, 1997.
10. F.K. Hwang, D.S. Richards, and P. Winter. *The Steiner Tree Problem*, volume 53 of *Annals of Discrete Mathematics*. North–Holland, 1992.
11. G. Kortsarz. On the hardness of approximating spanners. In *Proc. International Workshop Approximation Algorithms for Combinatorial Optimization, APPROX'98*, pages 135–146. Lecture Notes in Computer Science, vol. 1444, Springer, 1998.
12. D. Peleg and A.A. Schaeffer. Graph spanners. *J. of Graph Theory*, 13:99–116, 1989.
13. D. Peleg and J.D. Ullman. An optimal synchronizer for the hypercube. In *Proc. 6th ACM Symp. Principles of Distributed Computing, Vancouver*, pages 77–85, 1987.
14. G.N. Rouskas and I. Baldine. Multicast routing with end-to-end delay and delay variation constraints. *IEEE J. Selected Areas in Communications*, 15(3):346–356, 1997.

Minimal Size of Piggybacked Information for Tracking Causality: A Graph-Based Characterization

Jean Michel Hélary[1] and Giovanna Melideo[2,3]

[1] IRISA, Campus de Beaulieu, 35042 Rennes Cedex, France. helary@irisa.fr
[2] Dipartimento di Informatica e Sistemistica, Università "La Sapienza",
Via Salaria 113, 00198 Roma, Italy. melideo@dis.uniroma1.it
[3] Dipartimento di Matematica ed Applicazioni, Universitá di L'Aquila, Via Vetoio,
67100 L'Aquila, Italy.

Abstract. A fundamental problem in distributed computing consists in tracking causal dependencies between relevant events occurring during the computation, named *observable events*. Several methods have been proposed so far in order to track these dependencies on line. They require to propagate information among processes participating in the computation, by piggybacking additional control data to the computation messages. All these methods have to face the problem of the size of piggybacked information that could become prohibitive. However, bounding the size of piggybacked information may lead to the irremediable loss of causal dependencies, if the set of observable events is not correctly chosen. The challenge is to determine the minimal size of piggybacked information, in function of a given set of observable events, allowing to track all causal dependencies. This paper provides an answer to this previously open problem. This answer is based on the construction of a weighted graph modelizing the given computation with its observable events. Although the minimal value can be known only when all the computation is known, it can be used off line to perform *a posteriori* analysis of a computation.

1 Introduction

Causality is fundamental to many situations occurring in distributed computations, such as distributed debugging, crash recovery, concurrency measure, causal ordering in communication protocols, just to cite a few. As realistic distributed computations are asynchronous (because processes do not have access to a common clock and communication delays are unpredictable), the notion of real-time is not pertinent. Reasoning based on the causal structure of a system provides a better level of abstraction [11]. The notion of causal dependency between events has been formally defined by Lamport [9], endowing the set of events of a distributed computation with a partially ordered set structure $\widehat{E} = (E, \rightarrow_c)$ (also called the causal dependency graph). Since then, there has been a lot of work

U. Brandes and D. Wagner (Eds.): WG 2000, LNCS 1928, pp. 218–229, 2000.

aiming to design methods that track causal dependencies during the computation. All these methods are based on timestamps associated with events and on the piggybacking of information on messages used to update the timestamps. Timestamps are integer vectors of size n, where n is the number of processes involved in the computation.

In this paper, we are interested in *observation* methods which consider an additional observer process and a subset O of relevant events (called *observable events*). The observer gets the value of each observable event's timestamp and, from these values, builds an observed dependency graph tracking causal dependencies between observed events. With respect to a given distributed computation and a given set O, an observation method is *consistent* if it tracks *only* causal dependencies between events in O, and it is *complete* if it tracks *all* the causal dependencies between events in O. Several consistent observation methods have been proposed.

The *transitive dependency method*, proposed independently by Fidge [4] and Mattern [10] is complete whatever the choice of O (when O is empty, it means that $\widehat{E}$ can be constructed on line in a decentralized way, without the need of an observer). However, the size of information to be piggybacked is prohibitive, since it amounts to n integer values[1]. In the *direct dependency method*, proposed by Fowler and Zwaenepoel [5], the size of piggybacked information is reduced to one integer. But, if O is not correctly chosen, this method is not complete [8] (the original paper [5] implicitly assumes $O = E$, which ensures the completeness).

The *k-dependency method*, introduced by [1] is a generalization of the direct dependency method, where the size of piggybacked information is bounded by k integer values (k can be any integer ranging from 1 to n). But, except when $k = n$, its completeness requires some conditions on O, depending on the given value k. Yet, the only known condition on O, called NIVI and introduced in [6], refers to the case $k = 1$ and is only a sufficient condition. That is, no weaker sufficient condition has been stated when $k > 1$ (except $k = n$, where no condition is required) and, moreover, no *necessary* condition has been found so far.

At the operational level, two approaches have been proposed to ensure completeness of observation methods. In the first one [8], the set O is dynamically adapted to the size k. More precisely, the NIVI condition is enforced by directing processes to define additional observable events when this condition is about to be violated. Clearly, since the NIVI condition is only sufficient and designed for the case $k = 1$, it can be too strong and forces too many observable events. In the second approach, the size k is dynamically adapted to the given set O. This approach is illustrated by the *incremental transitive method* [8] which dynamically reduces the size of piggybacked information by removing some redundancies. But this method does not guarantee that the size of piggybacked informations will not grow up to n integer values, depending on O.

The aim of this paper is to provide a characterization of sets O ensuring completeness of observation methods with a bounded size k of piggybacked in-

---

[1] Following [3] this size is necessary if one requires complete methods whatever the set O.

formation. Equivalently, the main result is a characterization of the *minimal* value of k to be used in order to ensure completeness with respect to a given distributed computation and a given set of observable events. This characterization essentially shows that, the more observable events in the computation, the lesser k. In some sense, there is a tradeoff between the amount of information that must be propagated *within* the computation (piggybacked information) and the number of events that must *output* the information towards the observer process. To our knowledge, such a characterization (and the compromise which it reveals) had never been put forward, and is helpful for a better understanding of causality tracking methods.

The paper is made of three sections. Section 2 presents the computation model and causality tracking methods. Section 3 is devoted to the characterization result, and Section 4 concludes the paper.

2   Computation Model

2.1   Distributed Computation

A distributed computation is modeled by a finite set of n sequential processes $\{P_1, \dots, P_n\}$ communicating solely by exchanging messages. Each ordered pair of processes is connected by a reliable directed logical channel. Transmission delays are unpredictable but finite.

The execution of each process P_i produces a totally ordered set of *events* E_i. An event may be either *internal* or it may involve communication (*send* or *receive* event). We denote as $I_i \subseteq E_i$, $S_i \subseteq E_i$ and $R_i \subseteq E_i$ the set of internal, send and receive events produced by process P_i and as E the set of all the events, i.e. $E = \cup_{i=1}^n E_i$.

Following Lamport [9], we say that an event e *locally precedes* e', denoted as $e \to_l e'$, if e immediately precedes e' on the same process. Moreover, $e \to_m e'$ if e is the sending event of a message and e' is the corresponding receive event. Then we can see a distributed computation as a directed graph $\widehat{E} = (E, \to_c)$, where $\to_c$ is the well known *causality relation* (or Lamport's happened-before relation) defined as the transitive closure of the relation $\to_l \cup \to_m$. It is well known that $\widehat{E}$ is a strict finitary partial order. Two events e and e' are concurrent if $\neg(e \to_c e')$ and $\neg(e' \to_c e)$.

Notations. Given a directed graph $G = (X, \to_r)$ we will denote its transitive closure as $G^+ = (X, \to_r^+)$, its transitive reduction as $G^- = (X, \to_r^-)$, and

$$\forall x \in X \; : \; \downarrow_r (x) = \{x\} \cup \{y \in X \mid y \to_r^+ x\}$$

In particular, when considering a distributed computation $\widehat{E}$, we have $\widehat{E}^+ = \widehat{E}$, and the *causal past* of any $e \in E$ is $\downarrow_c (e) = \{e\} \cup \{e' \in E \mid e' \to_c e\}$.

2.2   Tracking Dependency between Observable Events

To design efficient distributed algorithms one needs to track causal dependencies during the computation. Nevertheless, detecting causal relations or concurrency between all internal and communication events is not desirable, so analyzing a distributed computation requires to define precisely a set of relevant events, called *observable events* and denoted by O (similarly, $O_i = O \cap E_i$). If I denotes the set of internal events of E, we assume $O \subseteq I$, that is only internal events can be observable. Note that this is not a restriction, because if a communication event must be "observed", it is equivalent to create an internal observable event immediately following the communication event. At the considered abstraction level, the distributed computation is characterized by the directed graph $\widehat{O} = (O, \rightarrow_o)$ defined as the subgraph of $\widehat{E}$ spanned by O, *i.e*, for every $e, e' \in O$, $e \rightarrow_o e' \Leftrightarrow e \rightarrow_c e'$.

Distributed computations can be visualized using *space-time diagrams* (see Figure 1.(a)). Without loss of generality, we do not represent non-observable internal events. Each observable event is depicted as a dot located on a process line and the x-th observable event generated on process P_i is denoted as $e_{i,x}$. Moreover, each message is represented by an interprocess arrow. Actually, a time diagram is a graphical representation of the transitive reduction $\widehat{O}^-$. In fact, $e \rightarrow_o e'$ if and only if there is a path starting at e and ending at e'.

Following [12], the causal past $\downarrow_o (e)$ of an observable event can be represented by an n-dimensional vector of integers $V(e)$ (where n is the number of processes), called *transitive dependency vector* and characterized by

$$\forall j \ : \ V(e)[j] = |\downarrow_o (e) \cap O_j| \, , \tag{1}$$

which implies [12]

$$V(e) = \max_{e' \in \downarrow_o(e)} V(e'), \tag{2}$$

or, equivalently, if P_i is the process producing e':

$$V(e')[i] \leq V(e)[i] \Leftrightarrow e' \rightarrow_o e. \tag{3}$$

Methods used to track causal dependencies between observable events consist in associating *n-dimensional vector timestamps* with observable events and in piggybacking control informations upon outgoing messages used to update timestamps. Moreover, an additional process, called *observer process*, receives the timestamps associated with observable events. Its role is to reconstruct $\widehat{O}$, or, equivalently, to compute exact values $V(e)$ for each observable event, just by analyzing the received timestamps. More precisely, the observer process receives pairs $(e, T(e))$ where events e are observable and $T(e)$ is some information control associated with e, approximating $V(e)$. Using these information, it builds an *observed dependency graph*, denoted ODG and defined as follows. Its vertices are events e received by the observer, labeled with $T(e)$. There is an edge from e to

e', denoted $e \rightarrow_{ODG} e'$ iff $T(e)[i] \le T(e')[i]$ (P_i being the process producing e). If the method is consistent, property (3) ensures $e' \rightarrow_{ODG} e \Rightarrow e' \rightarrow_o e$. Then, it has to backward explore the graph ODG in order to build the transitive closure ODG^+ [2,12] and computes, for every vertex e:

$$T_{obs}(e) = \max_{e' \in \downarrow_{ODG}(e)} T(e').$$

If ODG^+ is isomorphic to $\widehat{O}$ then $T_{obs}(e) = V(e)$.

Definition 1. *The coding of a set of observable events is* complete *when ODG^+ is isomorphic to $\widehat{O}$, that is the coding allows the observer to detect the correct dependencies between all observable events.*

2.3   Examples of Dependency Tracking Methods

To manage timestamps, each process P_i has a local n-dimensional integer vector V_i, initialized to zero, where $V_i[j] = x$ means $e_{j,x}$ causally precedes the current event of P_i. Different methods correspond to different sizes of piggybacking informations, ranging from 1 to n, which can be statically or dynamically defined. In the rest of the paper, a pair $(j, V_i[j])$ will be called an *event identifier* on P_i. In particular, the pair $(i, V_i[i])$ is called *direct dependency*.

The *k-dependency method* introduced by [1] actually represents a general scheme for dependency tracking based on piggybacking of information whose size, represented by the parameter k, is statically defined. Let $k \in \{1, \ldots, n\}$, and $T_k(e)$ be the timestamp associated with the event e by piggybacking control informations of size k. Rules for updating local vectors and assigning timestamps are the following:

R1 upon observation of an event e on P_i: $V_i[i] := V_i[i] + 1$; $T_k(e) := V_i$;

R2 upon sending a message m, P_i piggybacks on this one a list $m.L$ containing the direct dependency and $k - 1$ event identifiers selected according to some strategy;

R3 upon receiving a message m with the list $m.L$, $V_i[j] := max(V_i[j], V[j])$, $\forall j : (j, V[j]) \in m.L$.

When $k = n$, the previous scheme reduces to the known *transitive dependency method* [4,10], and $T_n(e)$ is really the vector $V(e)$. In this case, *whatever the choice of* $O \subseteq I$, ODG is isomorphic to $\widehat{O}$. In particular, it is already transitively closed. On the contrary, when $k < n$, the ODG is not necessarily transitively closed, as depicted in Figure 1, and, more important, if the set of observable events O is not correctly chosen, ODG^+ is not necessarily isomorphic to $\widehat{O}$ (see Figure 2): some transitive information can be irremediably lost if it was not previously recorded by an observable event, that is if it has not been previously sent to the observer process. So, the *completeness* can be impacted [8].

As an example, Figure 1(b) shows the ODG produced by the 1-dependency method depicted Figure 1(a). Each vertex is labeled with the corresponding

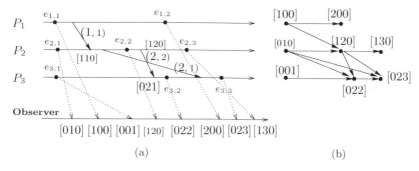

Fig. 1. A distributed computation using the 1-Dependency Vector method (a) and the related ODG (b).

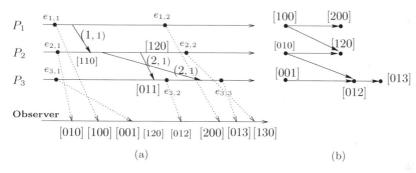

Fig. 2. A distributed computation using 1-Dependency method (a) and the related "incomplete" ODG (b).

timestamp T_1. Here, ODG^+ is isomorphic to $\widehat{O}$: if we consider the event $e_{3,2}$, we have $T_{obs}(e_{3,2}) = max_{e' \in \downarrow_{ODG}(e_{3,2})} T_1(e') = [1, 2, 2] = V(e_{3,2})$. Figure 2 shows the same when the observed event $e_{2,2}$ has been eliminated. In this case, $ODG^+ \subset \widehat{O}$. In fact, $T_{obs}(e_{3,2}) = max_{e' \in \downarrow_{ODG}(e_{3,2})} T_k(e') = [0, 1, 2]$ instead of the correct value $V(e_{3,2}) = [1, 1, 2]$. This means that the causal dependency $e_{1,1} \to_o e_{3,2}$ cannot be retrieved from the ODG. This is obvious by observing the figure 2: there is no path from $e_{1,1}$ to $e_{3,2}$ in the ODG.

In order to ensure the completeness of coding, two dual approaches have been proposed: one consists in "adapting" the set O to the given static value k, the other consists in dynamically adapting the value of k to the given set O.

Adapting O. In [6,8], a sufficient condition on O ensuring the completeness of the 1-dependency method has been stated. This condition, called NIVI, stipulates that after a receive event there must be an observable event before a send event (on the same process) occurs. At the operational level, processes have to define *forced* observable events when this condition is about to be violated. If the NIVI condition on O is satisfied (or forced), the k-dependency method is complete, for any given $k \geq 1$. In fact, even if in the local vector V_i there are more than k

different dependency informations, piggybacking the direct dependency is sufficient to ensure the completeness. The other event identifiers have already been
recorded in some observable event. They are redundant and their piggybacking
only improves the performance of the method (in terms of graph availability).

Adapting the size of piggybacked information. Jard and Jourdan [8] proposed the
incremental transitive method, which is *adaptive* with respect to the choice of O.
In this case the rules for updating local vectors V_i and assigning timestamps are
the following:

R1' upon observation of an event e on P_i: $V_i[i] := V_i[i] + 1$; $T(e) := V_i$;
$\forall j \neq i$ **do** $V_i[j] := 0$ (reset step)
R2' upon sending a message m to P_j, a list $m.L$ is piggybacked towards P_j
containing all pairs $(j, V_i[j])$ such that $V_i[j] \neq 0$;
R3' upon reception from P_j with the list $m.L$, $\forall j : (j, V[j]) \in m.L$ **do** $V_i[j] :=$
$max(V_i[j], V[j])$, .

With respect to the k-dependency method, there are two main differences: first,
the size of piggybacked information is variable (although it can grow up to
n). Second, all local information is propagated while no observable event occurs.
When such event e occurs, the information is recorded in the timestamp $T(e)$ and
sent to the observer process, so the propagation can stop (reset step). This allows
a process to "forget" the information received before the last observable event.
We remark that if the reset step is not performed, a process cannot distinguish
the "new" information received since the last observable event from the "old"
ones already transmitted to the observer. It can be argued that when the NIVI
condition is not satisfied, the reset step is necessary.

In this paper we are interested in finding the minimal size k of piggybacked
information in order to guarantee the completeness of the coding with respect to
a given distributed computation and a given set O. So, the methods considered
must not introduce additional observable events and thus must execute a reset
step when an event is observed on a process. The next section investigates this
relationship.

3   Characterizing the Minimal Size of Piggybacked Information

3.1   Construction of a Weighted Covering Graph

Given a distributed computation $\widehat{E}$ and a given set $O \subseteq I$ of observable events,
the *minimal* value of k ensuring a complete coding of $\widehat{O}$ when the size of piggybacked information is limited to k will be characterized. This characterization
is based on the construction of a weighted graph from which this value can be
easily computed.

The idea is based on the following observation. Let us consider a process
P_i where a sequence of l consecutive *receive* events occurs, each message being

received from a different sender. If the first event following this sequence is a *send* event, it means that the sent message must propagate l causal dependency informations towards the observable event that will follow *receive*(m). If, on the contrary, the first event following this sequence is observable, then all causal dependency information brought by these l messages is transferred towards the observer, and consequently has not to be propagated on the messages that will be sent by P_i after the observable event. Thus, every message sent appears to be the messenger of a certain amount of information, depending on how many causal information is stored on the sending process. However, this view is too pessimistic, because it does not take into account the fact that some information could be transferred from an observable event to another via several alternative paths in the distributed computation.

A careful analysis of this situation leads to the construction of a weighted graph on the set O, whose maximum edge weight is equal to the required minimal value of k. The construction of this graph is carried out in three steps: (1) a *propagation* graph, built on O and additional vertices modelizing the sequences of receive events followed by a send event, (2) a weighted *covering* graph, whose transitive closure is isomorphic to $\widehat{O}$, obtained by suppressing additional vertices from the propagation graph and computing weights accordingly, and (3) the transitive reduction of the weighted covering graph, whose aim is to remove redundant information transmission.

Propagation Graph. Let $PG = (V, \rightarrow_{pg})$ defined as follows. $V = O \cup P$, where an O-vertex is generated by each observable event, and a P-vertex is generated every time that a non empty sequence of consecutive receive events is followed by a send event (on the same process). Given two vertices X and Y generated on the same process P_i, we say that $X = pred(Y)$ (or equivalently $Y = succ(X)$) if the events generating X and Y occur in P_i in that order, and there is no vertex between X and Y. With every message m of the computation are associated two vertices $\overleftarrow{m}$ and $\overrightarrow{m}$ in V. $\overleftarrow{m}$ is the vertex corresponding to the last non send event that locally precedes *send*(m), and $\overrightarrow{m}$ is the vertex corresponding to the first non receive event that locally follows *receive*(m), that is $\overrightarrow{m}$ is an O-vertex if *receive*(m) is followed by an observable event, a P-vertex otherwise. There is an edge $X \rightarrow_{pg} Y$ iff $X = pred(Y)$ or there exists a message m such that $X = \overleftarrow{m}$ and $Y = \overrightarrow{m}$. O-vertices are labeled like their corresponding observable event, whereas $p_{i,x,h}$ denotes the h-th P-vertex on process P_i, between O-vertices $e_{i,x}$ and $e_{i,x+1}$.

As an example, in Figure 3 is shown the propagation graph (b) associated with the distributed computation depicted in (a). O-vertices are depicted as rounded rectangles, whereas P-vertices are depicted as sharp rectangles.

Reduction Procedure to Weighted Covering Graph. The weighted covering graph $WCG = (O, \rightarrow_{wcg})$ is a partial representation of $\widehat{O}$ where an edge from e to e' represents a path in the time diagram starting form e and ending to e' without observable event between e and e'. At the end of its construction, the weight

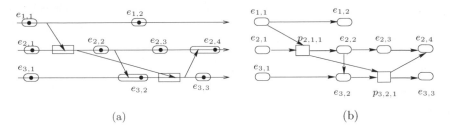

Fig. 3. A distributed computation (a) and its associated propagation graph (b).

of each edge from e to e' will represent the minimum size of the piggybacked information which guarantees that the information on e is propagated towards e' without any observable events after leaving e. This graph is built from the propagation graph by iteratively removing its P-vertices and replacing the removed edges by weighted ones linking only O-vertices. Rules for building WCG are described as follows. Initially, $WCG = PG$ with all weights equal to 1. For each P-vertex X whose all predecessors are O-vertices:

- for each Y such that $Y \rightarrow_{wcg} X$,
  if $\neg(Y \rightarrow_{wcg} succ(X))$
  then create an edge $Y \rightarrow_{wcg} succ(X)$ with $w(Y, succ(X)) = w(Y, X)$,
  else $w(Y, succ(X)) = \min(w(Y, succ(X)), w(Y, X))$.
- for each Z such that $X \rightarrow_{wcg} Z$, create an edge $prec(X) \rightarrow_{wcg} Z$ with $w(prec(X), Z) = 1$.
- for each $Y \neq prec(X)$, $Z \neq succ(X)$ such that $Y \rightarrow_{wcg} X \rightarrow_{wcg} Z$,
  if $\neg(Y \rightarrow_{wcg} Z)$
  then create an edge $Y \rightarrow_{wcg} Z$ with $w(Y, Z) = |\{y|y \rightarrow_{wcg} X\}|$,
  else $w(Y, Z) := \min(w(Y, Z), |\{y|y \rightarrow_{wcg} X\}|)$.
- remove X from WCG.

The construction stops when there are no more P-vertices in WCG. It is easy to see that, by construction, $WCG^+ = \widehat{O}$.

In the figure 4 is depicted an execution of the reduction's procedure, the propagation graph (a) being the input and the weighted covering graph (c) the output obtained after eliminating two P-vertices $p_{2,1,1}$ and $p_{3,2,1}$, as shown in (b) and (c), respectively.

Transitive Reduction of the Weighted Covering Graph. The Weighted Covering Graph obtained in the previous section can exhibit different paths between two observable events. Clearly, if a path is "doubled" by an edge, this indicates that the information transmitted along the edge is also transmitted by the path. So, removing the edge does not modify the possibilities of information transmission represented by the rest of WCG. This well-known procedure, called transitive reduction, can be carried out on WCG, to obtain the graph WCG^-. This graph allows to state the desired characterization.

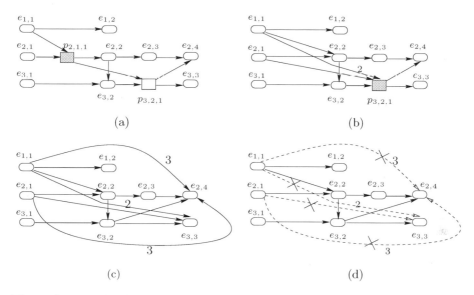

Fig. 4. An example of execution of the reduction procedure (a,b,c) on the propagation graph associated with the distributed computation depicted in Figure 3.

3.2   The Characterization

Theorem 1 (Minimal piggybacking size). *Let $\widehat{E} = (E, \rightarrow)$ be a distributed computation, and O be a set of observable events. The value k necessary and sufficient to be used in an observation method in order to obtain a complete coding of the sub-order $\widehat{O}$ is the maximum weight in the transitive reduction WCG^- of the weighted covering graph, that is:*

$$w_{max} = \max\{w(e, e') \mid e \rightarrow^-_{wcg} e'\}.$$

Proof. The proof is made of two parts: necessity and sufficiency.

Necessity. We show that, if $k < w_{max}$ then no coding can be complete. For that purpose, it is sufficient to exhibit a scenario where there exists an observable event e such that $T_{obs}(e) \neq V(e)$, where $T(e)$ is the timestamp of e in this coding. The figure 5 describes a computation with observable events (a) and its associated graph WCG^- (b) built according to the previous procedures. From WCG^- follows that $w_{max} = 3$. Now, let's consider a coding allowing only $k = 2$ event identifiers per message. Figure 6 shows the values piggybacked on messages (a), the resulting timestamps and the ODG graph (b). From this follows that $T_{obs}(e_{4,2}) = [0, 1, 1, 2] \neq V(e_{4,2}) = [1, 1, 1, 2]$. Clearly, the causal dependency $e_{1,1} \rightarrow_o e_{4,2}$ cannot be retrieved. Remark that another choice of vector entries to piggyback on the message m (in bold on Figure 6-a) would have led to the loss of causal dependency $e_{2,1} \rightarrow_o e_{4,2}$ instead.

Sufficiency. We show that, if $k \geq w_{max}$, it is possible to obtain a complete coding. To this purpose, we define a protocol that ensures a complete coding

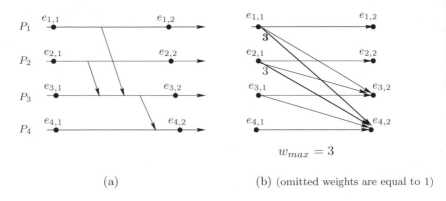

(a)   (b) (omitted weights are equal to 1)

Fig. 5. A computation and its associated WCG^-.

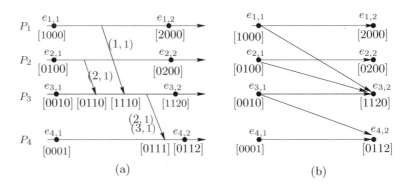

(a)   (b)

Fig. 6. Loss of causal dependencies.

under this assumption. This protocol is based on the k-dependency method with a reset step. It is based on the following rules (see rules R1' to R3' in Section 2.3):

R1" upon observation of an event e on P_i: $V_i[i] := V_i[i] + 1$; $T(e) := V_i$; $\forall j \neq i$: $V_i[j] := 0$; record $(e, T(e))$ on the observer (reset step)

R2" upon sending a message m, P_i piggybacks on this one a list $m.L$ containing the pair $(i, V_i[i])$ and at most $k - 1$ pairs $(j, V_i[j])$ such that $V_i[j] \neq 0$;

R3" upon receiving a message with the list $m.L$, for each j such that $(j, V[j]) \in m.L$, $V_i[j] := max(V_i[j], V[j])$.

This protocol provides a complete coding. Due to limitations, the detail of this proof is omitted. The complete proof can be found in [7]. It consists in proving by induction on the poset $\widehat{O}$ that if $k \geq w_{max}$, then $\forall e \in O : T_{obs(e)} = V(e)$.   □

As an example, Figure 4.(d), representing the transitive reduction of the weighted covering graph (c), shows that $w_{max} = 1$ in such a case. Previous property affirms that it is necessary and sufficient to piggyback upon messages only one integer in order to ensure the completeness of the coding of $\widehat{O}$.

4 Conclusion

Graph modelization is a powerful tool for analyzing properties of distributed computations. This technique has been used in this paper in order to determine the minimal size of piggybacked information ensuring the completeness of causal dependency tracking between a given set of observable events. This requires to associate a graph with the computation and to perform operations transforming this graph into an transitively reduced weighted graph (WCG). This transformation is similar to a transitive closure followed by a transitive reduction. This contribution is important, because, to our knowledge, no completeness characterization had been presented before. Such a characterization provides a better understanding of causality tracking techniques. Although the weighted graph WCG can be completely determined only when all the computation is known, it can be used off line to perform *a posteriori* analysis of a computation. A future work would be to find completeness conditions, less restrictive than the previously known NIVI condition, that could be tested on line. This result constitutes a theoretical basis for such an investigation.

References

1. R. Baldoni, M. Mechelli and G. Melideo. A General Scheme for Dependency Tracking in Distributed Computations, *Technical Report no. 17-99*, Dipartimento di Informatica e Sistemistica, Roma, 1999.
2. P. Baldy, H. Dicky, R. Medina, M. Morvan and J.-M. Vilarem. Efficient Reconstruction of the Causal Relationship in Distributed Systems. *Proc. 1st Canada-France Conference on Parallel and Distributed Computing*, LNCS #805, pp. 101-113, 1994.
3. B. Charron-Bost. Concerning the size of logical clocks in distributed systems, *Information Processing Letters*, 39, 11–16, 1991.
4. G. Fidge. Timestamps in message passing system that preserve the partial ordering, *Proc. 11th Australian Computer Science Conf.*, 55-66, 1988.
5. J. Fowler and W. Zwanepoel. Causal Distributed Breakpoints, *Proc. 10th IEEE Int. Conf. on Distributed Computing Systems*, pp.134-141, 1990.
6. E. Fromentin, C. Jard, G.-V. Jourdan and M. Raynal. On-the-fly Analysis of Distributed Computations, *Information Processing Letters*, 54:267-274, 1995.
7. J.M. Hélary and G. Melideo. Minimal Size of Piggybacked Information for Tracking Causality: a Graph-Based Characterization, *Research Report # 1300*, IRISA, Université de Rennes 1, February 2000. http://www.irisa.fr/EXTERNE/bibli/pi/1300/1300.html.
8. C. Jard and G.V. Jourdan. Incremental Transitive Dependency Tracking in distributed computations. Parallel Processing Letters 63, 427-435, 1996.
9. L. Lamport. Time, clocks, and the ordering of events in a distributed system, *Communications of the ACM*, **217**, 558-564, 1978.
10. F. Mattern. Virtual time and global states of distributed systems, M. Cosnard and P. Quinton eds., *Parallel and Distributed Algorithms* 215-226, 1988.
11. P. Panangaden, K. Taylor. *Concurrent Common Knowledge: Defining Agreement for Asynchronous Systems*, Distributed Computing 6(2), 73-93, 1992.
12. R. Schwarz and F. Mattern. *Detecting causal relations in distributed computations: in search of the holy grail*, Distributed Computing 7(3), 149–174, 1994.

The Expressive Power and Complexity of Dynamic Process Graphs

Andreas Jakoby[1,2], Maciej Liśkiewicz[1,3], and Rüdiger Reischuk[1]

[1] Institut für Theoretische Informatik, Medizinische Universität zu Lübeck,
Germany. {jakoby,liskiewi,reischuk}@informatik.mu-luebeck.de
[2] Department of Computer Science, University of Toronto, Canada.
jakoby@cs.toronto.edu
[3] Instytut Informatyki, Uniwersytet Wrocławski, Poland.

Abstract. A model for parallel and distributed programs, the *dynamic process graph,* is investigated under graph-theoretic and complexity aspects. Such graphs are capable of representing all possible executions of a parallel or distributed program in a very compact way. The size of this representation is small – in many cases only logarithmic with respect to the size of any execution of the program. An important feature of this model is that the encoded executions are directed acyclic graphs with a *regular* structure, which is typical of parallel programs, and that it embeds constructors for parallel programs, synchronization mechanisms as well as conditional branches.

In a previous paper we have analysed the expressive power of the general model and various restrictions. Furthermore, from an algorithmic point of view it is important to decide whether a given dynamic process graph can be executed correctly and to estimate the minimal deadline given enough parallelism. Our model takes into account communication delays between processors when exchanging data.

In this paper we study a variant with output restriction. It is appropriate in many situations, but its expressive power has not been known exactly. First, we investigate structural properties of the executions of such dynamic process graph s $\mathcal{G}$. A natural graph-theoretic conjecture that executions must always split into components isomorphic to subgraphs of $\mathcal{G}$ turns out to be wrong. We are able to establish a weaker property. This implies a quadratic bound on the maximal deadline in contrast to the general case, where the execution time may be exponential. However, we show that the problem to determine the minimal deadline is still intractable, namely this problem is $\mathcal{NEXPTIME}$-complete as is the general case. The lower bound is obtained by showing that this kind of dynamic process graph s can represent certain Boolean formulas in a highly succint way.

1 Introduction

Large parallel or distributed computations tend to have a lot of regularities. For example, the same instruction sequence may be executed by many processors in

U. Brandes and D. Wagner (Eds.): WG 2000, LNCS 1928, pp. 230–242, 2000.

parallel. To describe the elementary steps and the logical dependencies among them one can use graphs, often called *data flow graphs*. One would like to keep this description as compact as possible, for example not to unfold parallelism if this is not necessary. For this purpose, we have introduced the model of dynamic process graphs in [3]. For a more detailed motivation the reader may check there.

The novel property of this graph model are different modes for the input and output behaviour that can be assigned to each node, which models a process of the computation. This allows us to model basic primitives for writing parallel programs, like *fork* and *join*. The two modes are called ALT and PAR. To represent such parallel programs in a natural way, we introduce dynamic process graphs, which are generalizations of standard precedence graphs. A dynamic process graph is an acyclic graph $G = (V, E)$ with two sets of labels $I(v), O(v) \in \{\mathsf{PAR}, \mathsf{ALT}\}$ attached to the nodes $v \in V$. Nodes represent tasks, edges dependencies between tasks. A complete formal definition will be given in the next section. The label $I(v)$ describes the *input mode* of task v. If $I(v) = \mathsf{ALT}$ then to execute v at least one of the direct predecessor tasks has to be completed. $I(v) = \mathsf{PAR}$ requires that executions of all direct predecessors of v have to be completed before v can start. If task v gets executed then according to the *output mode* $O(v)$ one of its direct successors in case $O(v) = \mathsf{ALT}$ (resp. all of them in case $O(v) = \mathsf{PAR}$) has to be executed as well.

If one restricts the mode of nodes to PAR only, both for input and output, then this will be equivalent to ordinary data flow graphs. Using both modes, however, the representation of parallel programs by dynamic process graphs can provide an exponential compaction compared to the length of the actual execution sequences. Given a dynamic process graph, the first question that arises is whether it describes a legal parallel program. If yes then one would like to find an efficient execution of the program specified in such a compact form. We assume here that enough parallelism is available, so that the question turns into the problem to execute the program as fast as possible.

Dynamic process graphs and Boolean circuits are somehow related. We have shown that such graphs can be used to model computations of a circuit. This has then been used to prove that indeed dynamic process graphs that can use arbitrary combinations of modes provide a very compact nontrivial representation. To find an optimal schedule, which is $\mathcal{NP}$-complete for ordinary graphs, turns out to be $\mathcal{NEXPTIME}$-complete. A similar complexity jump has been observed for classical graph problems in [1,5,4]. These papers have shown that simple graph properties become $\mathcal{NP}$-complete when the graph is represented in a particular succinct form using generating circuits or a hierarchical decomposition. Under the same representation graph properties that are ordinarily $\mathcal{NP}$-complete, like HAMILTON CYCLE, 3-COLORABILITY, CLIQUE (with half the graph size), etc., become $\mathcal{NEXPTIME}$-complete.

If we put restrictions on the modes of a dynamic process graph its execution becomes easier. We have given a precise classification in [3] for all possible combinations, except the most interesting case where the output mode is restricted to PAR, but the input mode may use both alternatives. Such graphs are still able to model, for example, the two natural ways in which objects of an object-oriented languages can be activated: the total case, where all input parameters have to be specified before activation, and the partial case, where an object fires for any specified parameter once.

For dynamic process graphs with PAR output mode we could show that finding an optimal schedule is at least $\mathcal{BH}_2$-hard. There was no matching upper bound, only the $\mathcal{NEXPTIME}$ upper bound of the unrestricted case. Other combinations turn out to be less powerful. Restricting the input mode to ALT the problem becomes $\mathcal{P}$-complete, and also fixing the output mode the problem gets $\mathcal{NL}$-complete.

As main results of this paper we show the following two complexity bounds. A linear upper bound on the length of executing dynamic process graphs with PAR output mode, whereas in the unrestricted case this bound may be exponential. Thus, restricting the expressive power of this graph representation decreases the time complexity of parallel and distributed programs drastically. On the other hand, we prove that computing an optimal execution in a setting that takes communication delays into account remains hard – this problem is still $\mathcal{NEXPTIME}$-complete.

In the next section, we will give a formal definition of this graph model. In section 3 we exhibit structural properties of executing dynamic process graphs with PAR output mode restriction, which will imply the linear upper bound. Section 4 considers the scheduling problem and gives a general outline how the $\mathcal{NEXPTIME}$-hardness result can be shown. Due to space limitations, a bunch of technical details have to be omitted.

2 Dynamic Process Graphs and Runs

Given a DAG $G = (V, E)$ with node set V and edge set E, for $v \in V$ let $\mathbf{pred}(v):= \{u_1, \ldots, u_p\}$ denote the set of its direct predecessors, and $\mathbf{succ}(v):= \{w_1, \ldots, w_r\}$ its direct successors. $\mathbf{pred}^*(v)$ is the set of all anchestors of v.

Definition 1. *A **dynamic process graph**, DPG for short, $\mathcal{G} = (V, E, I, O)$ consists of a DAG (directed acyclic graph) with node set V and edges E and two node labellings $I, O : V \to \{\mathsf{ALT}, \mathsf{PAR}\}$. $V = \{v_1, v_2, \ldots, v_n\}$ represents a set of **processes** and E dependencies among them. I and O describe **input**, (resp. **output**) modes of the v_i.*

*A finite DAG $H_\mathcal{G} = (W, F)$ is a **run** of $\mathcal{G}$ iff the following conditions are fulfilled:*

1. *The set W is partitioned into subsets $W(v_1) \ \dot\cup\ W(v_2) \ \dot\cup\ \ldots \ \dot\cup\ W(v_n)$. The nodes in $W(v_i)$ are **execution instances** of the process v_i, and will be called **tasks**.*
2. *Each source node of $\mathcal{G}$, which represents a starting operation of the program modelled by $\mathcal{G}$, has exactly one execution instance in $H_\mathcal{G}$.*
3. *For every $v \in V$ with $pred(v) = \{u_1, \ldots, u_p\}$ and $succ(v) = \{w_1, \ldots, w_r\}$ and every execution instance $x \in W(v)$ it holds:*

- if $I(v) = $ ALT *then* x *has a unique predecessor* $y \in W(u_i)$ *for some* $i \leq p$;
- if $I(v) = $ PAR *then* $pred(x) = \{y_1, \dots, y_p\}$ *with* $y_i \in W(u_i)$ *for each* $i \leq p$;
- if $O(v) = $ ALT *then* x *has a unique successor* $z \in W(w_j)$ *for some* $j \leq r$;
- if $O(v) = $ PAR *then* $succ(x) = \{z_1, \dots, z_r\}$ *with* $z_j \in W(w_j)$ *for each* $j < r$.

We call a DPG $\mathcal{G}$ **executable** *iff there exists a run for it.*
Given $\mathcal{G}$ with run $H_{\mathcal{G}} = (W, F)$, for each edge (u, v) in $\mathcal{G}$ we define $F(u, v) :=$
$\{(y, z) \in F \mid y \in W(u) \text{ and } z \in W(v)\}.$

Throughout the paper we will illustrate the ALT-mode by a white box and the PAR-mode by a black box. To represent a node of a DPG with input and output mode e.g. ALT, resp. PAR we will use two boxes: white at the top representing the input mode and the black box at bottom for output mode. For a source or a node with indegree 1 the input mode obviously is inessential. Hence we will ignore such a label, and similarly the output label in case of a sink or a node with outdegree 1.

Fig. 1 gives an example of a DPG and two runs of it. DPGs can be used to specify parallel programs in a compact way. A run corresponds to an actual execution of the program. Observe that a run can be smaller than its defining DPG (see [3] Fig. 6). More typically, however, a run will be larger than the DPG itself since the PAR-constructor allows process duplications. The runs in Fig 1 illustrate this property. The following lemma gives an upper bound on the blow-up, resp. the possible compaction ratio of DPGs.

Lemma 1 ([3]). *Let $\mathcal{G} = (V, E, I, O)$ be a DPG and $H_{\mathcal{G}} = (W, F)$ be a corresponding run. Then $|W| \leq 2^{|V|-1}$, and this general upper bound is best possible.*

Certain DPGs thus have processes with exponential many execution instances.

The motivating question for our study of DPGs is *how efficiently a compactly specified parallel program Π can be executed.* Note that this is a crucial problem during compile-time. Furthermore, we like to construct an optimal schedule for the execution tasks of Π (compare e.g. [2]). Lemma 1 implies an upper bound for the execution time of such a compactly presented program. However, this bound may be much too pessimistic for massive parallelism. With respect to their size, DPGs may require exponential many execution instances for some of its processes. However, if enough processors are available it may be possible to construct runs that use only linear time by executing many tasks in parallel.

Definition 2. *Let $\mathcal{G} = (V, E, I, O)$ be a DPG and $H = (W, F)$ be a run of $\mathcal{G}$. A* **subrun** $\mathcal{R}(y)$ *of H is a subgraph that is induced by one of its sink nodes y, i.e. it consists of a sink y and all nodes in* **pred**$^*(y)$ *together with all their connections in F. We call a subrun $\mathcal{R}(y)$ of H* **k-overlapping** *iff for every node $v \in V$ it holds $|\mathcal{R}(y) \cap W(v)| \leq k + 1$. A run H is k-overlapping if all its subruns are k-overlapping. We call H* **non-overlapping** *if it is 0-overlapping.*

Note that non-overlapping is a slightly stronger condition than the requirement that each subrun has to be isomorphic to a subgraph of $\mathcal{G}$. The isomorphism requirement seems to be a very natural condition. Fig. 1, however, shows that it does not have to hold, even for DPGs with output restriction PAR. The two execution instances of the sink w induce two different subruns. For the run illustrated in (b) each subrun is non-overlapping, it is isomorphic to a subgraph of $\mathcal{G}$. In (c) each subrun contains both execution instances of process t, that means it is 1-overlapping.

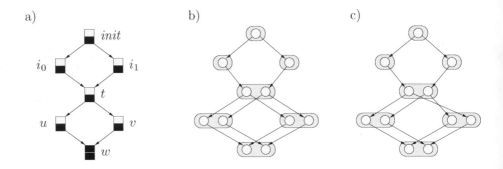

Fig. 1. *(a) A DPG with 2 nonisomorphic runs: process t has 2 execution instances, and the same holds for all its successors; (b) a non-overlapping run, (c) a run with overlap at $W(t)$. Recall, that in a schematic representation of node v of a DPG a white box pictures the* ALT-*mode and a black box – the* PAR-*mode.*

The maximal size of a subrun of H gives a better upper time bound for executing H than just the size of H because all subruns can be executed independently in parallel. Moreover, if a DPG $\mathcal{G}$ has a non-overlapping run then its processes can be executed in linear time with respect to the size of $\mathcal{G}$.

Next, we give a formal definition for scheduling DPGs. Let δ be a parameter describing the communication delay of the system. For the lower bounds it suffices to treat δ as a constant independent of the processes between which data is exchanged. Furthermore, it suffices to assume unit-execution time for all processes. Our algorithms achieving matching upper bounds can even handle variable delays and execution times.

To schedule a task graph $H = (W, F)$ with delay δ we assume that an unbounded number of processors can be used. In each unit-time interval a processor P can execute one single task w. In order to schedule w at time t each direct predecessor of w must have already been executed – either by P itself in previous time intervals or by some other processor by time interval $t - 1 - \delta$ such that the result of this predecessor can arrive at P on time. Scheduling tasks graphs in the presence of communication delays has first been considered in [6].

Definition 3. *A schedule S for a DPG $\mathcal{G}$ with delay δ is a schedule of a run $H = (W, F)$. Let $T(S)$ denote the duration of S, i.e. the point of time when S has executed all tasks, and $\boldsymbol{T_{opt}(\mathcal{G}, \delta)} := \min_{S \text{ schedule for } (\mathcal{G}, \delta)} T(S)$.*

For lower bounds on the complexity of this problem, instead of the optimization version to compute $T_{opt}(\mathcal{G}, \delta)$ it will suffice to consider the decision problem.

Definition 4 (DYNAMIC PROCESS GRAPH SCHEDULE (DPGS)).
Given a DPG $\mathcal{G}$ with communication delay δ and a deadline T^*, does $T_{opt}(\mathcal{G}, \delta) \leq T^*$ hold?

3   The Structure of Runs

Lemma 1 has shown that the size of a run can be exponential with respect to the size of its underlying DPG. But this does not imply that an optimal parallel execution time of a run requires exponential length, too, because each of its subruns can be small. In this section we investigate the problem what is the *minimal-overlapping* run for a given DPG $\mathcal{G}$. A simple thought shows that if $\mathcal{G} = (V, E, I, O)$ possesses a non-overlapping run then independently of the communication delay one can achieve the bound $T_{opt}(\mathcal{G}, \delta) \leq |V|$.

Lemma 2. *Let $\mathcal{G}$ be a DPG with input mode restriction, either $I \equiv$ ALT, or $I \equiv$ PAR. Then every run of $\mathcal{G}$ is non-overlapping.*

The *proof* of this claim is easy: In case $I \equiv$ ALT each subrun is a simple path from a sink back to a single source. If $I \equiv$ PAR then one can show by a topological induction starting with the sources that no process can have more than one execution instance.

The non-overlapping property does not hold anymore when the input mode may vary – even if one restricts the output mode. In case of $O \equiv$ ALT it is easy to construct a DPG $\mathcal{G}$ such that each run of $\mathcal{G}$ is not non-overlapping. However, we can show:

Lemma 3. *Let $\mathcal{G} = (V, E, I, O)$ be a DPG with κ sources and $O \equiv$ ALT. Then $\mathcal{G}$ has a $(\kappa - 1)$-overlapping run.*

This claim can be shown by a flow-argument. The requirement for the sources to have exactly one execution instance generates an initial flow of size κ from the sources to their direct successors. The ALT output mode implies that the flow does not increase. On the contrary, an internal node v with input mode PAR will decrease the flow by $|\text{pred}(v)| - 1$. The flow leaving a node corresponds exactly to the number of execution instances this process requires.

Thus, in case of ALT output mode runs have a quadratic increase at most. In the unrestricted case, this property does not hold anymore as the following result shows.

Lemma 4. *There exists a family $\mathcal{G}_1, \mathcal{G}_2, \mathcal{G}_3, \ldots$ of DPGs $\mathcal{G}_k = (V_k, E_k, I_k, O_k)$ with $|V_k| = 2k + 1$ such that every run H_k of $\mathcal{G}_k$ has a unique subrun $\mathcal{R}$ – which is $(2^{(|V_k|-1)/2-1} - 1)$-overlapping.*

The DPGs are constructed as follows. $\mathcal{G}_k$ consists of $2k+1$ nodes $V_k := \{ v_i \mid 1 \le i \le 2k+1 \}$ such that the first $k+1$ nodes form a complete DAG as well as the last $k+1$ nodes:

$$E_k := \{ (v_i, v_j) \mid 1 \le i < j \le k+1 \} \cup \{ (v_i, v_j) \mid k+1 \le i < j \le 2k+1 \}.$$

Furthermore, define the modes as follows: $I(v_i)$ is ALT for $i \le k+1$ and PAR for $i > k+1$. Moreover $O(v_i)$ is PAR for $i < k+1$ and ALT otherwise. Now, if the delay δ_k for $\mathcal{G}_k$ is chosen large enough – $\delta_k = |H_k| = 3 \cdot 2^{k-1}$ suffices – then it does not pay to utilize more than 1 processor, since H_k has only one sink.

Corollary 1. *For the family* $\mathcal{G}_1, \mathcal{G}_2, \ldots$ *defined above it holds* $T_{opt}(\mathcal{G}_k, \delta) \ge 3 \cdot 2^{(|V|-1)/2-1}$ *if the communication delay is sufficiently large.*

Now, let us investigate the most interesting restricted case, namely DPGs with $O \equiv$ PAR, in the following called **PAR-output DPGs**. As one can see in Fig. 1 such DPGs may have overlapping as well as non-overlapping runs. Hence, a natural question is whether for any executable PAR-output DPG $\mathcal{G}$ one can always find a non-overlapping run. The example in Fig. 2 gives a negative answer to this question. However, the situation cannot be too bad, we can show an exponential improvement of the trivial upper bound.

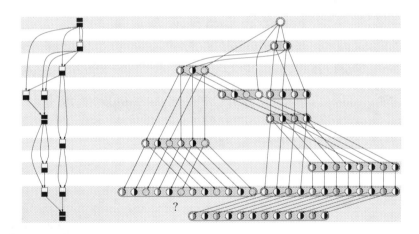

Fig. 2. An executable PAR-output DPG (left) with no non-overlapping run.

Theorem 1. *Every executable PAR-output DPG* $\mathcal{G} = (V, E, I, O)$ *has a* k-*overlapping run where* $k := \lfloor \log_2 \max_{v \in V} |W(v)| \rfloor$.

A k-overlapping run can be executed in parallel: for every subrun a unique processor completely processes all of its execution instances. This requires at most $|V| \cdot (k+1)$ many steps. Together with Lemma 1 we thus get

Corollary 2. *If a PAR-output DPG* $\mathcal{G} = (V, E, I, O)$ *can be executed at all it can be executed in time* $O(|V|^2)$.

Below we sketch the idea how to prove Theorem 1. Assume that $\mathcal{G}$ is an executable PAR-output DPG with nodes $V = \{v_1, \ldots, v_n\}$. For a run $H_{\mathcal{G}}$ of $\mathcal{G}$ define the **characteristic vector** as the sequence $(|W(v_1)|, |W(v_2)|, \ldots, |W(v_n)|)$ and $\gamma(\mathcal{G}) := \lfloor \log_2 \max_{v \in V} |W(v)| \rfloor$. Note that on any path $v_{i_1}, \ldots, v_{i_d}$ from a source to a sink the sequence of multiplicities $|W(v_{i_j})|$ must be increasing. Furthermore, we have proven that for PAR-output DPGs $\mathcal{G}$ all runs have exactly the same characteristic vector [3], thus $\gamma(\mathcal{G})$ is independent of a particular run and hence well defined. We will use five steps to transform $\mathcal{G}$ into a graph $\tilde{\mathcal{G}}$

$$\mathcal{G} \xrightarrow{\text{(A)}} \mathcal{G}_A \xrightarrow{\text{(B)}} \mathcal{G}_B \xrightarrow{\text{(C)}} \mathcal{G}_C \xrightarrow{\text{(D)}} \mathcal{G}_D \xrightarrow{\text{(E)}} \mathcal{G}_E = \tilde{\mathcal{G}}$$

in such a way that the intermediate DPGs fulfill the following increasing list of properties (that means $\mathcal{G}_A$ fulfills (A), then $\mathcal{G}_B$ fulfills (A) and (B), etc):

(A) every node with input mode ALT has indegree exactly two;
(B) every node with input mode PAR has only direct predecessors and successors with input mode ALT;
(C) the sources and sinks have input mode PAR: on every path from a source to a sink the input mode alternates between PAR and ALT;
(D) for every value $k \leq \overline{k} := \max_{v \in \mathcal{G}} |W(v)|$ there exists a unique PAR-input node $\boldsymbol{u_k}$ with $|W(u_k)| = k$;
(E) for all triples (i, j, k) with $i + j = k \leq \overline{k}$ considering the unique nodes u_i, u_j, u_k according to (D), there exists a unique node $v_{i,j,k}$ in $\mathcal{G}_E$ with input mode ALT such that $(u_i, v_{i,j,k}), (u_j, v_{i,j,k}), (v_{i,j,k}, u_k)$ are edges in $\mathcal{G}_E$.

Property (D) and the monotonicity of the multiplicities $|W(u_k)|$ imply that the sink is unique and achieves maximal multiplicity $\overline{k}$.

These transformations expand as well as shrink subgraphs of $\mathcal{G}$ in order to achieve this unique layered structure. Although the size of $\tilde{\mathcal{G}}$ can be exponential in the size of $\mathcal{G}$, we can guarantee that the γ-values do not change, in particular $\gamma(\mathcal{G}) = \gamma(\tilde{\mathcal{G}})$. Furthermore, the property that $\gamma(\mathcal{G})$-overlapping runs exist is preserved as well. We call a DPG fulfilling property (A) to (E) a **complete PAR-output DPG** $\mathcal{C}_k$, where k denotes the multiplicity of the sink of $\mathcal{C}_k$.

Lemma 5. *Every complete* PAR-*output DPG* $\mathcal{C}_k$ *has a* $\gamma(\mathcal{C}_k)$-*overlapping run.*

One might be tempted to guess that a run with minimal overlap serves as the basis for a schedule of minimal length. But this is not correct. Overlapping may be crucial to design optimal executions of DPGs as it is shown by the following result. A proof can be found in the full paper, where we construct a family of DPGs exhibiting this behaviour.

Theorem 2. *There exist* PAR-*output DPGs* $\mathcal{G}$ *with an appropriate communication delay* δ *such that no non-overlapping run of* $\mathcal{G}$ *can achieve the minimal schedule length* $T_{\mathrm{opt}}(\mathcal{G}, \delta)$.

4   The Complexity of Scheduling

The previous section has shown a linear bound on the maximal execution time for PAR-output DPGs. On the other hand, execution times of unrestricted DPGs may be exponential. In this section we prove that the computational complexity to determine the minimal execution time for PAR-output DPGs is still very difficult, namely the scheduling problem for such DPGs is $\mathcal{NEXPTIME}$-complete as in the general case.

The $\mathcal{NEXPTIME}$-completeness of the scheduling problem in the unrestricted case has been proved in [3]. Furthermore, we have shown that for any variant with input or output mode restrictions, except the case of PAR-output DPGs, the computational complexity of the scheduling problem decreases drastically, namely it can be solved in $\mathcal{NP}$, in some cases even with less effort. These upper bounds have been matched with corresponding lower bounds.

For scheduling PAR-output DPGs we were able to prove $\mathcal{BH}_2$-hardness. This means that the scheduling problem is probably more difficult for PAR-output DPGs than for the other restrictions. In this section we improve the lower bound showing that the problem is $\mathcal{NEXPTIME}$-hard. Since an optimal schedule of PAR-output DPGs can always be found in $\mathcal{NEXPTIME}$ as in the unrestricted case the lower bound is best possible. The scheduling problem is thus $\mathcal{NEXPTIME}$-complete.

Theorem 3. *Scheduling* PAR-*output DPGs with constant communication delay is $\mathcal{NEXPTIME}$-hard.*

The hardness result is obtained by showing that PAR-output DPGs can represent certain Boolean formulas in a highly succinct way. A sketch of this construction will be given in the rest of this section. For the reduction we will use the following $\mathcal{NEXPTIME}$-complete problem [5]:

Definition 5 (SUCCINCT-3SAT). *As input we are given a Boolean circuit over the standard AND, OR, NOT-basis that succinctly codes a Boolean formula in conjunctive normal form with the additional property that each clause has exactly three literals and each literal appears exactly three times. Suppose that the encoded formula consists of n variables and m clauses. On input $(0, i, k)$ with $i \in \{0, \dots, n - 1\}$ and $k \in \{1, 2, 3\}$ (appropriately coded in binary), the coding circuit returns the index of the clause where the literal $\neg x_i$ appears the k-th time. On input $(1, i, k)$ it returns the index of the clause where x_i appears for the k-th time. On input $(2, j, k)$ with $j \in \{0, \dots, m - 1\}$ and $k \in \{1, 2, 3\}$, it returns the k-th literal of the j-th clause. The problem is to decide whether the encoded formula is satisfiable.*

4.1   Boolean Circuits and Dynamic Process Graphs

In [3] the first step of the $\mathcal{NEXPTIME}$-hardness proof for the general case made a transformation of a given Boolean circuit B into a DPG $\mathcal{G}_B$ such that runs of $\mathcal{G}_B$ simulate computations of B on arbitrary input assignments. To encode an

arbitrary Boolean circuit B by a DPG $\mathcal{G}_B$ we code each gate of B separately
by its own subgraph and draw edges from the output nodes of each gate to the
appropriate input nodes (for details of this construction see [3]). The simulation
of a gate-subgraph is based on testing the different possible input constellations.
For this simulation the ALT-output mode is essential and it cannot be simulated
using only the PAR-output mode.

In this subsection we describe a different technique to *transform* a Boolean
circuit into a DPG that requires only nodes with output mode PAR. Let B have
n input variables. To construct a PAR-output DPG $\mathcal{G}_B$ we proceed in two steps:

1. Construct an equivalent circuit B' that contains only AND and OR gates
   (of fanin two), NOT gates are superfluous since B' receives as input also the
   negation of each variable. B' computes the same function as B and has the
   additional properties:
   a) the lengths of all paths from an input to an output are equal,
   b) on every input output path the AND and OR gates alternate.
2. From B' generate $\mathcal{G}_B$ as follows: Let $G = (V, E)$ be the graph of B', then
   $\mathcal{G}_B := (V, E, I, O)$ with $O \equiv$ PAR. For $u \in V$ we choose $I(u) :=$ ALT iff the
   corresponding gate of u in B' is an OR -gate, else $I(u) :=$ PAR.

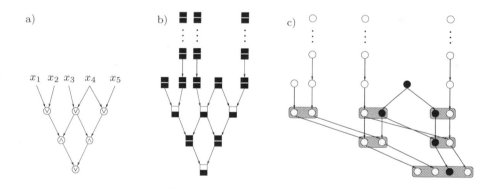

Fig. 3. *A Boolean circuit B, its DPG $\mathcal{G}_{B,\alpha}$, and a run for an input $x_1 = x_4 = 1$ and
$x_2 = x_3 = x_5 = 0$.*

One can show that $\mathcal{G}_B$ is executable and that for any communication delay
δ it holds: $T_{opt}(\mathcal{G}_B, \delta) \le |B'|$.

To understand the functionality of $\mathcal{G}_B$ we will now modify $\mathcal{G}_B$, such that the
run of the resulting DPG simulates a computation of B on an arbitrary input
vector $\alpha \in \{0,1\}^n$.

Define the line DPG $\mathcal{L}$ of length $k \ge 1$ as a DPG consisting of k vertices
$v_1, v_2, \ldots, v_k$ and $k-1$ edges: $(v_1, v_2), \ldots, (v_{k-1}, v_k)$. The input and output
modes are inessential. For completeness we set $I(v_i) = O(v_i) =$ PAR. Now let ℓ

denote the number of nodes of $\mathcal{G}_B$ and let $k := 2\ell$. We add n lines $\mathcal{L}_1, \ldots, \mathcal{L}_n$ of length k to $\mathcal{G}_B$ such that $\mathcal{L}_i$ corresponds to the variable x_i as follows: If $x_i = 1$ then we connect the sink node $v_{k,i}$ of $\mathcal{L}_i$ to the input node of $\mathcal{G}_B$ that corresponds to the literal $\overline{x}_i$ in B'. If $x_i = 0$ we connect $v_{k,i}$ to the input node of $\mathcal{G}_B$ that corresponds to the literal x_i in B'. The resulting DPG will be denoted by $\boldsymbol{\mathcal{G}_{B,\alpha}}$. An example for such a DPG $\mathcal{G}_{B,\alpha}$ is shown in Fig. 3(b).

Lemma 6. *Let $y_1, \ldots, y_m$ be the output nodes of circuit B, let α be an input assignmet, and let $\beta \in \{0,1\}^m$ be the output vector of B for input α. Then for every run H of $\mathcal{G}_{B,\alpha}$ and for any $i \in [1..m]$ the following holds:*

$$\beta(i) = 0 \;\Rightarrow\; \forall w \in W(y_i): \text{ the subrun induced by } w \text{ has size} > 2\ell.$$

Moreover there exists such a run H_α for $\mathcal{G}_{B,\alpha}$ that for any sequence $i_1 < i_2 < \ldots < i_k$, with $\beta(i_j) = 1$, one can chose tasks $w_{i_j} \in W(y_{i_j})$ in H_α such that the subgraph induced by the tasks w_{i_j} and their predecessors has size bounded by ℓ, i.e. $|\bigcup_{1 \le j \le k} \mathcal{R}(w_{i_k})| \le \ell$.

Hence to test some output gates $y_{i_1}, \ldots, y_{i_k}$ of B for a particular input assignment α one can check for the run H_α whether for some tasks of $y_{i_1}, \ldots, y_{i_k}$ the corresponding subgraph has size at most ℓ.

4.2   The Reduction

Let B be an input of the SUCCINCT-3SAT problem, that means a circuit coding a Boolean 3CNF formula $\mathcal{F}$. We may assume that $\mathcal{F}$ has exactly $m = 2^d$ clauses. Then there will be $n = 2^{d-1}$ variables.

The first step of our reduction is to construct two circuits A_1 and A_2. The first circuit A_1 verifies the syntax of the formula encoded by B. Speaking more formally, A_1 with the input j returns 1 iff the given circuit B returns j on exactly three inputs $(\boldsymbol{k},\ \boldsymbol{B(2,j,k')},\ \boldsymbol{k''})$ for all $k \in \{0,1\}$ and $k', k'' \in \{1,2,3\}$. Otherwise A_1 returns 0.

The second circuit A_2 will be used to check whether the encoded formula is satisfiable. A_2 with input (k,j,i), is defined as follows. $A_2(k,j,i)$ returns $(\boldsymbol{1,\beta})$ if x_i^β is the k-th literal of the j-th clause, where x_i^0 denotes $\neg x_i$ and x_i^1 means x_i. In any other case, A_2 returns $(\boldsymbol{0,0})$. Obviously, these circuits can easily be constructed for a given B. Let us denote the DPG that corresponds to circuit A_i by $\mathcal{A}_i$. We will call $\mathcal{A}_1$ the **syntax-verifier**, and $\mathcal{A}_2$ the **index-verifier** graph. Because of lack of space only the syntax-verifier will be described in the following.

Let us call the sources of $\mathcal{A}_1$ $\mathbf{j_{t1}, j_{f1}, j_{t2}, j_{f2}, \ldots, j_{td}, j_{fd}}$. They encode the integer j as follows: If the ℓ-th digit of the binary representation of j is 1 then in a run for $\mathcal{A}_1$ that represents j the node $\mathbf{j_{t\ell}}$ has an execution instance that can be computed within a certain time, while the execution instance of $\mathbf{j_{f\ell}}$ requires slightly more time. If the ℓ-th digit of j is 0 then $\mathbf{j_{f\ell}}$ can be computed fast and $\mathbf{j_{t\ell}}$ requires the larger amount of time. The unique instance of either $\mathbf{j_{t\ell}}$ or $\mathbf{j_{f\ell}}$ that can be computed fast will be denoted by $t(\ell)$.

Our first goal is to construct a DPG that represents the input values of j. More precisely, the run of this DPG represents all possible 2^d values for j. The crucial point of the whole construction is then to couple this value genarator DPG with $\mathcal{A}_1$ in an appropriate way.

To generate the binary representation of j we will use a DPG called a **counting graph** $\mathcal{TC}_{k,\ell_1,\ell_2}$ with parameters k, ℓ_1 and ℓ_2. Their main building blocks are subgraphs $\mathcal{C}_\ell := (V, E, I, O)$ with $V = \{u_i, v_{0,i}, v_{1,i} | 1 \leq i \leq \ell\} \cup \{u_0\}$, $E := \{(u_{i-1}, v_{j,i}), (v_{j,i}, u_i) | 1 \leq i \leq \ell, j \in \{0,1\}\}$, $O \equiv$ PAR, and $I \equiv$ ALT. Note that for each $\mathcal{C}_\ell$ if considered separately the sink process u_ℓ has 2^ℓ execution nodes.

Using these subgraphs one can achieve the following property which will be important for the reduction. $\mathcal{TC}_{k,\ell_1,\ell_2}$ has $2k$ sink nodes, which will be denoted by $y_1^0, y_1^1, \dots, y_k^0, y_k^1$. For every run of $\mathcal{TC}_{k,\ell_1,\ell_2}$ it holds that each y_i^j has exactly 2^k execution tasks and for exactly 2^{k-1} of these tasks each corresponding subrun has size $S_t := k(\ell_1 + 2) + k^2 + 4$. For the remaining 2^{k-1} tasks each subrun has size $S_f := S_t + (k - 1) \cdot (\ell_2 + 1) + 2$. This means that for the suitable values ℓ_1 and ℓ_2, exactly 2^{k-1} subruns have small size whereas the remaining subruns are much bigger. Furthermore, any task t_i^β of a node y_i^β is associated with a task $t_i^{1-\beta}$ of a node $y_i^{1-\beta}$ such that $|\mathcal{R}(t_i^\beta)| = S_t$ iff $|\mathcal{R}(t_i^{1-\beta})| = S_f$. Finally, the construction guarantees that any processor can only compute one task for any pair y_i^0, y_i^1.

Drawing edges $(y_i^0, j_{fi}), (y_i^1, j_{t1})$ from the sinks of the counting graph to the sources of $\mathcal{A}_1$ we obtain a DPG which has similar properties as the DPG $\mathcal{G}_{B,\alpha}$ discussed in the previuous subsection. Note that the pair y_i^0, y_i^1 of the counting graph specifies the value of $t(i)$. Using Lemma 6 one can prove that for any input j the circuit $\mathcal{A}_1$ returns the value 1 iff for any value j (represented by the particular tasks t_i^0, t_i^1) there exists a subrun for the sink of $\mathcal{A}_1$ of moderate size, i.e. $\leq k(\ell_1 + 2) + k(k^2 + 4) + |\mathcal{A}_1|$.

The DPG for checking whether the encoded formula is satisfiable is technically more involved. This graph has to check whether the input values i, j, and k fit together (which is something like a local test). Furthermore, it requires a mechanism to choose the value of the variables x_i and decide if these values satisfy each clause C_j (which is something like a global test). The details can be found in the full paper.

5   Conclusion

We have shown that the computational complexity to determine the optimal schedule length T_{opt} for PAR-output DPGs is quite high, this problem is complete for $\mathcal{NEXPTIME}$. On the other hand, we have shown that the maximal size of a subrun of a minimal-overlapping run H gives a good approximation for T_{opt}. However, this bound is not tight and an interesting problem is to give some other *natural* structural properties of runs which allow better approximations.

Another interesting question is how efficiently a minimal-overlapping run can be constructed for a given PAR-output DPG $\mathcal{G}$. In this paper we have shown that all such runs have the same size. Can we find a minimal-overlapping run H in linear time with respect to the size of H? Note that Lemma 1 and Lemma 2 in [3] imply that for certain DPGs the number of different runs can be exponential with respect to the size of H. Therefore, a brute-force searching will be inefficient.

References

1. H. Galperin, A. Wigderson, Succinct Representations of Graphs, Information and Control 56, 1983, 183-198.
2. S. Ha, E. Lee, *Compile-time Scheduling and Assignment of Data-flow Program Graphs with Data-dependent Iteration*, IEEE Trans. Computers 40, 1991, 1225-1238.
3. A. Jakoby, M. Liśkiewicz, R. Reischuk, *Scheduling Dynamic Graphs*, in Proc. 16th STACS'99, LNCS 1563, Springer-Verlag, 1999, 383-392; for a complete version see TR A-00-02, Universität Lübeck, 2000.
4. T. Lengauer, K. Wagner, *The Correlation between the Complexities of the Non-hierarchical and Hierarchical Versions of Graph Problems*, J. CSS 44, 1992, 63-93.
5. C. Papadimitriou and M. Yannakakis, *A Note on Succinct Representations of Graphs*, Information and Control 71, 1986, 181-185.
6. C. Papadimitriou and M. Yannakakis, *Towards an Architecture-Independent Analysis of Parallel Algorithms*, Proc. 20. STOC, 1988, 510-513, see also SIAM J. Comput. 19, 1990, 322-328.

Bandwidth of Split and Circular Permutation Graphs[*]

Ton Kloks[1], Dieter Kratsch[2], Yvan Le Borgne[3,**], and Haiko Müller[4]

[1] Vrije Universiteit Amsterdam, Department of Mathematics and Computer Science,
1081 HV Amsterdam, The Netherlands. `kloks@cs.vu.nl`
[2] Université de Metz, Laboratoire d'Informatique Théorique et Appliquée,
57045 Metz Cedex 01, France. `kratsch@lrim.univ-metz.fr`
[3] Laboratoire Bordelais de Recherche en Informatique, 351 cours de la Libération,
33 405 Talence Cedex, France. `borgne@labri.u-bordeaux.fr`
[4] Friedrich-Schiller-Universität Jena, Fakultät für Mathematik und Informatik,
07740 Jena, Germany. `hm@minet.uni-jena.de`

Abstract. The BANDWIDTH minimization problem on graphs of some special graph classes is studied and the following results are obtained. The problem remains NP-complete when restricted to splitgraphs. There is a linear time algorithm to compute the exact bandwidth of a subclass of splitgraphs called hedgehogs. There is an efficient algorithm to approximate the bandwidth of circular permutation graphs within a factor of four.

1 Introduction

The BANDWIDTH problem for a graph is that of labeling its vertices with distinct integers so that the maximum difference across an edge is minimized. One of the original motives in graph algorithms was finding the most efficient way to represent the adjacency matrix of a graph. The problem is namely equivalent with finding a permutation of the vertices such that the non-zeros lie in a small band around the diagonal. For matrix problems in general this problem is of interest since it can clearly be used to accelerate matrix operations like multiplications and inversions. The problem was formulated for graphs in the beginning of the sixties. Heuristics for the bandwidth problem for graphs exist already for a long time, however their worst-case performance cannot be guaranteed (for an overview and further references see [3,4]).

Nowadays, the BANDWIDTH problem is standing in the midst of all kinds of new research areas because it has become a sort of test-case in the area of complexity analysis, approximation algorithms, and research into special classes of graphs. This is apparent by the long list of publications on the subject.

The problem is NP-complete [18] and it remains NP-complete even for trees [7,16]. Concerning fixed parameter complexity, the problem is $W[t]$-hard for

---

[*] This work was supported by Netherlands Organization for Scientific Research (NWO).
[**] Supported by Ecole Normale Supérieure de Lyon.

every t (see, e.g. [5]). This is seen as strong evidence that to test if a graph has bandwidth at most k is not fixed parameter tractable. On the positive side, there is an algorithm to test if a graph has bandwidth at most k with running time $O(n^k)$ [8].

The BANDWIDTH minimization problem is also in the center of renewed interest in the area of approximation algorithms. We mention some recent results. There is no polynomial time algorithm to approximate the bandwidth of a graph within a constant factor, unless P=NP [22]. An approximation algorithm for the bandwidth problem has been obtained using semi-definite programming [2]. Furthermore a polylogarithmic approximation algorithm has been given [6]. Much research has already been done on the approximation of the bandwidth for graphs in special classes of graphs. We mention a few of them. It has been shown that there is no polynomial time constant factor approximation algorithm for trees unless P=NP [22]. Approximation algorithms for special classes of trees have been given in [9,10]. For AT-free graphs a factor two approximation was obtained [13]. For split graphs a factor two approximation is known [23]. It has been shown that there is no polynomial time approximation algorithm for circular arc graphs with a factor less than two unless P=NP [22]. On the other hand, an efficient algorithm to approximate the bandwidth of circular arc graphs within a factor two is known [15].

Unfortunately only a few typically small graph classes are known to have a polynomial time exact bandwidth algorithm, among them caterpillars with hairs of length at most two [1], chain graphs [14], interval graphs [11] and theta graphs [20].

In this paper we consider the complexity of the BANDWIDTH problem for different graph classes. We show that the problem remains NP-complete when restricted to splitgraphs. For a subclass of splitgraphs, called hedgehogs, we were able to find a linear time exact algorithm. As an easy consequence, we obtain a linear time 2-approximation algorithm for chordal graphs of diameter at most three (including all split graphs). Finally we use the above-mentioned approximation algorithm for circular arc graphs to obtain the first constant factor approximation algorithm for circular permutation graphs.

2   The NP-Completeness for Splitgraphs

In this section we show that the BANDWIDTH problem remains NP-complete when restricted to splitgraphs.

Definition 1. *A* layout *L of G is a 1-1 mapping from V onto $\{1, ..., |V|\}$. The* width *$b(G, L)$ of L is zero, if G has no edges, and otherwise*

$$b(G, L) = \max\{|L(u) - L(v)| : \{u, v\} \in E\}.$$

The bandwidth *of G is*

$$\mathrm{bw}(G) = \min\{b(G, L) : L \text{ is a layout of } G\}.$$

We call a layout L a k-*layout* if its width equals k. We say that a layout L of G is *optimal* if $b(G, L) = \mathrm{bw}(G)$.

The following lower bound for the bandwidth is easy to obtain, see e.g. [3].

Lemma 1. *For every graph* $G = (V, E)$ *we have*

$$\mathrm{bw}(G) \geq \max_{W \subseteq V} \frac{|W| - 1}{\mathrm{diam}(G[W])},$$

especially

$$\mathrm{bw}(G) \geq \max\left\{ \frac{|V| - 1}{\mathrm{diam}(G)}, \frac{\Delta(G)}{2}, \omega(G) - 1 \right\}.$$

Proof. Let L be an optimal layout of G and $W \subseteq V$. We consider the vertices $w, z \in W$ at leftmost and rightmost position in L among all vertices in W. By the pigeonhole principle there is an edge $\{x, y\}$ on every path from w to z in G such that $|L(x) - L(y)| \geq (|W| - 1)/(\mathrm{diam}(G[W]))$, where $\mathrm{diam}(G)$ denotes the diameter of G.

Especially we choose W to be the whole vertex set V, the neighbourhood $N(v)$ of a vertex v of maximum degree augmented by v itself ($|W| = \Delta(G) + 1$) and a maximum clique of G ($|W| = \omega(G)$). □

Definition 2. *A* splitgraph $G = (C, I, E)$ *is a graph with vertex set* $C \cup I$ *and edge set* E *such that* C *is a clique and* I *an independent set in* G.

The complexity of the BANDWIDTH problem for splitgraphs was open for a long time (see, e.g., [15,23]). To prove its NP-completeness we use a reduction from the BANDWIDTH problem on cobipartite graphs, which is known to be NP-complete (see [13]). Recall that *cobipartite graphs* are just the complements of bipartite graphs.

Lemma 2. *For every cobipartite graph* $G = (X, Y, E)$ *there is an optimal layout* L *such that* $L(x) < L(y)$ *for every* $x \in X$ *and* $y \in Y$.

Proof. By Lemma 1 we know $|X|, |Y| \leq \mathrm{bw}(G) + 1$ because X and Y form cliques of G. We add vertices to X and Y without neighbours in the other color class, such that the new sets X' and Y' satisfy $|X'| = |Y'| = \mathrm{bw}(G) + 1$ and call this new graph G'. Then $\mathrm{bw}(G) = \mathrm{tw}(G) = \mathrm{tw}(G') = \mathrm{bw}(G')$, because for cobipartite graphs bandwidth and treewidth are equal [19] and every tree-decomposition of G can easily extended to a tree-decomposition of G'.

The new graph G' is cobipartite. In every optimal layout of G' the vertices in X' appear consecutively because $|X'| = \mathrm{bw}(G') + 1$ as well as the vertices in Y' do. Hence we obtain a layout L of G with $L(x) < L(y)$ for every $x \in X$ and $y \in Y$ if we restrict an optimal layout of G' with the analogous property. □

Theorem 1. *The* BANDWIDTH *problem is NP-complete for splitgraphs.*

Proof. Consider a cobipartite graph $G = (X, Y, E)$ with color classes X and Y and integer k. By Lemma 1 we may assume $k \geq |X| - 1$ and $|X| \geq |Y|$. First choose sets A and B of additional vertices such that $|A| = |X|$ and $|B| = k + 1 - |X|$. Starting from G we form a splitgraph H with clique $B \cup X$ and independent set $A \cup Y$. Therefore we add all edges $\{a, b\}$ with $a \in A$ and $b \in B$ and remove the edges from $A \cup Y$. The edges of G between X and Y remain unchanged.

Let L be a k-layout of G such that $L(x) < L(y)$ for all $x \in X$ and $y \in Y$. Then a k-layout of H starts with vertices of A, followed by the vertices of B and then followed by the layout L of G.

Next consider a k-layout of H. By the cardinalities of A and B all vertices in the clique $B \cup X$ are consecutive as well as all vertices of $A \cup B$. Without loss of generality, the layout is of the form A, B, X, Y. The restriction to X and Y is a k-layout for G because $k > |X| \geq |Y|$, i.e., the edges inside Y missing in H have at most the length of the longest edge inside X. □

3   Bandwidth of Hedgehogs

In this section we consider a subclass of splitgraphs. We present a formula for the bandwidth of hedgehogs and describe corresponding optimal layouts.

Definition 3. *A splitgraph $G = (C, I, E)$ is said to be a* hedgehog *if every vertex of I has degree one.*

Let $G = (C, I, E)$ be a hedgehog. An edge $e = \{c, x\}$ of G is called a *hair* if $c \in C$ and $x \in I$. The *root* (marked by ○ in forthcoming figures) of the hair e is c and the *tip* (marked by ●) of e is x. Clearly every vertex in I is a tip, but not all vertices in C need to be roots. A vertex $p \in C$ that is not incident with a hair is called *pore* (□). For every vertex $c \in C$ we define $M(c) = I \cap N(c)$, $M[c] = \{c\} \cup M(c)$. Throughout this section let $n = |C \cup I|$.

An hedgehog is called *trivial* if it has at most one root. Since every hedgehog is connected, trivial hedgehogs have diameter at most 2 and nontrivial hedgehogs have diameter 3. It is not hard to see that $\mathrm{bw}(G) = \max\left\{|C| - 1, \left\lceil \frac{1}{2}\Delta(G)\right\rceil\right\}$ for trivial hedgehogs. (Notice that $\Delta(G) = n - 1$ for trivial hedgehogs.)

Definition 4. *A layout $L : C \cup I \to \{1, 2, \ldots, n\}$ of a hedgehog $G = (C, I, E)$ is* normal *if $L(a) < L(c) \iff L(x) < L(z)$ for every pair of hairs $\{a, x\}$, $\{c, z\}$ with $a \neq c$ and $x, z \in I$.*

Lemma 3. *For every hedgehog there is an optimal normal layout.*

Proof. Let L be an optimal layout of a hedgehog $G = (C, I, E)$ that is not normal. Then there is a pair of hairs $\{a, x\}$ and $\{c, z\}$, $x, z \in I$, with $L(a) < L(c)$ and $L(x) > L(z)$. Changing positions of x and z we do not increase the width of the layout, but we decrease the number of such pairs of hairs. □

Let L be a normal layout of $G = (C, I, E)$. For each pair of integers s, t with $1 \leq s \leq t \leq n$, we call $\{v : s \leq L(v) \leq t\}$ an *interval* of L. Then for every root $c \in C$, the set $M(c)$ either forms an interval of L or the smallest interval containing $M(c)$ consists of $M(c)$ and some vertices of C, but not other vertices of I.

Now let L be a fixed normal layout of a hedgehog $G = (C, I, E)$. We partition the tips into *left tips* ($\blacktriangleleft$), these are vertices $x \in I$ with $L(x) < \min\{L(c) : c \in C\}$, *middle tips* ($\blacktriangle$), these are vertices $x \in I$ with $\min\{L(c) : c \in C\} < L(x) < \max\{L(c) : c \in C\}$ and *right tips* ($\blacktriangleright$), these are vertices $x \in I$ with $\max\{L(c) : c \in C\} < L(x)$. Similarly we define *left roots* ($\triangleleft$), these are roots adjacent to left tips only, *middle roots* ($\triangle$), these are roots adjacent to middle tips only and *right roots* ($\triangleright$), these are roots adjacent to right tips only. A vertex $c \in C$ that is not a left, middle or right root is either a *pore* (if $M(c) = \varnothing$) or a *parting*. A *left parting* ($<$) is adjacent to left and middle tips, but not to right tips. A *right parting* ($>$) is adjacent to right and middle tips, but not to left tips. Finally, a *centre parting* ($\diamondsuit$) is adjacent to at least one left tip and one right tip. (See e.g. Fig. 1.)

Lemma 4. *Let L be a normal layout of a hedgehog $G = (C, I, E)$. Then there is at most one left parting, at most one centre parting and at most one right parting. If either a left or a right parting exists then there is no centre parting. If there is a centre parting then there is neither a left nor a right parting.*

Proof. Otherwise we have hairs $\{a, x\}$ and $\{c, z\}$ incident to partings a and c such that $L(a) < L(c) < L(z) < L(x)$ or $L(z) < L(x) < L(a) < L(c)$ contradicting the fact that L is normal. □

Definition 5. *A normal layout $L : C \cup I \to \{1, 2, \ldots, n\}$ of a hedgehog $G = (C, I, E)$ is said to be a* sorted *layout if the vertices appear in the following order: in the leftmost positions (from the left) first all left tips, then all left roots, then a left parting (if any); in the rightmost positions (from the right) first all right tips, then all right roots, then a right parting (if any).*

Remark 1. It is not hard to verify that every hedgehog has an optimal sorted layout (that we shall obtain as an immediate consequence of the construction of the optimal layouts in the proof of Theorem 2).

Lemma 5. *Let L be a sorted layout of a hedgehog $G = (C, I, E)$. Let ℓ be the number of left tips in L and let r be the number of right tips in L. If L has no centre parting then $b(G, L) = \max\{\ell, r, n - \ell - r - 1\}$. Otherwise let $c_m \in C$ be a centre parting and let d be the maximum of $L(c_m) - \min\{L(x) : x \in M(c_m)\}$ and $\max\{L(x) : x \in M(c_m)\} - L(c_m)$. Then $b(G, L) = \max\{\ell, r, n - \ell - r - 1, d\}$.*

Furthermore $L(c_m) - \min\{L(x) : x \in M(c)\} \leq \ell$ if the centre parting is next to the rightmost left root and $\max\{L(x) : x \in M(c_m)\} \leq r$ if the centre parting is next to the leftmost right root in L.

Proof. Clearly an edge violating any statement of the lemma must be a hair.

First we consider a vertex $c \in C$ which is either a left root, a left parting or a centre parting such that no pore is left of c in L. Let $\{c, t\}$ be any hair such that t is a left tip. Let $\{a, z\}$ be the hair with $L(z) = 1$ and $L(a) = \ell + 1$. Since each vertex v of C with $L(v) \in \{\ell + 1, \ldots, L(c)\}$ is adjacent to at least one left tip we may conclude $L(c) - L(t) \leq \ell$.

Analogously we obtain $L(t') - L(c') \leq r$ for each hair $\{c', t'\}$ with t' right tip. Together this implies all statements of the lemma. $\square$

Now we are ready to present the formula for the bandwidth of hedgehogs.

Theorem 2. *Let $G = (C, I, E)$ be a hedgehog. Then*

$$\mathrm{bw}(G) = \max \left\{ |C| - 1, \left\lceil \tfrac{1}{2}\Delta(G) \right\rceil, \left\lceil \tfrac{1}{3}(|C| + |I| - 1) \right\rceil \right\}.$$

Proof. The formula holds for trivial hedgehogs, thus we may assume that G is a nontrivial hedgehog. Hence G has diameter 3.

The lower bound follows from Lemma 1.

To establish the upper bound we shall distinguish three cases depending on which of the three values in the formula is maximum. Nevertheless there is a general way of obtaining an optimal *sorted layout* that we shall describe first. (See Fig. 1 for an illustration.)

To obtain a layout of I, place the sets $M(c)$ for all roots $c \in C$ as interval I_c side by side where the order of the intervals is *arbitrary*. We shall cut the obtained I-*layout* (between two consecutive tips) at two positions that we call *left cut* and *right cut* (and that may coincide).

Then all tips left of the left cut will be the left tips and all tips right of the right cut will be the right tips of the final optimal layout. All other tips will be middle tips. Middle and all vertices of C will be placed between the left cut and the right cut. Notice that there is a centre parting c_m in L iff left and right cut both divide the same interval I_{c_m}.

Since we construct a sorted layout L, all left roots, left partings, right partings and right roots and their position in L is fixed by the I-layout and the two cuts. Notice that the positions of middle roots, middle tips and pores are of no importance for the width of L by Lemma 5. It remains to fix the position of the centre parting in L if there is one. The feasible positions for c_m are right of all left tips and all left roots and left of all right roots and all right tips. We shall show that we can always place the centre parting in a feasible position of L such that L is optimal.

Now we distinguish three cases. For each case we show how to choose the left and the right cut and that the constructed layout L has the requested width and is optimal.

Case A: $\left\lceil \tfrac{1}{2}\Delta(G) \right\rceil \geq \max \left\{ |C| - 1, \left\lceil \tfrac{1}{3}(n - 1) \right\rceil \right\}$

We choose the left and the right cut of the I-layout such that the final layout L has $\ell = \left\lceil \tfrac{1}{2}\Delta(G) \right\rceil$ left tips and $r = \left\lceil \tfrac{1}{2}\Delta(G) \right\rceil$ right tips if $2 \left\lceil \tfrac{1}{2}\Delta(G) \right\rceil \leq |I|$ (i.e. there are enough tips). Otherwise we choose $\ell = \left\lceil \tfrac{1}{2}|I| \right\rceil$ and $r = \left\lfloor \tfrac{1}{2}|I| \right\rfloor$. Thus in both cases $\left\lceil \frac{1}{2}\Delta(G) \right\rceil \geq n - \ell - r - 1$ either since $\left\lceil \frac{1}{2}\Delta(G) \right\rceil \geq \left\lceil \frac{1}{3}(n-1) \right\rceil$ or since $\left\lceil \frac{1}{2}\Delta(G) \right\rceil \geq |C| - 1$. Thus, if L has no centre parting then $b(G, L) \leq \left\lceil \frac{1}{2}\Delta(G) \right\rceil$ by Lemma 5. Consequently L is optimal.

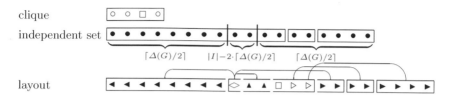

Fig. 1. Case A

Suppose c_m is the centre parting of L, i.e. the left and the right cut divide the interval I_{c_m}. Placing c_m in the leftmost feasible position the distance to its leftmost neighbor in L is at most $\ell \leq \left\lceil \frac{1}{2}\Delta(G) \right\rceil$ either since L is sorted or, if there is no left root, since $L(c_m) = \ell + 1$. Analogously, placing c_m in the rightmost feasible position the distance to its rightmost neighbor in L is at most $r \leq \left\lceil \frac{1}{2}\Delta(G) \right\rceil$ either since $L(c_m) = n - r$ (i.e. no right root) or since the upper bound r of the distance is guaranteed since L is a sorted layout.

Consequently there is a feasible position of c_m such that the distance to its leftmost neighbor and the distance to its rightmost neighbor in L are both at most $\left\lceil \frac{1}{2}\Delta(G) \right\rceil$. Placing c_m in such a position we obtain a layout L of width at most $\left\lceil \frac{1}{2}\Delta(G) \right\rceil$. Consequently L is optimal.

Case B: $|C| - 1 \geq \max \left\{ \left\lceil \frac{1}{2}\Delta(G) \right\rceil, \left\lceil \frac{1}{3}(n-1) \right\rceil \right\}$

We choose the left and the right cut of the I-layout such that the final layout L has $\ell = \left\lceil \frac{1}{2}|I| \right\rceil$ left tips and $r = \left\lfloor \frac{1}{2}|I| \right\rfloor$ right tips. Clearly $n - \ell - r - 1 = |C| - 1$. Notice that $|C| - 1 \geq \left\lceil \frac{1}{3}(|C| + |I| - 1) \right\rceil$ implies $\left\lceil \frac{1}{2}|I| \right\rceil \leq |C| - 1$. Consequently if L has no centre parting then $b(G, L) \leq |C| - 1$ by Lemma 5 and thus L is optimal.

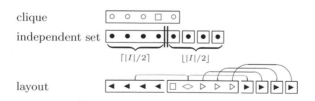

Fig. 2. Case B

Suppose c_m is the centre parting of L. Analogously to Case A, placing c_m in the leftmost feasible position the distance to its leftmost neighbor in L is at most $\ell \leq |C| - 1$ since $L(c_m) = \ell + 1$ or since L is sorted. Similarly, placing c_m in the rightmost feasible position the distance to its rightmost neighbor in L is at most $r \leq |C| - 1$. Consequently there is a feasible position of c_m such that its distance to its leftmost neighbor and its distance to its rightmost neighbor are both at most $\lceil \frac{1}{2}\Delta(G) \rceil \leq |C| - 1$. We place c_m in such a position and obtain an optimal layout L.

Case C: $\quad \lceil \frac{1}{3}(n-1) \rceil \geq \max\left\{|C| - 1, \lceil \frac{1}{2}\Delta(G) \rceil\right\}$

We choose the left and the right cut of the I-layout such that the final layout L has $\ell = \lceil \frac{1}{3}(n-1) \rceil$ left tips and $r = \lceil \frac{1}{3}(n-1) \rceil$ right tips. If there should not be enough tips we choose either one, say ℓ, or both, ℓ and r, equal to $\lfloor \frac{1}{3}(n-1) \rfloor$. Consequently $n - \ell - r - 1 \leq \lceil \frac{1}{3}(n-1) \rceil$.

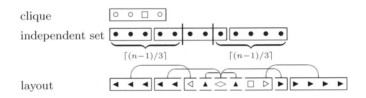

Fig. 3. Case C

Thus, if L has no centre parting then $b(G, L) \leq \lceil \frac{1}{3}(n-1) \rceil$ by Lemma 5 and L is optimal. If there is a centre parting c_m then analogously to the cases A and B we obtain that there is a feasible position of c_m such that the distance to its leftmost neighbor is at most ℓ and the distance to its rightmost neighbor is at most r. We place c_m in such a position and obtain an optimal layout L. $\qquad \square$

Corollary 1. *There is a linear time algorithm for hedgehogs that computes the bandwidth and an optimal layout.*

4 Consequences for Bandwidth Approximation

The following tool for the design of bandwidth approximation algorithms for special graph classes has been given in [13].

Lemma 6. *Let G and H be graphs with $G \subseteq H \subseteq G^d$ or $H \subseteq G \subseteq H^d$ for an integer $d \geq 1$, and let L be an optimal layout for H, i.e., $b(H, L) = \mathrm{bw}(H)$. Then L approximates the bandwidth of G by a factor of d, i.e., $b(G, L) \leq d \cdot \mathrm{bw}(G)$.*

Proof. First we assume $G \subseteq H \subseteq G^d$. Then

$$
\begin{aligned}
b(G, L) &\leq b(H, L), &&\text{since } G \subseteq H \\
&= \mathrm{bw}(H), &&\text{since } L \text{ is optimal for } H \\
&\leq \mathrm{bw}(G^d), &&\text{since } H \subseteq G^d \\
&\leq d \cdot \mathrm{bw}(G).
\end{aligned}
$$

Now let $H \subseteq G \subseteq H^d$. Then

$$
\begin{aligned}
b(G, L) &\leq b(H^d, L), &&\text{since } G \subseteq H^d \\
&\leq d \cdot b(H, L), \\
&= d \cdot \mathrm{bw}(H), &&\text{since } L \text{ is optimal for } H \\
&\leq d \cdot \mathrm{bw}(G), &&\text{since } H \subseteq G.
\end{aligned}
$$

This proves the lemma. □

In a general fashion its consequences might be formulated as follows.

Lemma 7. *Let $\mathcal{G}$ and $\mathcal{H}$ be graph classes and $d \geq 1$ an integer. Assume there is an $s(n)$ algorithm to approximate the bandwidth of graphs in class $\mathcal{H}$ within a factor of c.*

Suppose there is either a $t(n)$ algorithm to compute for a given graph $G \in \mathcal{G}$ a graph $H \in \mathcal{H}$ such that $G \subseteq H \subseteq G^d$, or there is a $t(n)$ algorithm to compute for a given graph $G \in \mathcal{G}$ a graph $H \in \mathcal{H}$ such that $H \subseteq G \subseteq H^d$.

Then there is an $s(n)+t(n)$ algorithm to approximate the bandwidth of graphs in class $\mathcal{G}$ within a factor of $c \cdot d$.

Let $G = (C, I, E)$ be any splitgraph. A hedgehog $H = (C, I, F)$ with the same clique C and same independent set I *spans* G if $F \subseteq E$. A spanning hedgehog of G is easy to find. Obviously we have $H \subseteq G \subseteq H^2$ for every spanning hedgehog H of G. By Lemma 6 and Lemma 7 respectively, combined with Corollary 1 we obtain

Corollary 2. *There is a linear time algorithm to approximate the bandwidth of splitgraphs with a factor of 2.*

We mention that a simple and different linear time 2-approximation algorithm for split graphs was obtained in [23].

Observing that any graph with a dominating clique, i.e. a clique C such that every vertex outside C has a neighbor in C, has spanning hedgehog, we can slightly extend our 2-approximation algorithm to chordal graphs having a dominating clique. Using a theorem of [12] stating that a chordal graph has a dominating clique iff it has diameter at most three, we obtain

Corollary 3. *There is a linear time algorithm to approximate the bandwidth of chordal graphs of diameter at most three within a factor of 2.*

5   Circular Permutation Graphs

In this section we use the technique of Lemmas 6 and 7 to obtain an efficient algorithm to approximate the bandwidth of circular permutation graphs within a factor of four.

A family $M = \{o_v : v \in V\}$ of subsets of a universal set U is an *intersection model* of the graph $G = (V, E)$ if $E = \{\{x, y\} : o_x \cap o_y \neq \varnothing\}$. A graph G is an *intersection graph* if it has an intersection model. In this section we consider classes of intersection graphs defined by special intersection models.

Interval graphs are the intersection graphs of intervals on the real line. The exact bandwidth of an interval graph is computable in polynomial time [11, 21]. Circular arc graphs are the intersection graphs of circular arcs on a circle. Obviously, every interval graph is a circular arc graph. The following result was proved in [15].

Theorem 3. *The bandwidth of a circular arc graph can be approximated to within a factor of four in $\mathcal{O}(n)$ time, and to within a factor of two in $\mathcal{O}(n \log^2 n)$ time.*

Permutation graphs are the intersection graphs of straight line segments between two parallel lines. Circular permutation graphs generalize permutation graphs in the same way as circular arc graphs generalize interval graphs.

We consider a finite family of spiral segments in the annual region between two concentric circles in the plane such that any two segments have at most one common point. The graphs defined by such intersection models form the class of *circular permutation graphs* [17].

In the following we consider a circular permutation graph $G = (V, E)$ that is not a permutation graph. For a fixed intersection model $\{s_v : v \in V\}$ of G we construct a circular arc graph $H = (V, F)$ such that $G \subseteq H \subseteq G^2$.

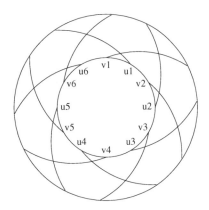 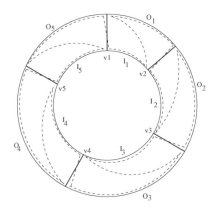

Fig. 4. The spiral segments of a chordless dominating cycle C_{12} in a circular permutation diagram.

Fig. 5. The folded circular arc model is indicated by the dashed line in the circular permutation diagram.

Therefore we fix a chordless dominating cycle $(v_1, u_1, v_2, u_2, \ldots, v_l, u_l)$ of G. The segments s_{v_i} define a partition of the annular region into l sections. The border of such an section consists of the segment s_{v_i}, the circular arc I_i on the inner circle, the segment $s_{v_{i+1}}$ and the circular arc O_i on the outer circle. Note that either for all $i = 1, \ldots, l$, the endpoints of the segment s_{u_i} belong to I_{i+1}

and O_{i-1}, or for all $i = 1, \ldots, l$, the endpoints of the segment s_{u_i} belong to I_{i-1} and O_{i+1}. Without loss of generality we assume henceforth the first case.

We form a circle C from the arcs $I_1, O_1, I_2, O_2, \ldots I_l$ and O_l in this circular order. Let $I_i O_i$ and $O_i I_{i+1}$ denote the common points on C of the arcs I_i and O_i and of the arcs O_i and I_{i+1}, respectively. For every vertex $v \in V$ we define a circular arc a_v on C as follows: The endpoints of a_v are the endpoints of the segment s_v. If $v = v_i$ for an index i, then the arc a_v contains exactly one point, namely $O_{i-1} I_i$. If $v \in N(v_i)$ then a_v contains $O_{i-1} I_i$ and for $v \in N[u_i]$ the arc a_v contains the point $I_i O_i$.

Let $H = (V, F)$ be the circular arc graph defined by the intersection model $\{a_v : v \in V\}$.

Lemma 8. $G \subseteq H \subseteq G^2$.

Proof. First we consider an edge $\{x, y\} \in E$. Let i be the index such that the common point of s_x and s_y belongs to the section between s_{v_i} and $s_{v_{i+1}}$ (including s_{v_i}). If $x \in N[v_i]$ and $y \in N[v_i]$ then $\{O_{i-1} I_i\} \in a_x \cap a_y$. If $x \notin N[v_i]$ and $y \notin N[v_i]$ then $\{I_i O_i\} \in a_x \cap a_y$. Hence we may assume $x \in N[v_i]$ and $y \notin N[v_i]$. This implies that a_x and a_y have a point of I_i in common. Hence, in all cases we have $\{x, y\} \in F$.

Now let $\{x, y\} \in F$. Both a_x and a_y contain a point $I_i O_i$ or $O_i I_{i+1}$ for a suitable index i. For $I_i O_i \in a_x \cap a_y$ we have $u_i \in N[x] \cap N[y]$. $O_i I_{i+1} \in a_x \cap a_y$ implies $u_i \in N[x] \cap N[y]$. $I_i O_i \in a_x \setminus a_y$ and $O_i I_{i+1} \in a_y \setminus a_x$ is impossible because $\{x, y\} \in F$. Consequently, x and y are adjacent in G or have a common neighbour. □

Theorem 4. *The bandwidth of a circular permutation graph can be approximated to within a factor of four in $\mathcal{O}(n \log^2 n)$ time.*

Proof. By Lemmas 6, 7 and 8 and Theorem 3. □

References

1. Assmann, S. F., G. W. Peck, M. M. Sysło and J. Zak, The bandwidth of caterpillars with hairs of length 1 and 2, *SIAM J. Algebraic Discrete Methods* **2** (1981), pp. 387–393.
2. Blum, A., G. Konjevod, R. Ravi and S. Vempala, Semi-definite relaxations for minimum bandwidth and other vertex-ordering problems, *Proceedings of the Thirtieth Annual ACM Symposium on Theory of Computing* (Dallas, TX, 1998), pp. 111–105.
3. Chinn, P. Z., J. Chvátalová, A. K. Dewdney and N. E. Gibbs, The bandwidth problem for graphs and matrices—a survey, *Journal of Graph Theory* **6** (1982), pp. 223–254.
4. Chung, F. R. K., Labelings of graphs, *Selected topics in graph theory* **3** (1988), pp. 151–168.
5. Downey, R. G. and M. R. Fellows, *Parameterized complexity*, Springer–Verlag New York, 1998.

6. Feige, U., Approximating the bandwidth via volume respecting embeddings, *Proceedings of the Thirtieth Annual ACM Symposium on Theory of Computing* (Dallas, TX, 1998).

7. Garey, M. R., R. L. Graham, D. S. Johnson and D. E. Knuth, Complexity results for bandwidth minimization, *SIAM Journal on Applied Mathematics* **34** (1978), pp. 477–495.

8. Gurari, E. M. and I. H. Sudborough, Improved dynamic programming algorithms for bandwidth minimization and the MinCut linear arrangement problem, *J. Algorithms* **5** (1984), pp. 531–546.

9. Haralambides, J. and F. Makedon, Approximation algorithms for the bandwidth minimization problem for a large class of trees, *Theory Comput. Syst.* **30** (1997), pp. 67–90.

10. Haralambides, J., F. Makedon and F. and B. Monien, Bandwidth minimization: An approximation algorithm for caterpillars, *Math. Syst. Theory* **24** (1991), pp. 169–177.

11. Kleitman, D.J. and R.V. Vohra, Computing the bandwidth of interval graphs, *SIAM Journal on Discrete Mathematics* **3** (1990), pp. 373–375.

12. Kratsch, D., P. Damaschke and A. Lubiw, Dominating cliques in chordal graphs, *Discrete Mathematics* **128** (1994), pp. 269–276.

13. Kloks, T., D. Kratsch and H. Müller, Approximating the bandwidth for asteroidal triple-free graphs, *Journal of Algorithms* **32** (1999), pp. 41–57.

14. Kloks, T., D. Kratsch, and H. Müller, Bandwidth of chain graphs, *Information Processing Letters* **68** (1998), pp. 313–315.

15. Kratsch, D. and L. Stewart, Approximating bandwidth by mixing layouts of interval graphs, Proceedings of STACS'99, LNCS 1563, pp. 248–258.

16. Monien, B., The bandwidth minimization problem for caterpillars with hair length 3 is NP-complete, *SIAM Journal on Algebraic and Discrete Methods* **7** (1986), pp. 505–512.

17. Rotem, D. and J. Urrutia, Circular permutation graphs, *Networks* **12** (1982), pp. 429–437.

18. Papadimitriou, C.H., The NP-completeness of the bandwidth minimization problem, *Computing* **16** (1976), pp. 263–270.

19. Parra, A., *structural and algorithmic aspects of chordal graph embeddings*, PhD–thesis, Technische Universität Berlin, 1996.

20. Peck, G. W. and Aditya Shastri, Bandwidth of theta graphs with short paths, *Discrete Mathematics* **103** (1992), pp. 177–187.

21. Sprague, A. P., An $O(n \log n)$ algorithm for bandwidth of interval graphs, *SIAM Journal on Discrete Mathematics* **7** (1994), pp. 213–220.

22. Unger, W., The complexity of the approximation of the bandwidth problem, *Proceedings of the 39th Annual Symposium on Foundations of Computer Science* (Palo Alto, California, 1998), pp. 82–91.

23. Venkatesan, G., U. Rotics, M. S. Madanlal, J. A. Makowsky, and C. Pandu Rangan, Restrictions of minimum spanner problems, *Information and Computation* **136** (1997), pp. 143–164.

Recognizing Graphs without Asteroidal Triples
(Extended Abstract)

Ekkehard Köhler*

Fields Institute Toronto, Canada, and Fachbereich Mathematik, Technische
Universität Berlin, Strasse des 17. Juni 136, Berlin, Germany.
ekoehler@Math.TU-Berlin.DE

Abstract. We consider the problem of recognizing AT-free graphs. Although there is a simple $O(n^3)$ algorithm, no faster method for solving this problem had been known. Here we give three different algorithms which have a better time complexity for graphs which are sparse or have a sparse complement; in particular we give algorithms which recognize AT-free graphs in $O(n\overline{m} + n^2)$, $O(\overline{m}^{3/2} + n^2)$, and $O(n^{2.82} + nm)$. In addition we give a new characterization of graphs with bounded asteroidal number by the help of the knotting graph, a combinatorial structure which was introduced by Gallai for considering comparability graphs.

1 Introduction

An *asteroidal triple* or, briefly, an *AT* of a given graph G is a set of three independent vertices such that there is path between each pair of these vertices that does not contain any vertex of the neighborhood of the third. Consequently, a graph G is called *asteroidal triple-free* or *AT-free* if there is no asteroidal triple in G and it is called *coAT-free* if $\overline{G}$ is AT-free.

Almost forty years ago, Lekkerkerker and Boland [13] defined the concept of an asteroidal triple for the first time. They used it for the investigation of intersection graphs corresponding to intervals of the real line—the interval graphs—and proved the well known characterization, that a graph G is an interval graph if and only if it is chordal and AT-free. Already in this early paper Lekkerkerker and Boland consider the problem of deciding whether a given graph contains an asteroidal triple. In fact, they gave a simple $O(n^3)$ algorithm—we call it STRAIGHTFORWARD ALGORITHM in the following—and used it for recognizing interval graphs. Of course by now there are much faster algorithms for recognizing interval graphs. However, for deciding whether a given graph contains an asteroidal triple no faster algorithm had been known.

In this paper we study the recognition problem of AT-free graphs from different perspectives and present three different recognition algorithms for AT-free

---

* The author would like to thank the Fields Institute for Research in Mathematical Sciences in Toronto, and NSERC Canada for financial assistance under Derek Corneil's grant.

graphs. At first we examine the above-mentioned STRAIGHTFORWARD ALGO-
RITHM a bit closer and design an algorithm that runs in $O(n\overline{m} + n^2)$, where
$\overline{m}$ is the number of non-edges of the input graph G. For the second recognition algorithm we use an algorithm for listing all triangles of a given graph and
achieve a time bound of $O(\overline{m}^{3/2} + n^2)$. Finally, in the last section we present
the KNOTTING GRAPH ALGORITHM. It makes use of a characterization of AT-
free graphs by the help of the knotting graph and recognizes AT-free graphs in
$O(n\,m + n^{2.82})$.

Since AT-free graphs are defined as a generalization of interval graphs it
seems to be plausible to try similar methods for recognizing AT-free graphs as
proved useful for recognizing interval graphs. However, for different reasons non
of the fast interval graph recognition algorithms seems to help for our purpose:

Booth and Lueker [2] designed the first linear time recognition algorithm for
interval graphs making use of the vertex–maximal clique matrix of the input
graph. A vertex–maximal clique matrix is a $0 - 1$ matrix M, such that each row
of the matrix corresponds to a vertex of the graph and each column corresponds
to a maximal clique of the graph and an entry m_{ij} is 1 if and only if vertex i
is contained in the maximal clique j. Fulkerson et al. [6] showed, that a graph
G is an interval graph if and only if the vertex–maximal clique matrix M of G
has the consecutive ones property for rows, i.e., if there is a permutation of the
columns of M, such that no ones in a single row are separated by zeroes in that
same row. For AT-free graphs we do not have such a strong characterization.
Of course, there is a close relationship between AT-free graphs and interval
graphs, since every interval graph is AT-free and every minimal triangulation
of an AT-free graph is an interval graph [15]. By Parra's characterization of
minimal triangulations (see [16]) there is a one-one correspondence between the
minimal triangulations of a given graph G and the inclusion maximal sets of
pairwise parallel minimal separators of G. For AT-free graphs, this implies that
for each set of pairwise parallel minimal separators the corresponding vertex–
minimal separator matrix has the consecutive ones property for rows. However,
this does not give a characterization of AT-free graphs yet and, even worse, the
number of minimal separators of an AT-free graph can be quite large—it can
even be exponential in the number of vertices of G. Hence, it is not very likely
that one can use a method which is similar to the consecutive ones testing for
the recognition of AT-free graphs.

The algorithm of Booth and Lueker can be interpreted also as one that makes
use of the geometric model of interval graphs, since a vertex–maximal clique
matrix that has the consecutive ones property for rows provides an interval
model of the corresponding graph. Again, for AT-free graphs such an approach
is not applicable, since there is no known geometric model for AT-free graphs.

A different method for recognizing interval graphs, was recently suggested
by Corneil et al. [3] by applying a simple four-sweep LBFS-algorithm for this
problem. They make especially use of the characterization of interval graphs to
be chordal and AT-free, where the first of these properties was known to be
checkable using LBFS in linear time before (see [17]). The key property that

Corneil et al. make use of, is the existence of a so-called *interval ordering*, i.e., a linear ordering $v_1, \ldots, v_n$ of the vertices of the graph with the property that for each edge $(v_i, v_j) \in E$ with $i < j$, all vertices v_k with $i < k < j$ are adjacent to v_j. This ordering characterizes interval graphs (see [5]). A couple of nice properties of LBFS are known for AT-free graphs as well (see [4]) and several researchers have considered the problem of recognizing AT-free graphs by the help of LBFS or at least similar methods like LBFS. However, up to now no fast algorithm using this approach is known. One of the main reasons for the difficulty of any such method is, that there is no known linear ordering that characterizes graphs without asteroidal triples.

Thus, none of the mentioned approaches seems to be applicable for recognizing AT-free graphs. In spite of this dejecting observation there are, in fact, efficient methods to handle AT-free graphs as you will see in the following sections. But all three presented recognition algorithms for AT-free graphs that we give here do not achieve a linear time bound. Recently, Spinrad showed, that it is rather unlikely to find a much faster algorithm to recognize AT-free graphs. He gave a construction for comparing the complexity of the recognition of asteroidal triple-free graphs to the complexity of finding an independent triple and showed: Given an algorithm that recognizes AT-free graphs in $O(f(n,m))$ then there is an $O(f(n,m) + n^2)$ algorithm for deciding whether a given graph contains an independent triple. (Due to space restrictions we cannot give this construction here but refer the reader to [18].) Hence it is unlikely to find algorithms for recognizing AT-free graphs which are much faster than the known algorithms for finding triangles. By a similar argument as used in Spinrad's construction, Hempel and Kratsch [10] showed that already the recognition of claw-free AT-free graphs is as hard as finding an independent triple in a given graph. The fastest known algorithm for finding triangles in a graph is matrix multiplication with a time bound of $O(n^\alpha)$, with $\alpha < 2.376$. Hempel and Kratsch also gave an $O(n^\alpha)$ algorithm for recognizing this restricted subclass of AT-free graphs.

In the following we will denote the set of all neighbors of a vertex v in a graph G by $N_G(v)$ and $N_G[v] = N_G(v) \cup \{v\}$ (if no ambiguities are possible we omit the subscript). For a set of vertices S of a graph G we denote by $G[S]$ the graph induces in G by the vertices of S and we denote with $G - S$ the graph $G[V \setminus S]$. For a graph G we use n for the number of vertices and m for the number of edges; $\overline{m}$ is used for the number of edges of $\overline{G}$, the complementary graph of G.

Because of space restrictions some of the results in this extended abstract are stated without proof; the interested reader can find the omitted parts in [12].

2   Straightforward Algorithm and Its Improvement

There are several characterizations for asteroidal triple-free graphs and, as we will examine in the course of this paper, some of them are more useful for the recognition of this graph class than others. A very simple characterization is already given by a slight alteration of the definition of AT-free graphs. For a graph G and vertices v, w of G, let $C^v(w)$ be the connected component of $G - \mathrm{N}[v]$ containing vertex w.

Observation 1. *An independent triple u, v, w of a graph G forms an asteroidal triple of G if and only if $C^v(u) = C^v(w)$ and $C^u(v) = C^u(w)$ and $C^w(v) = C^w(u)$ holds.*

If such a triple $\{u, v, w\}$ exists we know that u and w are in the same connected component of $G - \mathrm{N}[v]$, in other words there is an u, w-path, that avoids the neighborhood of v. Analogously, there is a v, w-path avoiding the neighborhood of u and an u, v-path avoiding the neighborhood of w. Thus, u, v, w is an asteroidal triple of G.

This straightforward characterization immediately implies a STRAIGHTFORWARD ALGORITHM; it was first suggested by Lekkerkerker and Boland in [13] when they constructed an $O(n^4)$ algorithm for recognizing interval graphs. In a first step, for each vertex v the connected components of $G - \mathrm{N}[v]$ are determined and each is assigned a different label. Then, in the second step, for each triple of vertices it is checked whether the condition of Observation 1 is fulfilled.

For implementing this algorithm one can use a simple data structure called *component structure* which we define here for later use. For a given graph $G = (V, E)$ with n vertices, the *component structure* of G is an $n \times n$ matrix C, where each column and each row of G corresponds to a vertex of G. For each vertex v the matrix entry c_{vw} is 0, if $w \in \mathrm{N}[v]$, otherwise c_{vw} is set to the label of the connected component of $G - \mathrm{N}[v]$ containing w.

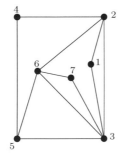

	1	2	3	4	5	6	7
1	0	0	0	a	a	a	a
2	0	0	0	0	b	0	c
3	0	0	0	d	0	0	0
4	e	0	e	0	0	e	e
5	f	f	0	0	0	0	g
6	h	0	0	i	0	0	0
7	k	k	0	k	k	0	0

Fig. 1. Example of a graph together with its component structure

Example 1. Let G be the 7-vertex graph of Figure 1. The corresponding component structure is given in the table. The labels of the connected components are the letters a to k. Note that a component that occurs for different vertices gets different labels; e.g. the component containing only vertex 4 occurs both for vertex 3 and vertex 6 and thus has both label d and label i.

The first part of the STRAIGHTFORWARD ALGORITHM can be implemented to run in $O(nm)$ by applying an $O(n + m)$ breadth-first search to $G - \mathrm{N}[v]$

for each vertex v of G. Using the component structure, the second part of the algorithm can easily be implemented to run in $O(n^3)$, by checking for each triple u, v, w of G whether $c_{uv} = c_{uw}$, $c_{vu} = c_{vw}$ and $c_{wu} = c_{wv}$. Hence, AT-free graphs can be recognized in $O(n^3)$ using the STRAIGHTFORWARD ALGORITHM.

Obviously, the "bottleneck" of the complexity of the STRAIGHTFORWARD ALGORITHM seems to be the checking of all possible triples of the given graph, and it is a reasonable question to ask, whether all those tests are indeed required. Actually, it is not really necessary to do this for *all* triples of the graph, because, when searching for asteroidal triples, we are interested only in independent triples. Thus, as a first step, it is sufficient to check for all non-edges (v, w) of G, whether there is a vertex u such that the triple u, v, w is an asteroidal triple. Obviously, there are $O(n\,\overline{m})$ those triples in G, where $\overline{m}$ is the number of non-edges of G. A look back to the complexity of the first part of the STRAIGHTFORWARD ALGORITHM tells us that we did not really get a better time complexity yet, since $O(n\,m) + O(n\,\overline{m})$ is still $O(n^3)$. However, also this first part of the algorithm can be altered to run in $O(n\,\overline{m})$, leading to an $O(n^2 + n\,\overline{m})$ algorithm, as we show in the following.

The method that we use for this improvement is a BFS conducted on $\overline{G}$. McConnell [14] observed, that one can implement a BFS in such a way that it constructs the BFS-layers for $\overline{G}$ and runs in $O(n + m)$ time, where m is the number of edges of G—not of $\overline{G}$ (see also [11]).

To achieve a BFS ordering on the complement by the help of partition refinement, we just have to change the ordering of T_1 and T_2, i.e., T_2 is placed in front of T_1 in the ordering of the sets of the partition.

Putting together the BFS on $\overline{G}$ and the improved checking of triples, leads to the COMPLEMENT ALGORITHM.

Theorem 1. *Recognition of AT-free graphs, using the* COMPLEMENT ALGORITHM, *takes* $O(n^2 + n\,\overline{m})$ *time, where* $\overline{m}$ *is the number of non-edges of* G.

Proof. Computing the complement of a graph G can, of course, be done in $O(n^2)$. To compute the component structure, by the help of the complement-BFS takes $O(n + \overline{m})$ for each vertex; hence, for all vertices it can be done in $O(n\,\overline{m})$. Finally, checking for each non-edges (v, w) and each vertex u, whether u, v, w form an AT, takes $O(n\,\overline{m})$ as well. $\square$

3   Triangle Algorithm

Before stating the next algorithm for recognizing AT-free graphs we first consider a different problem, the problem of finding triangles in a given graph. As we will see later, it turns out to be closely related to our recognition problem. Here we have to distinguish between the problem of deciding whether a given graph contains a triangle and the problem of finding one or all triangles of G. For the beginning we are interested in the second of these problems, i.e., we want to list all triangles of a given graph G. For this we make use of an observation made by Gabow, concerning the number of different triangles in a graph G, i.e., the number of triangles of G, such that each pair of those triangles differs in at least one vertex.

Lemma 1 (Gabow [7]). *Given a graph G with m edges, there are at most $O(m^{3/2})$ different triangles.*

For listing all triangles of G in $O(m^{3/2})$ time, we can use the following algorithm. First the vertices are ordered, according to their degree. Then for each vertex v_i a list L_{v_i} of all neighbors of v_i with equal or larger degree that occur after v_i in the ordering is created. Within this list the vertices are ordered according to the ordering that was determined in the first step. In the last part of the algorithm for each vertex v_i the corresponding list L_{v_i} is traversed and for each vertex v_j in L_{v_i} the lists L_{v_i} and L_{v_j} are compared. For each vertex v_k that is contained in both lists the algorithm outputs the corresponding triangle v_i, v_j, v_k.

Lemma 2. *For a given graph G a complete list of all different triangles of G can be determined in $O(m^{3/2} + n)$.*

Similarly as in Lemma 1, one can bound the number of different triangles of a graph G also by the help of $\Delta(G)$, the maximum vertex degree of G.

Lemma 3. *In a graph G with m edges and maximum degree $\Delta(G)$ there are at most $O(\Delta(G)\,m)$ different triangles.*

Of course, this bound can also be used for the algorithm, that lists all triangles of G.

Corollary 1. *For a given graph G a complete list of all different triangles of G can be determined in $O(\Delta(G)\,m)$.*

In Section 2, exploring the structure of the complement of a graph G seemed to be helpful for deciding whether G contains asteroidal triples. In the following algorithm—the TRIANGLE ALGORITHM—we consider again $\overline{G}$, this time using the set of its triangles.

Theorem 2. *Recognition of AT-free graphs, using the TRIANGLE ALGORITHM (Algorithm 1), takes $O(\overline{m}^{3/2} + n^2)$ time, where n is the number of vertices and $\overline{m}$ is the number of non-edges of G.*

Using the results of Lemma 1 and Lemma 3, we can also give a recognition algorithm that runs in $O(\Delta(\overline{G})\,\overline{m} + n^2)$, where $\Delta(\overline{G})$ is the maximum degree of $\overline{G}$. Hence we have the following corollary.

Corollary 2. *Recognition of AT-free graphs, using the TRIANGLE ALGORITHM, takes $O(n^2 + \min\{\overline{m}^{1/2}, \Delta(\overline{G})\}\,\overline{m})$ time, where n is the number of vertices, $\overline{m}$ is the number of non-edges of G and $\Delta(\overline{G})$ the maximum degree of $\overline{G}$.*

Algorithm 1: TRIANGLE ALGORITHM

---

begin

> Compute $\overline{G}$;
> Determine a list L of all different triangles of $\overline{G}$;
> **for** $v \in G$ **do**
> > find all edges of $\overline{G}[N_{\overline{G}}(v)]$ using L;
> > compute connected components of $\overline{G[N_{\overline{G}}(v)]} = G - N[v]$;
> > store labels of components in component structure C;
>
> **end**
> **for** *each triangle u, v, w of L* **do**
> > check whether u, v, w is an AT;
>
> **end**

end

Remark 1. As part of the TRIANGLE ALGORITHM we used an algorithm that lists all triangles of the complement of G in a *certain time* and then runs in the order of the size of the set of triangles (plus $O(n^2)$ for initializing the component structure). Thus, if we are given the set of all independent triples of G and this set has size $|T|$, then we can check in $O(|T| + n^2)$ time whether G is AT-free. If we take, for example, a graph G such that $\overline{G}$ is planar, we can find the list of triangles of $\overline{G}$ in $O(n^2)$, implying that we can decide in $O(n^2)$ whether G is AT-free.

The time bounds of both the TRIANGLE ALGORITHM and the COMPLEMENT ALGORITHM are measured in the size of $\overline{G}$. Another way to look at these algorithms is to consider them to be algorithms to recognize coAT-free graphs. In other words, we have an $O(nm)$, an $O(m^{3/2} + n^2)$ and an $O(\Delta(G)\, m + n^2)$ algorithm which decides the problem: Given a graph G, is G a coAT-free graph?

For recognizing AT-free graphs the n^2 term for the above algorithms was not avoidable, since both the edges and the non-edges of G were used during the algorithms and for every graph G, either m or $\overline{m}$ is in the order of n^2. For recognizing coAT-free graphs we can do better. All we have to consider are the edges of G whereas the non-edges of G are not of any interest. The way the algorithm is given above, we need $O(n^2)$ for initializing the component structure C. We can get rid of this by the following method: Instead of storing the information about the connected components of the neighborhood in the component structure, we store this information "within" the edges of the graph. For a vertex v we store the label of the connected component of $\overline{G}[N_G[v]]$ containing some vertex u together with the edge (v, u). Obviously, for every edge only two labels are stored. To check, whether there is an asteroidal triple in $\overline{G}$ we just have to check for each triangle of G, whether the edges of the triangle have pairwise the same label for their common incident vertex. Consequently, we have the following theorem.

Theorem 3. *For coAT-free graphs the recognition problem can be solved in $O(\min\{m^{1/2}, \Delta(G)\}\, m)$ time, where m is the number of edges and $\Delta(G)$ the maximum degree of G.*

4   The Knotting Graph

When studying AT-free graphs one learns to appreciate the strong relationship between AT-free and comparability graphs. It was proved by Golumbic, Monma, and Trotter in their 1984 paper on tolerance graphs [9], that a graph G is a cocomparability graph then G contains no asteroidal triple. In fact, they were not the first to realize this relationship. A closer look at the paper of Gallai [8] on transitively orientable graphs reveals, that he already achieved this result almost twenty years earlier. In fact, he proves a much stronger result. To state his theorem we have to define one more concept.

Definition 1. *The sequence* $\sigma = x_1, P_1, x_2, P_2, x_3, \ldots, x_{2k+1}, P_{2k+1}, x_1, (k \geq 1)$ *is said to be a $(2k+1)$-asteroid of a graph G, if $x_1, \ldots, x_{2k+1}$ are different vertices of G and P_i are x_i, x_{i+1}-paths of G, such that for each x_i $(1 \leq i \leq 2k+1)$ there is no neighbor of x_i on the path P_{i+k} where the paths P_α and P_β are assumed to be equal for $\alpha \equiv \beta$ (mod $2k + 1$).*

Not surprisingly, every 3-asteroid contains an asteroidal triple, as one can derive from the definition. Examples of $(2k + 1)$-asteroids for $k \geq 2$ are all complements of chord-less $2k+1$ cycles for $k \geq 2$. A complete list of all *irreducible* $(2k + 1)$-asteroids for $k \geq 2$ which do not contain any asteroid of smaller length was given in [8]. Now we are able to state the result of Gallai.

Theorem 4 (Gallai [8]). *A graph G is a comparability graph if and only if $\overline{G}$ does not contain a $(2k + 1)$-asteroid for $k \geq 1$.*

For the proof of his result Gallai makes use of a certain structure called the *knotting graph* (*Verknüpfungsgraph*) of G.

Definition 2. *For a given graph $G = (V, E)$ the corresponding* knotting graph *is given by* $K[G] = (V_K, E_K)$ *where V_K and E_K are defined as follows. For each vertex v of G there are copies $v_1, v_2, \ldots, v_{i_v}$ in V_K, where i_v is the number of connected components of $\overline{N(v)}$, the complement of the graph induced by $N(v)$. For each edge (v, w) of E there is an edge (v_i, w_j) in E_K, where w is contained in the ith connected component of $\overline{N(v)}$ and v is contained in the jth connected component of $\overline{N(w)}$.*

Example 2. In Figure 2 one can see a graph G together with its knotting graph. Here small dots in the knotting graph that are drawn closely together indicate that they are copies for the same original vertex of the graph.

For the purpose of characterizing comparability graphs, i.e., the class of graphs which have a transitive orientation, the knotting graph has special importance, as shown in the following theorem.

Theorem 5 (Gallai [8]). *A graph G is transitively orientable if and only if $K[G]$ is bipartite.*

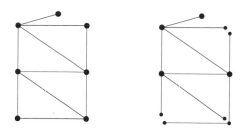

Fig. 2. Example of a graph, together with its knotting graph.

What makes this graph interesting for us is the close connection between the asteroidal sets of a graph G (a generalization of asteroidal triples) and the cliques of the knotting graph of $\overline{G}$. Before we can state this result we first have to define the concept of an asteroidal set and the asteroidal number of a graph G.

Definition 3. *For a given graph G, an independent set of vertices S is called asteroidal set if for each $x \in S$ the set $S - \{x\}$ is in one connected component of the graph $G - N[x]$. The asteroidal number of a graph G is defined as the maximum cardinality of an asteroidal set of G, and is denoted by $\mathrm{an}(G)$.*

Theorem 6. *Let G be a graph, then $\mathrm{an}(G) = \omega(\mathrm{K}[\overline{G}])$.*

Proof. Let $\mathrm{an}(G) = k$ and let $A = \{a_1, \ldots, a_k\}$ be an asteroidal set of G. By the definition of asteroidal sets the vertices of A are pairwise independent. Consequently, A induces a clique in $\overline{G}$ and for each $a_i \in A$ the set $A \setminus \{a_i\}$ is contained in the neighborhood of a_i in $\overline{G}$. Since A is an asteroidal set, for each $a_j, a_k \in A \setminus \{a_i\}$ ($j \neq k$) there is an a_j, a_k-path in G that avoids the neighborhood of a_i. Therefore a_j and a_k are in the same connected component of $G - N[a_i]$. By the definition of the knotting graph this implies that the knotting graph edges corresponding to the edges $(a_i, a_j), (a_i, a_k)$ of $\overline{G}$ are incident to the same copy of a_i in the knotting graph. Since this is true for all pairs of vertices in $A \setminus \{a_i\}$, all edges corresponding to edges from vertices of $A \setminus \{a_i\}$ to a_i in $\overline{G}$, are incident to the same copy of a_i in the knotting graph. Consequently, there is a k-clique in $\mathrm{K}[\overline{G}]$ formed by copies of vertices of A.

Now suppose there is a k-clique in the knotting graph $\mathrm{K}[\overline{G}]$. Since there is a 1-1 correspondence between the edges of $\overline{G}$ and the edges of $\mathrm{K}[\overline{G}]$ there is a set $A = \{a_1, \ldots, a_k\}$ of k vertices of G corresponding to the vertices of the clique in $\mathrm{K}[\overline{G}]$. By the definition of the knotting graph, for each vertex $a_i \in A$ the vertices of $A \setminus \{a_i\}$ are contained in the same connected component of $G - N[a_i]$. Consequently, A is an asteroidal set of G. ☐

For AT-free graphs we can draw the following corollary.

Corollary 3. *A graph G is asteroidal triple-free if and only if $\mathrm{K}[\overline{G}]$ is triangle-free.*

Because of the close relationship between an AT-free graph G and the knotting graph $K[\overline{G}]$ of the complementary graph $\overline{G}$, we will sometimes call $K[\overline{G}]$ the knotting graph *corresponding* to the AT-free graph G.

At this point we would like to mention some questions that arise from considering $(2k + 1)$-asteroids. As we have seen, AT-free graphs are exactly those graphs, that do not contain a 3-asteroid. On the other hand, cocomparability graphs are those graphs that do not contain *any* $(2k + 1)$-asteroid. It seems to be an interesting question to consider those graph classes, that are defined by forbidding only certain kinds of asteroids. For example one could develop a whole hierarchy of graph classes that are superclasses of cocomparability graphs and subclasses of AT-free graphs.

Another way of generalizing this concept is to consider the definition of asteroids again. This definition has the unsatisfying property that it exists only for odd numbers. Thus, it seems to be natural to ask whether one can state a reasonable definition also for even asteroids. The definition of Gallai implies, that an asteroid of a given graph G corresponds to a cycle in the knotting graph of $\overline{G}$. This can be used for our purpose in the following way. We define a k-asteroid as a sequence $\sigma = x_1, \ldots, x_k$ $(k \geq 3)$ of different vertices of G, such that for each vertex x_i there is a path between x_{i-1} and x_{i+1} in $G - N[x_i]$. This, in fact, does cover the definition of odd asteroids, although the new numbering is different to the one used in the definition of Gallai. We leave it as an open question to characterize the graphs that are characterized by forbidding certain (not necessarily odd) asteroids.

There are a couple of other properties of the knotting graph, especially with respect to AT-free graphs. For further results the reader is refered to [12].

5   The Knotting Graph Algorithm

Before we can state the KNOTTING GRAPH ALGORITHM, we first consider the relationship between the knotting graph, corresponding to a graph G and the component structure of this graph.

Lemma 4. *Let $G = (V, E)$ be a graph and C the corresponding component structure, then $K[\overline{G}] = (V_K, E_K)$ is the knotting graph of $\overline{G}$, with*

$$V_K = \{c_{vw} : \exists v, w \in V\} \setminus \{0\}$$
$$E_K = \{(c_{vw}, c_{wv}) : \exists v, w \in V \text{ such that } c_{vw} \neq 0 \text{ and } c_{wv} \neq 0\}.$$

Now we are ready to present the KNOTTING GRAPH ALGORITHM (see Algorithm 2).

As shown in Section 2, the component structure of G can be computed in $O(n\,m)$ time. By Lemma 4, V_K and E_K, the vertex and edge set of $K[\overline{G}]$, can be determined by the help of the component structure. To compute V_K one has to scan through all rows r of the component structure C and for each new label one inserts a new copy of the vertex corresponding to row r. Hence, finding V_K takes $O(n^2)$. The edge set E_K can be computed by checking for each possible

---

Algorithm 2: Knotting Graph Algorithm

---

begin

    for $v \in G$ **do**

        $H \leftarrow G - N[v]$;

        compute connected components of H, using BFS;

        store labels of components in component structure $C(i_v, \cdot)$;

    end

    $V_K = \{c_{vw} : \exists v, w \in V \text{ such that } c_{vw} \neq 0\}$;

    $E_K = \{(c_{vw}, c_{wv}) : \exists v, w \in G \text{ such that } c_{vw} \neq 0 \text{ and } c_{wv} \neq 0\}$;

    check whether $K[\overline{G}]$ contains a triangle;

end

---

row-column pair r, c of C whether both c_{rc} and c_{cr} are non-zero. Since there are n^2 those pairs for our $n \times n$ matrix C, this can be done in $O(n^2)$ as well.

The difficult part of the algorithm is finding triangles in $K[\overline{G}]$. The fastest known algorithm for this problem is matrix multiplication, which runs in $O(n^\alpha)$. Unfortunately, we cannot use this algorithm for our problem. The reason is, that the number of vertices of $K[\overline{G}]$ can be considerably larger than the number of vertices of G. There are examples of graphs, showing that there can be $\Omega(n^2)$ vertices in $K[\overline{G}]$, where n is the number of vertices of G. One information that we do have about the number of vertices of $K[\overline{G}]$ is, that there are not more than twice as many vertices as edges in $K[\overline{G}]$, i.e., $|V_K| \leq 2|E_K|$ (an exception is, of course, the case that there are universal vertices in G, but for finding triangles in $K[\overline{G}]$ this case is not of interest). Hence, if we apply an algorithm, that runs in $O(f(|E_K|))$, it is in fact an $O(f(n^2))$ algorithm, since the number of edges of $K[\overline{G}]$ is equal to the number of edges of $\overline{G}$. For the case that the input graph is sparse, Alon et al. [1] suggested an improvement of the matrix multiplication algorithm for finding a triangle. The complexity of this algorithm is $O(m^{\frac{2\alpha}{\alpha+1}})$; for $\alpha < 2.376$ this is $O(m^{1.41})$. Consequently, if we apply the triangle algorithm for sparse graphs of Alon et al., we can decide in $O(\overline{m}^{\frac{2\alpha}{\alpha+1}}) = O(n^{\frac{4\alpha}{\alpha+1}}) = O(m\,n + n^{2.815})$ for $\alpha < 2.376$ whether $K[\overline{G}]$ contains a triangle. Thus we have shown the following Theorem.

Theorem 7. *Recognition of AT-free graphs using the* KNOTTING GRAPH ALGORITHM *takes* $O(m\,n + n^{2.815})$ *time.*

Remark 2. It is an open question, whether there is a certificate for a graph to be AT-free, which can be checked in less than $O(n^3)$. The knotting graph does provide some kind of "partial certificate", since it can indeed be checked in less than $O(n^3)$ for containing a triangle. Of course, for being a proper certificate one should be able to check also within this time bound whether the knotting graph is correctly determined. This can currently be done in $O(nm)$, implying that we have a fast checkable certificate for sparse graphs.

References

1. N. ALON, R. YUSTER, AND U. ZWICK, *Finding and counting given length cycles*, Algorithmica, 17 (1997), pp. 209–223.
2. S. BOOTH AND G. LUEKER, *Testing for the consecutive ones property, interval graphs, and planarity using PQ-tree algorithms*, J. Comput. System Sci., 13 (1976), pp. 335–379.
3. D. G. CORNEIL, S. OLARIU, AND L. STEWART, *The ultimate interval graph recognition algorithm?*, in Proceedings of the 9th annual ACM-SIAM symposium on discrete algorithms, San Francisco, CA, USA, 1998, pp. 175–180.
4. D. G. CORNEIL, S. OLARIU, AND L. STEWART, *Linear time algorithms for dominating pairs in asteroidal triple-free graphs*, SIAM J. Comput., 28 (1999), pp. 1284–1297.
5. P. DAMASCHKE, *Forbidden ordered subgraphs*, in Topics in Combinatorics and Graph Theory, R. Bodendiek and R. Henn, eds., Physica-Verlag Heidelberg, 1990, pp. 219–229.
6. D. FULKERSON AND O. GROSS, *Incidence matrices and interval graphs*, Pacific J. Math., 15 (1965), pp. 835–855.
7. H. GABOW. private communication, 1994.
8. T. GALLAI, *Transitiv orientierbare Graphen*, Acta Math. Acd. Sci. Hungar., 18 (1967), pp. 25–66.
9. M. C. GOLUMBIC, C. L. MONMA, AND W. T. TROTTER, *Tolerance graphs*, Discrete Appl. Math., 9 (1984), pp. 157–170.
10. H. HEMPEL AND D. KRATSCH, *On claw-free asteroidal triple-free graphs*, in Proceedings of WG'99, P. Widmayer, ed., vol. LNCS 1665, 1999, pp. 377–390.
11. H. ITO AND M. YOKOYAMA, *Linear time algorithms for graph search and connectivity determination on complement graphs*, Inform. Process. Lett., 66 (1998), pp. 209–213.
12. E. G. KÖHLER, *Graphs without Asteroidal Triples*, PhD thesis, Technische Universität Berlin, Cuvillier Verlag Göttingen, 1999.
13. C. LEKKERKERKER AND J. BOLAND, *Representation of a finite graph by a set of intervals on the real line*, Fund. Math., 51 (1962), pp. 45–64.
14. R. MCCONNELL. private communication, 1996.
15. R. H. MÖHRING, *Triangulating graphs without asteroidal triples*, Discrete Appl. Math., 64 (1996), pp. 281–287.
16. A. PARRA, *Structural and Algorithmic Aspects of Chordal Graph Embeddings*, PhD thesis, Technical University Berlin, 1996.
17. D. J. ROSE, R. E. TARJAN, AND G. S. LUEKER, *Algorithmic aspects of vertex elimination on graphs*, SIAM J. Comput., 5 (1976), pp. 266–283.
18. J. P. SPINRAD, *manuscript* http://www.vuse.vanderbilt.edu/ spin/research .html.

Budget Constrained Minimum Cost Connected Medians

Goran Konjevod[1,*], Sven O. Krumke[2,**], and Madhav Marathe[3,***]

[1] Department of Mathematical Sciences, Carnegie Mellon University, Pittsburgh,
PA 15213-3890 and Los Alamos National Laboratory, USA.
konjevod@andrew.cmu.edu
[2] Konrad-Zuse-Zentrum für Informationstechnik Berlin, Department Optimization,
Takustr. 7, 14195 Berlin-Dahlem, Germany. krumke@zib.de
[3] Los Alamos National Laboratory, P.O. Box 1663, MS B265, Los Alamos,
NM 87545, USA. marathe@lanl.gov

Abstract. Several practical instances of network design problems require the network to satisfy multiple constraints. In this paper, we address the *Budget Constrained Connected Median Problem*: We are given an undirected graph $G = (V, E)$ with two different edge-weight functions c (modeling the construction or communication cost) and d (modeling the service distance), and a bound B on the total service distance. The goal is to find a subtree T of G with minimum c-cost $c(T)$ subject to the constraint that the sum of the service distances of all the remaining nodes $v \in V \setminus T$ to their closest neighbor in T does not exceed the specified budget B. This problem has applications in optical network design and the efficient maintenance of distributed databases.

We formulate this problem as bicriteria network design problem, and present bicriteria approximation algorithms. We also prove lower bounds on the approximability of the problem that demonstrate that our performance ratios are close to best possible.

1 Introduction and Overview

The problem of interfacing optic and electronic networks has become an important problem in telecommunication network design [20,21]. As an example, consider the following problem: Given a set of sites in a network, we wish to select a subset of the sites at which to place optoelectronic switches and routers. The backbone sites should be connected together using fiber-optic links in a minimum cost tree, while the end users are connected to the backbone via direct links. The major requirement is that the total access cost for the users be within a specified bound, whereas the construction cost of the backbone network should be minimized.

* Supported by an NSF CAREER grant CCR-9625297 and DOE Contract W-7405-ENG-36
** Research supported by the German Science Foundation (DFG, grant Gr 883/5-3).
*** Research supported by the Department of Energy under Contract W-7405-ENG-36.

U. Brandes and D. Wagner (Eds.): WG 2000, LNCS 1928, pp. 267–278, 2000.
© Springer-Verlag Berlin Heidelberg 2000

Problems of similar nature arise in the efficient maintenance of distributed databases [22,4,3,15,7]. Other applications of the *Budget Constrained Connected Median Problem* studied in this paper include location theory and manufacturing logistics (see [20,21] and the references cited therein).

The above problems can be cast in a graph theoretic framework as follows: Given an undirected graph $G = (V, E)$ with two different edge-weight functions c (modeling the construction cost of the backbone/inter-database links) and d (modeling the service distance), the goal is to find a subtree T of G with minimum c-cost $c(T)$ subject to the constraint that the sum of the service distances of all the remaining nodes $v \in V \setminus T$ to their closest neighbor in T does not exceed a specified budget B.

We study the approximability of the *Budget Constrained Connected Median Problem*. This paper is organized as follows. In Section 2 we formally define the problem under study and the notion of bicriteria approximation. Section 3 contains a brief summary of the main results in the paper and a discussion of related work. In Section 4 we prove hardness results. Section 5 contains a fully polynomial approximation scheme on trees. An approximation algorithm for the general case is presented in Section 6.

2   Problem Definition and Preliminaries

Throughout the paper $G = (V, E)$ denotes a finite connected undirected graph with $n := |V|$ vertices and $m := |E|$ edges. The *Budget Constrained Connected Median Problem* (BCCMED) problem considered in this paper is defined as follows:

Definition 1 (Budget Constrained Connected Median Problem). *An instance consists of an undirected graph $G = (V, E)$ with two different edge-cost functions c (modeling the construction or communication cost) and d (modeling the service distance), and a bound B on the total service distance. The problem is to find a subtree T of G of minimum cost $c(T) := \sum_{e \in T} c(e)$ subject to the constraint that the total service distance of each of the vertices from V is at most B, that is,*

$$median_d(T) := \sum_{v \in V} \min_{u \in T} dist_d(v, u) \leq B.$$

(where $dist_d(v, u)$ denotes the shortest path distance between vertices v and u with respect to the edge-cost function d).

The problem BCCMED can be formulated within the framework developed in [18,13]. A generic bicriteria network design problem, (A, B, S), is defined by identifying two minimization objectives, – A and B, – from a set of possible objectives, and specifying a membership requirement in a class of subgraphs, – S. The problem specifies a budget value on the first objective, A, and seeks to find a network having minimum possible value for the second objective, B, such that this network is within the budget on the first objective A. The solution network must belong to the subgraph-class S. In this framework BCCMED is stated as (total d-service distance, total c-edge cost, subtree). : the budgeted objective A is the total service distance median$_d(T)$ with respect to the edge weights specified by d, the cost-minimization objective B is the total c-cost of the edges in the solution subgraph which is required to be a subtree of the original network.

Definition 2 (Bicriteria Approximation Algorithm). *A polynomial time algorithm for a bicriteria problem (A, B, S) is said to have* performance (α, β), *if it has the following property: For any instance of (A, B, S), the algorithm produces a solution from the subgraph class S for which the value of objective A is at most α times the specified budget and the value of objective B is at most β times the minimum value of a solution from S that satisfies the budget constraint.*

Notice that a "standard" c-approximation algorithm is a $(1, c)$-bicriteria approximation algorithm. A family $\{A_\epsilon\}_\epsilon$ of approximation algorithms, is called a *fully polynomial approximation scheme* or *FPAS*, if algorithm A_ϵ is a $(1, 1 + \epsilon)$-approximation algorithm and its running time is polynomial in the size of the input and $1/\epsilon$.

3   Summary of Results and Related Work

In this paper, we study the complexity and approximability of the problem BCCMED. Our main results include the following:

1. BCCMED is NP-hard even on trees. This result continues to hold even if the edge-weight functions c and d are identical. We strengthen this hardness result to obtain strong NP-hardness results for bipartite graphs.
2. We strengthen the above hardness results for general graphs further and show that unless NP $\subseteq$ DTIME($N^{\log \log N}$), there can be no polynomial time approximation algorithm for BCCMED with a performance $(1, (1/20 - \epsilon) \ln n)$, where n denotes teh number of vertices in the input graphs.

Our hardness results are complemented by the following approximation results:

1. There exists a FPAS for BCCMED on trees.
2. For any fixed $\epsilon > 0$ there exists a $(1 + \epsilon, (1 + 1/\epsilon)\mathcal{O}(\log^3 n \log \log n))$-approximation algorithm for BCCMED on general graphs.

3.1   Relationship to the Traveling Purchaser Problem

The BCCMED problem is closely related to a well studied variant of the classical traveling salesperson problem called the *Traveling Purchaser Problem* (see [20] and the references therein). In this problem we are given a bipartite graph $G =$

$(M \cup P, E)$, where M denotes a set of markets and P denotes the set of products. There is a (metric) cost c_{ij} to travel from market i to market j. An edge between market i and product p with cost d_{ip} denotes the cost of purchasing product p at market i. A *tour* consists of starting at a specified market visiting a subset of market nodes, thereby purchasing *all* the products and returning to the starting location. The cost of the tour is the sum of the travel costs used between markets and the cost of buying each of the products. The budgeted version of this problem as formulated by Ravi and Salman [20] aims at finding a minimum cost tour subject to a budget constraint on the purchasing costs.

It is easy to see that a (α, β)-approximation algorithm for the budgeted traveling purchaser problem implies a $(\alpha, 2\beta)$-approximation for BCCMED: just delete one edge of the tour to obtain a tree. Using the $(1+\epsilon, (1+1/\epsilon)\mathcal{O}(\log^3 m \log \log m))$-approximation algorithm from [20] we get a $(1+\epsilon, 2(1+1/\epsilon)\mathcal{O}(\log^3 m \log \log m))$-approximation for BCCMED. Our algorithm given in Section 6 uses the techniques from [20] directly and improves this result.

3.2   Related Work

Other service-constrained minimum cost network problems have been considered in [1,6,12,17,16]. These papers consider the variant that prescribes a budget on the service distance for each node not in the tree. The goal is to find a minimum length salesperson tour (or a tree as may be the case) so that all the (customer) nodes are strictly serviced. Restrictions of the problems to geometric instances were considered in [1,12,19]. Finally, the problem BCCMED can be seen as a generalization of the classical *k-Median Problem*, where we require the set of medians to be connected.

4   Hardness Results

Theorem 1. *The problem* BCCMED *is* NP-*hard even on trees. This result continues to hold even if we require the two cost functions c and d to coincide.*

Proof. We use a reduction from the PARTITION problem, which is well known to be NP-complete [10, Problem SP12]. Given a multiset of (not necessarily distinct) positive integers $\{a_1, \ldots, a_n\}$, the question is whether there exists a subset $U \subseteq \{1, \ldots, n\}$ such that $\sum_{i \in U} a_i = \sum_{i \notin U} a_i$.

Given any instance of PARTITION we construct a star-shaped graph G having $n+1$ nodes $\{x, x_1, \ldots, x_n\}$ and n edges (x, x_i), $i = 1, \ldots, n$. We define $c(x, x_i) := d(x, x_i) := a_i$. Let $D := \sum_{i=1}^{n} a_i$. We set the budget for the median cost of the tree to be $B := D/2$. It is easy to see that there exists a feasible tree T of cost $c(T)$ at most $D/2$ if and only if the instance of PARTITION has a solution. $\square$

Next, we prove our inapproximability results for general graphs. Before stating the hardness result we recall the definition of the MIN SET COVER problem [10, Problem SP5] and cite the hardness results from [9,2] about the hardness of approximating MIN SET COVER. An instance $(U, \mathcal{S})$ of MIN SET COVER consists of a finite set U of ground elements and a family $\mathcal{S}$ of subsets of U. The objective is to find a subcollection $\mathcal{C} \subseteq \mathcal{S}$ of minimum size $|\mathcal{S}|$ which contains all the ground elements.

Theorem 2 (Feige [9]). *Unless* NP $\subseteq$ DTIME($N^{\mathcal{O}(\log \log N)}$), *for any $\epsilon > 0$ there is no approximation algorithm for* MIN SET COVER *with a performance of $(1 - \epsilon) \ln |U|$, where U is the set of ground elements.* □

Theorem 3 (Arora and Sudan [2]). *There exists a constant $\eta > 0$ such that, unless* P= NP, *there is no approximation algorithm for* MIN SET COVER *with a performance of $\eta \ln |U|$, where U is the set of ground elements.* □

We are now ready to prove the result about the inapproximability of BCCMED on general graphs.

Theorem 4. *The problem* BCCMED *is strongly* NP-*hard even on bipartite graphs. If there exists an approximation algorithm for* BCCMED *on bipartite graphs with performance $\alpha(|V|) \in \mathcal{O}(\ln |V|)$, then there exists an approximation algorithm for* MIN SET COVER *with performance $2\alpha(2|U| + 2|\mathcal{S}|)$. All results continue to hold even if we require the two cost functions c and d to coincide.*

Proof. Let $(U, \mathcal{S})$ be an instance of MIN SET COVER. We assume without loss of generality that the minimum size set cover for this instance contains at least two sets (implying also that $|U| \geq 2$).

For each $k \in \{2, \ldots, n\}$ we construct an instance I_k of BCCMED as follows: The bipartite graph constructed for instance I_k has $|V_k| = 2|U| + 2|\mathcal{S}| + 2 - k \leq 2(|U| + |\mathcal{S}|)$ vertices. First construct the natural bipartite graph with node set $U \cup \mathcal{S}$. We add an edge between an element node $u \in U$ and a set node $S \in \mathcal{S}$ if and only if $u \in S$. We now add a root node r which is connected via edges to all the set nodes from $\mathcal{S}$. Finally, we add a set L_k of $|U| + |\mathcal{S}| - k + 1$ nodes which are connected to the root node via the edges (l, r), $l \in L_k$. Let $X := k\lceil \alpha(|V_k|) \rceil + 1$. The edges between element nodes and set nodes have weight X, all other edges have weight 1. The budget on the median cost for instance I_k is set to $B_k := |L_k| + X|U| + |\mathcal{S}| - k$.

As noted above, the bipartite graph constructed for instance I_k has $|V_k| = 2|U| + 2|\mathcal{S}| + 2 - k \leq 2(|U| + |\mathcal{S}|)$ vertices. Thus, $\alpha(|V_k|) \leq \alpha(2|U| + 2|\mathcal{S}|)$.

The main goal of the proof is to show that (i) if there exists a set cover of size k, then instance I_k of BCCMED has a solution with value at most k; (ii) any feasible solution for instance I_k of BCCMED with cost $C \leq \alpha(|V_k|)k$ can be used to obtain a set cover of size at most $2C$. Using these two properties of the reduction, we can show that any $\alpha(|V|)$-approximation to BCCMED transforms into a $\alpha(2|U| + 2|\mathcal{S}|)$-approximation for MIN SET COVER: Find the minimum value $k^* \in \{1, \ldots, n\}$ such that the hypothetical α-approximation algorithm A for BCCMED outputs a solution of cost at most $\alpha(|V_k|)k^*$ for instance I_{k^*}. By property (i) and the performance of A it follows that k^* is no greater than the optimum size set cover. By property (ii) we get a set cover of size at most $2\alpha(|V_{k^*}|)k^*$ which is at most $2\alpha(2|U| + 2|\mathcal{S}|)$ times the optimum size cover.

We first prove (i). Any set cover $\mathcal{C}$ of size k can be used to obtain a tree by choosing the subgraph induced by the set nodes corresponding to the sets in $\mathcal{C}$ and the root node r. Clearly, the cost of the tree is k. Since the sets in $\mathcal{C}$ form a cover, each element node is within distance of X from a vertex in the tree. Thus, the total median-cost of T is no more than $X|U| + |\mathcal{S}| - k + |L_k| = B_k$.

We now address (ii). Assume conversely, that T is a solution for I_k with value C, i.e., a tree with $\text{median}_d(T) \leq B_k$ and $c(T) = C \leq \alpha(|V_k|)k$. We first show that the root node r must be contained in the tree.

In fact, if this were not the case, then at least $|L_k| - 1$ nodes from L_k can not be in T either. Moreover, these at least $|L_k| - 1$ nodes are at distance at least two from any node in the tree. Moreover, since $X \geq C + 1$, the tree T can not contain any edge between set nodes and element nodes. Thus, T consists either of a single element node or T does not contain any element nodes. In the first case, the median cost of T is at least

$$2X(|U| - 1) + 3|L_k| = B_k + (|U| - 2)X + |U| + |\mathcal{S}| - k + 2 > B_k,$$

which contradicts that T is feasible. In the second case, (T does not contain element nodes) either T consists of a single node from L_k or does not contain any node from L_k. Thus, we get that the median cost of T is at least

$$X|U| + 2(|L_k| - 1) + 1 + 2 = X|U| + |L_k| + |U| + |\mathcal{S}| - k + 1 = B_k + 1 > B_k,$$

which is again a contradiction. (The additive terms in the above calculation stem from the fact that r is not in the tree and a set node or the remaining node from L_k is at distance at least two from the tree).

We now show that the collection $\mathcal{C}$ of set nodes spanned by the tree T can be used to obtain a cover of size at most $2C$. Let $\bar{U}_{\mathcal{C}} \subseteq U$ be the subset of element nodes not covered by the collection $\mathcal{C}$ of sets. For each element $u \in \bar{U}_{\mathcal{C}}$ its distance to any node in T is at least $X + 1$. The median cost of T thus satisfies:

$$\text{median}_d(T) \geq |L_k| + X|U| + |\bar{U}_{\mathcal{C}}| + |\mathcal{S}| - |T \cap (L_k \cup \mathcal{S})| = B_k + k + |\bar{U}_{\mathcal{C}}| - C. \quad (1)$$

On the other hand, since T is feasible, $\text{median}_d(T) \leq B_k$, and hence we get from (1) that $|\bar{U}_{\mathcal{C}}| \leq C - k < C$. In words, the number of elements left uncovered by the collection $\mathcal{C}$ of sets is at most the cost C of the tree T. Hence, we can augment $\mathcal{C}$ to a valid cover by adding at most $|\bar{U}_{\mathcal{C}}| \leq C$ sets. This leads to a set cover of size at most C.   □

The instances of MIN SET COVER used in [9] have the property that the number of sets is at most $|U|^5$, where U is the ground set (see [8] for an explicit computation of the number of sets used). Thus from Theorem 4 we get a lower bound for BCCMED of $(1/20 - \epsilon) \ln |V|$ (assuming that $\mathsf{NP} \not\subseteq \mathsf{DTIME}(N^{\log \log N})$). Since the number of sets in any instance of MIN SET COVER is bounded by 2^U, we can use the result from [2] to obtain a result under the weaker assumption that $\mathsf{P} \neq \mathsf{NP}$:

Theorem 5. *(i) Unless* NP $\subseteq$ DTIME($N^{\log\log N}$)*, for any $\epsilon > 0$ there can be no polynomial time approximation algorithm for* BCCMED *with a performance* $(1/20 - \epsilon)\ln n$.
(ii) Unless P$=$ NP*, for any $\epsilon > 0$ there is no approximation algorithm for* BCC-MED *with a performance of* $(1/8 - \epsilon)\ln\ln n$. □

5 Approximation Scheme on Trees

We first consider the problem BCCMED when restricted to trees. We present an FPAS for a slightly more general problem than BCCMED, called *generalized* BCCMED in the following: We are additionally given a subset $U \subseteq V$ of the vertex set and the budget constraint requires that the total service distance of all vertices in U (instead of V) does not exceed B.

Theorem 6. *There is a FPAS for the generalized* BCCMED *on trees with running time $\mathcal{O}(\log(nC)n^3/\epsilon^2)$, where C denotes the maximum c-weight of an edge in a given instance.*

Proof. Let $T = (V, E)$ be the tree given in the instance I of BCCMED. We root the tree at an arbitrary vertex $r \in V$. In the sequel we denote by T_v the subtree of T rooted at vertex $v \in V$. So $T_r = T$. Without loss of generality we can assume that r is contained in some optimal solution I (we can run our algorithm for all vertices as the root vertex). We can also assume without loss of generality that the rooted tree T is binary (since we can add zero cost edges and dummy nodes to turn it into a binary tree).

Let $T^* = (V^*, E^*)$ be an optimal solution for I which contains r. Denote by $\text{OPT} = c(T^*)$ its cost. Define $C := \max_{e \in E} c(e)$ and let $K \in [0, nC]$ be an integral value. The value K will act as "guess value" for the optimum cost in the final algorithm. Notice that the optimum cost is an integer between 0 and nC.

For a vertex $v \in V$ and an integer $k \in [0, K]$ we denote by $D[v, k]$ the minimum service cost of a tree $T^*_{v,k}$ servicing all nodes in U contained in the subtree T_v rooted at v and which has following properties: (1) $T^*_{v,k}$ contains v, and (2) $c(T^*_{v,k}) \leq k$. If no such tree exists, then we set $D[v, k] := +\infty$. Notice that

$$c(T^*) = \min\{\, k : D[r, k] \leq B \,\}.$$

Let $v \in V$ be arbitrary and let v_1, v_2 be its children in the rooted tree T. We show how to compute all the values $D[v, k]$, $1 \leq k \leq B$ given the values $D[v_i, \cdot]$, $i = 1, 2$.

For $i = 1, 2$ let $S_i := \sum_{w \in T_{v_i} \cap U} c(w, v)$. If v_i is not in the tree $T^*_{v,k}$ then none of the vertices in T_{v_i} can be contained in $T^*_{v,k}$. Let

$$X_k := S_1 + S_2$$

and

$$Y_k := \min\{\, D[v_1, k'] + D[v_2, k''] : k' + k'' = k - c(v, v_1) - c(v, v_2) \,\}.$$

Then we have that

$$D[v, k] = \min\{S_2 + D[v_1, k - c(v, v_1)], S_1 + D[v_2, k - c(v, v_1)], X_k, Y_k\}.$$

The first term in the last equation corresponds to the case that v_1 is in $T^*_{v,k}$ but not v_2. The second term is the symmetric case when v_2 is in the tree but not v_1. The third term concerns the case that none of v_1 and v_2 is in the tree. Finally, the fourth term models the case that both children are contained in $T^*_{v,k}$.

It is straightforward to see that this way all the values $D[v, k], 0 \le k \le K$ can be computed in $\mathcal{O}(K^2)$ time given the values for the children v_1 and v_2. Since the table values for each leaf of T can be computed in time $\mathcal{O}(K)$, the dynamic programming algorithm correctly finds an optimal solution within time $\mathcal{O}(nK^2)$.

Let $\epsilon > 0$ be a given accuracy requirement. Now consider the following test for a parameter $M \in [0, (n-1)C]$: First we scale all edge costs $c(e)$ in the graph by setting

$$c^M(e) := \left\lceil \frac{(n-1)c(e)}{M\epsilon} \right\rceil. \tag{2}$$

We then run the dynamic programming algorithm from above with the scaled edge costs and $K := (1 + 1/\epsilon)(n - 1)$. We call the test *successful* if the algorithm gives the information that $D[r, K] \le B$. Observe that the running time for one test is $\mathcal{O}(\frac{n^3}{\epsilon^2})$.

We now prove that the test is successful if $M \ge$ OPT. To this end we have to show that there exists a tree of cost at most K such that its service cost is at most B. Recall that T^* denotes an optimum solution. Since we have only scaled the c-weights, it follows that T^* is also a feasible solution for the scaled instance with service cost at most B. If $M \ge$ OPT we have

$$\sum_{e \in T^*} c^M(e) \le \sum_{e \in T^*} \left(\frac{(n-1)c(e)}{M\epsilon} + 1 \right) \le \frac{n-1}{\epsilon} + |T^*| \le \left(1 + \frac{1}{\epsilon}\right)(n-1).$$

Hence for $M \ge$ OPT, the test will be successful. We now use a binary search to find the minimum integer $M' \in [0, (n-1)C]$ such that the test described above succeeds. Our arguments from above show that the value M' found this way satisfies $M' \le$ OPT. Let T' be the corresponding tree found which has service cost no more than B. Then

$$\sum_{e \in T'} c(e) \le \frac{M'\epsilon}{n-1} \sum_{e \in T'} c^{M'}(e) \le \frac{M'\epsilon}{n-1}\left(1 + \frac{1}{\epsilon}\right)(n-1) \le (1+\epsilon)\text{OPT}. \tag{3}$$

Thus, the tree T' found by our algorithm has cost at most $1 + \epsilon$ times the optimum cost. The running time of the algorithm can be bounded as follows: We run $\mathcal{O}(\log(nC))$ tests on scaled instances, each of which needs time $\mathcal{O}(n^3/\epsilon^2)$ time. Thus, the total running time is $\mathcal{O}(\log(nC)n^3/\epsilon^2)$, which is bounded by a polynomial in the input size and $1/\epsilon$. □

6   Approximation Algorithm on General Graphs

In this section we use a Linear Programming relaxation in conjunction with filtering techniques (cf. [14]) to design an approximation algorithm. The techniques used in this section are similar to those given in [20] for the *Traveling Purchaser Problem*. The basic outline of our algorithm is as follows:

1. Formulate the problem as an *Integer Linear Program (IP)*.
2. Solve the Linear Programming relaxation (LP).
3. With the help of the optimal fractional solution define a *service-cluster* for each vertex. The goal is to service each vertex by one node from its cluster.
4. Solve a Group Steiner Tree problem on the clusters to obtain a tree.

Integer Linear Programming Formulation and Relaxation. In the following we assume again that there is one node r (the root) that must be included in the tree. This assumption is without loss of generality. Consider the following Integer Linear Program (IP) which we will show to be a relaxation of BCCMED. The meaning of the binary decision variables is as follows: $z_e = 1$ if and only if edge e is included in the tree; furthermore $x_{vw} = 1$ if and only if vertex w is serviced by v. The constraints (4) ensure that every vertex is serviced, constraint (5) enforces the budget-constraint on the service distance. Inequalities (6) are a relaxation of the connectivity and service requirements: For each vertex w and each subset S which does not contain the root r either w is serviced by a node in $V \setminus S$ (this is expressed by the first term) or there must be a an edge of T crossing the cut induced by S (this is expressed by the second term).

$$(\text{IP}) \qquad \min \sum_{e \in E} c(e) z_e$$

$$\sum_{v \in V} x_{vw} = 1 \qquad\qquad (w \in V) \qquad\qquad (4)$$

$$\sum_{v \in V} \sum_{w \in V} d(v,w) x_{vw} \leq D \qquad\qquad\qquad (5)$$

$$\sum_{v \notin S} x_{vw} + \sum_{v \in S, u \notin S} z_{vu} \geq 1 \qquad (w \in V, S \subset V, r \notin S) \quad (6)$$

$$z_e \in \{0,1\} \qquad\qquad (e \in E) \qquad\qquad (7)$$

$$x_{vw} \in \{0,1\} \qquad\qquad (v \in V, w \in V) \qquad\qquad (8)$$

The Linear Programming relaxation (LP) of (IP) is obtained by replacing the integrality constraints (7) and (8) by the constraints $z_e \in [0,1]$ ($e \in E$) and $x_{vw} \in [0,1]$ ($v \in V, w \in V$).

Lemma 1. *The relaxation (LP) of (IP) can be solved in polynomial time.*

Proof. We show that there is a polynomial time separation oracle for the constraints (6). Using the result from [11] implies the claim.

Suppose that (x, z) is a solution to be tested for satisfying the constraints (6) for a fixed w. We set up a complete graph with edge capacities z_{vu} $(u, v \in U)$. We then add a new node $\tilde{w}$ and edges $(\tilde{w}, v)$ of capacity x_{wv} for all $v \in V$. It is now easy to see that there exists a cut separating r and $\tilde{w}$ of capacity less than one if and only if constraints (6) are violated for w. □

Service Clusters and Group Steiner Tree Construction. Let $\epsilon > 0$. Denote by $(\hat{x}, \hat{z})$ the optimal fractional solution of (LP) and by $Z_{\text{LP}} := \sum_{e \in E} c(e)\hat{z}_e$ the optimal objective function value. For each vertex $w \in V$ define the value

$$D_w := \sum_{v \in V} d(v, w)\hat{x}_{vw}$$

and the subset (service cluster)

$$G_w(\epsilon) := \{ v \in V : d(v, w) \leq (1 + \epsilon)D_w \}.$$

The value D_w is the contribution of vertex w to the total service cost in the optimum fractional solution of the Linear Program. The set $G_w(\epsilon)$ consists of all those vertices that are "sufficiently close" to w.

Lemma 2. *For each $w \in V$ we have $\sum_{v \in G_w(\epsilon)} \hat{x}_{vw} \geq \epsilon/(1 + \epsilon)$.*

Proof. If the claim is false for $w \in V$ then we have $\sum_{v \notin G_w(\epsilon)} \hat{x}_{vw} > 1 - \epsilon/(1+\epsilon) = 1/(1 + \epsilon)$. Thus

$$D_w = \sum_{v \in V} d(v, w)\hat{x}_{vw} \geq \sum_{v \notin G_w(\epsilon)} d(v, w)\hat{x}_{vw} \geq (1 + \epsilon)D_w \sum_{v \notin G_w(\epsilon)} \hat{x}_{vw} > D_w.$$

This is a contradiction. Hence the claim must hold. □

The *Group Steiner Tree Problem* (GST) is defined as follows: Given a complete undirected graph $G = (V, E)$ with edge weights $c(e)$ $(e \in E)$ and a collection $G_1, \ldots, G_k$ of (not necessarily disjoint) subsets of V, find a subtree of G of minimum cost such that this tree contains at least one vertex from each of the groups $G_1, \ldots, G_k$. Charikar et al.[5] gave an approximation algorithm for GST with polylogarithmic performance guarantee. We will use this algorithm as a subroutine. Consider the instance of GST on the graph G given in the instance of BCCMED where the groups are the sets $G_w(\epsilon)$ $(w \in V)$, and the edge weights are the c-weights. This problem is formulated as an Integer Linear Program as follows:

$$(\text{GST}) \quad \min \sum_{e \in E} c(e)z_e$$

$$\sum_{v \in S, w \notin S} z_{vw} \geq 1 \quad (S \subset V, r \notin S, G_w(\epsilon) \subseteq S \text{ for some } w)$$

$$z_e \in \{0, 1\} \quad (e \in E)$$

The algorithm [5] finds a group Steiner tree of cost at most $\mathcal{O}(\log^3 m \log \log m)$ times $\sum_{e \in E} c(e) z_e^* = Z_{\text{LP-GST}}$, where z_e^* denotes the optimal fractional solution of the LP-relaxation (LP-GST) and $Z_{\text{LP-GST}}$ denotes its objective function value.

Lemma 3. *Denote by $Z_{LP\text{-}GST}$ and Z_{LP} the optimal values of the LP-relaxations of the Integer Linear Program (GST) and (IP), respectively. Then $Z_{LP\text{-}GST} \leq (1 + 1/\epsilon)Z_{LP}$.*

Proof. We show that the vector $\bar{z}$ defined by $\bar{z}_{vw} := (1 + 1/\epsilon)\hat{z}_{vw}$ is feasible for the LP-relaxation of (GST). This implies the claim of the lemma. To this end let S be an arbitrary subset such that $r \notin S$, $G_w(\epsilon) \subseteq S$ for some w and $\sum_{v \in S, w \notin S} z_{vw} < 1$. Since $(\hat{x}, \hat{z})$ is feasible for (LP), it satisfies constraint (6), i.e., $\sum_{v \notin S} \hat{x}_{vw} + \sum_{v \in S, w \notin S} \hat{z}_{vw} \geq 1$. Hence we get that

$$\sum_{v \in S, w \notin S} \hat{z}_{vw} \geq 1 - \sum_{v \notin S} \hat{x}_{vw} \geq 1 - \sum_{v \notin G_w(\epsilon)} \hat{x}_{vw} \qquad \text{(since } G_w(\epsilon) \subset S\text{)}$$

$$\geq 1 - \left(1 - \frac{\epsilon}{1 + \epsilon}\right) \qquad \text{(by Lemma 2)}$$

$$= \frac{\epsilon}{1 + \epsilon}.$$

Multiplying the above chain of inequalities by $1 + 1/\epsilon$ yields the claim. □

Hence we know that $Z_{\text{LP-GST}} \leq \left(1 + \frac{1}{\epsilon}\right) Z_{\text{LP}} \leq \left(1 + \frac{1}{\epsilon}\right)$ OPT. We can now use the algorithm from [5] to obtain a group Steiner tree. By the last chain of inequalities this tree is within a factor $(1 + 1/\epsilon)\mathcal{O}(\log^3 m \log \log m)$ of the optimal solution value for the instance of BCCMED while the budget constraint on the service distance is violated by a factor of at most $1 + \epsilon$:

Theorem 7. *For any fixed $\epsilon > 0$ there is a $(1 + \epsilon, (1 + 1/\epsilon)\mathcal{O}(\log^3 m \log \log m))$-approximation algorithm for BCCMED.* □

Acknowledgement. We thank Professor R. Ravi (Carnegie Mellon University) for his collaboration in early stages of this work.

References

1. E. M. Arkin and R. Hassin, *Approximation algorithms for the geometric covering salesman problem*, Discrete Applied Mathematics **55** (1994), 197–218.
2. S. Arora and M. Sudan, *Improved low-degree testing and its applications*, Proceedings of the 29th Annual ACM Symposium on the Theory of Computing, 1997, pp. 485–496.
3. B. Awerbuch, Y. Bartal, and A. Fiat, *Competitve distributed file allocation*, Proceedings of the 25th Annual ACM Symposium on the Theory of Computing, 1993, pp. 164–173.
4. Y. Bartal, A. Fiat, and Y. Rabani, *Competitive algorithms for distributed data management*, Journal of Computer and System Sciences **51** (1995), no. 3, 341–358.

5. M. Charikar, C. Chekuri, A. Goel, and S. Guha, *Rounding via tree: Deterministic approximation algorithms for group Steiner trees and k-median*, Proceedings of the 30th Annual ACM Symposium on the Theory of Computing, 1998, pp. 114–123.
6. J. T. Current and D. A. Schilling, *The covering salesman problem*, Transportation Science **23** (1989), 208–213.
7. L. W. Dowdy and D. V. Foster, *Comparative models of the file assignment problem*, ACM Computing Surveys **14** (1982), no. 2, 287–313.
8. S. Eidenbenz, Ch. Stamm, and P. Widmayer, *Positioning guards at fixed height above a terrain – an optimum inapproximability result*, Proceedings of the 6th Annual European Symposium on Algorithms, Lecture Notes in Computer Science, vol. 1461, Springer, 1998, pp. 187–198.
9. U. Feige, *A threshold of* ln n *for approximating set cover*, Proceedings of the 28th Annual ACM Symposium on the Theory of Computing, 1996, pp. 314–318.
10. M. R. Garey and D. S. Johnson, *Computers and intractability (a guide to the theory of NP-completeness)*, W.H. Freeman and Company, New York, 1979.
11. M. Grötschel, L. Lovász, and A. Schrijver, *Geometric algorithms and combinatorial optimization*, Springer-Verlag, Berlin Heidelberg, 1988.
12. K. Iwano, P. Raghavan, and H. Tamaki, *The traveling cameraman problem, with applications to automatic optical inspection*, Proceedings of the 5th International Symposium on Algorithms and Computation, Lecture Notes in Computer Science, vol. 834, Springer, 1994, pp. 29–37.
13. S. O. Krumke, *On the approximability of location and network design problems*, Ph.D. thesis, Lehrstuhl für Informatik I, Universität Würzburg, December 1996.
14. J. H. Lin and J. S. Vitter, *ε-approximations with minimum packing constraint violation*, Proceedings of the 24th Annual ACM Symposium on the Theory of Computing, 1992, pp. 771–781.
15. C. Lund, N. Reingold, J. Westbrook, and D. C. K. Yan, *On-line distributed data management*, Proceedings of the 2nd Annual European Symposium on Algorithms, Lecture Notes in Computer Science, vol. 855, Springer, 1994, pp. 202–214.
16. M. V. Marathe, R. Ravi, and R. Sundaram, *Service constrained network design problems*, Nordic Journal on Computing **3** (1996), no. 4, 367–387.
17. M. V. Marathe, R. Ravi, and R. Sundaram, *Improved results for service constrained network design problems*, Network Design: Connectivity and Facilities Location (P. M. Pardalos and D. Du, eds.), AMS-DIMACS Volume Series in Discrete Mathematics and Theoretical Computer Science, vol. 40, American Mathematical Society, 1998, pp. 269–276.
18. M. V. Marathe, R. Ravi, R. Sundaram, S. S. Ravi, D. J. Rosenkrantz, and H. B. Hunt III, *Bicriteria network design problems*, Journal of Algorithms **28** (1998), no. 1, 142–171.
19. C. Mata and J. B. Mitchell, *Approximation algorithms for geometric tour and network design problems*, Proceedings of the 11th Annual Symposium on Computational Geometry, ACM Press, June 1995, pp. 360–369.
20. R. Ravi and F. S. Salman, *Approximation algorithms for the traveling purchaser problem and itsvariants in network design*, Proceedings of the 7th Annual European Symposium on Algorithms, Lecture Notes in Computer Science, vol. 1643, Springer, 1999, pp. 29–40.
21. S. Voss, *Designing special communication networks with the traveling purchaser problem*, Proceedings of the FIRST ORSA Telecommuncation Conference, 1990, pp. 106–110.
22. O. Wolfson and A. Milo, *The multicast policy and its relationship to replicated data placement*, ACM Transactions on Database Systems **16** (1991), no. 1, 181–205.

Coloring Mixed Hypertrees

Daniel Král'[1,*], Jan Kratochvíl[2,**], Andrzej Proskurowski[3,***], and
Heinz-Jürgen Voss[4]

[1] Department of Applied Mathematics, Charles University, Malostranské nám. 25,
118 00 Prague, Czech Republic. kral@kam.ms.mff.cuni.cz
[2] Department of Applied Mathematics and Institute for Theoretical Computer
Science, Charles University, Malostranské nám. 25, 118 00 Prague, Czech Republic.
honza@kam.ms.mff.cuni.cz
[3] Department of Computer and Information Science, University of Oregon, Eugene,
USA. andrzej@cs.uoregon.edu
[4] Institute of Algebra, Technische Universität Dresden, Germany.
voss@math.tu-dresden.de

Abstract. A mixed hypergraph is a hypergraph with edges classified as
of type 1 or type 2. A vertex coloring is strict if no edge of type 1 is totally multicolored, and no edge of type 2 monochromatic. The chromatic
spectrum of a mixed hypergraph is the set of integers k for which there
exists a strict coloring using exactly k different colors. A mixed hypertree
is a mixed hypergraph in which every hyperedge induces a subtree of the
given underlying tree. We prove that mixed hypertrees have continuous
spectra (unlike general hypergraphs, whose spectra may contain gaps [cf.
Jiang et al.: The chromatic spectrum of mixed hypergraphs, submitted].
We prove that determining the upper chromatic number (the maximum
of the spectrum) of mixed hypertrees is NP-hard, and we identify several
polynomially solvable classes of instances of the problem.

1 Introduction

Definition 1. *A* mixed hypergraph *is a triple $H = (V, \mathcal{C}, \mathcal{D})$, where V is a set of
vertices and $\mathcal{C}, \mathcal{D}$ are sets of hyperedges (hyperedges are subsets of the vertex set).
A vertex coloring of a mixed hypergraph is* **strict** *if every edge $e \in \mathcal{C}$ contains
two vertices of the same color, and every edge $e \in \mathcal{D}$ contains two vertices of
different colors. A strict coloring that uses exactly k distinct colors is called a
strict k-coloring. The* chromatic spectrum *of H is the set $Sp(H)$ of integers k
such that H has a strict k-coloring. The spectrum of H is called* continuous *if it
is an interval (in the set of integers). We denote $\overline{\chi}(H) = \max Sp(H)$ the* upper
chromatic number *of H.*

---

[*] Research supported in part by Czech Research Grant GaČR 201/1999/0242.
[**] This author acknowledges partial support of Czech research grants GAUK 158/1999
and KONTAKT 338/99. Part of the research was carried on while visiting Technical
University Dresden in June 1999 and University of Oregon in the fall of 1999.
[***] Research supported in part by National Science Foundation grants NSF-INT-
9802416 and NSF-ANI-9977524.

U. Brandes and D. Wagner (Eds.): WG 2000, LNCS 1928, pp. 279–289, 2000.
© Springer-Verlag Berlin Heidelberg 2000

The notion of mixed hypergraphs was introduced by Voloshin [10]. The concept is steadily gaining more interest both for potential applications and interesting theoretical results. Applications to list colorings, integer programming, scheduling and molecular biology can be found in [11,9]. Mixed colorings of block designs of various types were considered in [3,7,6,8]. Other coloring problems studied from a different point of view can be rephrased in terms of mixed hypergraphs [5].

Solving a long-standing open problem, Jiang et al. [4] showed that there are mixed hypergraphs whose spectra have gaps, i.e., are not continuous. Moreover, for every finite set of integers greater than 1 there exists a mixed hypergraph whose chromatic spectrum coincides with the given set. They also showed that spectra of interval hypergraphs are continuous. (A hypergraph is *interval* if its vertex set allows a linear ordering such that every edge is an interval in this ordering.) In this paper, we extend the former result to the class of hypertrees.

Definition 2. *A mixed hypertree is a mixed hypergraph $H = (V, \mathcal{C}, \mathcal{D})$ such that there exists a tree T with vertex set $V(T) = V$ and such that every edge of H induces a subtree in T.*

Without loss of generality, we assume that every edge of our hypertree has size at least 2, and every $\mathcal{C}$-edge has size at least three (the endpoints of a $\mathcal{C}$-edge of size two must be colored by the same color and we can contract this edge). We also assume that the hypertree has at least one edge. Every 2-coloring of the underlying tree (i.e., a bipartition of T) is then a strict coloring of H, as every $\mathcal{D}$-edge contains two vertices adjacent in T (and hence colored by different colors), and every $\mathcal{C}$-edge contains at least 3 vertices (and hence at least 2 vertices of the same color by the pigeon-hole principle). Thus we have the following:

Observation 1. *For every hypertree H, 2 belongs to the chromatic spectrum of H and is its minimum member, unless H has no $\mathcal{D}$-edges.*

Our aim is to prove the following theorem:

Theorem 1. *The chromatic spectrum of any mixed hypertree is continuous, in particular $Sp(H) = [2, \overline{\chi}(H)]$ if H contains a $\mathcal{D}$-edge, and $Sp(H) = [1, \overline{\chi}(H)]$ otherwise.*

This theorem is proved in the next section. Later, we address the algorithmic questions. In general, the questions of deciding if a mixed hypergraph is strict k-colorable, or whether its upper chromatic number has a given value may be of different complexity; e.g., if H has no $\mathcal{C}$ edges then $\overline{\chi}(H) = n$ while it is NP-complete to decide whether $2 \in Sp(H)$ (the problem of bicolorability of ordinary hypergraphs). Due to Theorem 1, these two problems are equally difficult for mixed hypertrees:

Proposition 1. *The problem of determining the upper chromatic number of a mixed hypertree and the problem of deciding $k \in Sp(H)$ for an input integer k, are polynomially equivalent.*

Proof. Given $k \geq 2$, this k belongs to $Sp(H)$ if and only if $k \leq \overline{\chi}(H)$. On the other hand, the upper chromatic number can be found by deciding for every $k \leq n = |V|$ whether H is strict k-colorable. ☐

2   Chromatic Spectra of Mixed Hypertrees

First, we show that a minimal counterexample to Theorem 1 (if it existed) would have $\mathcal{D} = E(T)$, i.e. the edges of the underlying tree, $E(T)$, would be precisely the $\mathcal{D}$-edges of H.

Lemma 1. *Let H be a hypertree with disconnected spectrum and with minimum possible number of vertices. Then every tree edge xy is a subset of some $\mathcal{D}$-edge of H.*

Proof. Suppose H has 2 vertices, say x, y, adjacent in T which do not belong together to the same $\mathcal{D}$-edge of H. Let T_1 be the largest subtree of T which contains x but not y, and let T_2 be the largest subtree containing y but not x. Let H_i, $i = 1, 2$, be the hypertree induced by T_i, i.e., H_i contains exactly those edges of H that consist only from vertices of T_i. Both H_1 and H_2 are smaller than H and hence have continuous spectra.

Denote $c = \overline{\chi}(H)$ and consider a c-coloring f of H. Then f restricted to H_1 is a strict coloring and uses some $c_1 \leq c$ colors, and similarly, f restricted to H_2 uses some c_2 colors. The union of these two color sets is the set of c colors used on H, and hence

$$c_1 + c_2 \geq c.$$

For $i = 1, 2$ and for every t_i, $\min Sp(H_i) \leq t_i \leq c_i$, consider a t_i-coloring f_i of H_i. If necessary, rename and permute the colors so that $f_1(x) = f_2(y)$ and so that this is the only color that f_1 and f_2 have in common. Regard the union of f_1 and f_2 as coloring g of the entire H. This coloring uses $t_1 + t_2 - 1$ colors, and it is a strict coloring of H: every $\mathcal{D}$-edge of H lies entirely within H_1 or within H_2, and hence cannot be monochromatic; no $\mathcal{C}$-edge that lies totally within H_1 or within H_2 is totally multicolored since it was presented and hence well colored in H_1 or H_2, and all the remaining $\mathcal{C}$-edges are well colored because they contain both x and y (and $g(x) = g(y)$).

If we let t_1 and t_2 range over $Sp(H_1)$ and $Sp(H_2)$, respectively, we obtain strict t-colorings of H at least in the range $3 \leq t \leq c_1 + c_2 - 1$. Since $c_1 + c_2 - 1 \geq c - 1$, this covers the range $[3, c - 1]$. But H has a strict 2-coloring (Lemma 1), and a strict k-coloring by the definition of $c = \overline{\chi}(H)$. Hence H has a continuous spectrum, contradicting the assumption. ☐

We define the *size* of H as the sum of the sizes of its edges, i.e., $s(H) = \sum_{e \in \mathcal{C} \cup \mathcal{D}} |e|$.

Lemma 2. *Let H be a hypertree with disconnected spectrum, and of minimum possible size. Then all $\mathcal{D}$-edges have size 2.*

Proof. Suppose H has a $\mathcal{D}$-edge e of size greater than 2. Denote $c = \overline{\chi}(H)$, and fix a strict c-coloring f of H. The edge e contains 2 vertices, say x and y, such that $f(x) \neq f(y)$ and such that x and y are adjacent in the underlying tree T (otherwise e would be monochromatic in f). Consider $H' = (V, \mathcal{C}, (\mathcal{D} \setminus e) \cup \{\{x, y\}\})$, i.e., replace e by the pair x, y. The resulting hypergraph H' is a hypertree again, because xy is an edge of T. Since f is a strict coloring of H' as well, $c \in Sp(H')$. And since $s(H') < s(H)$, and H was a minimum hypertree with disconnected spectrum, the spectrum of H' is continuous, i.e., for every $t, 2 \leq t \leq c$, H' has a strict t-coloring, which is also a strict t-coloring of H. This contradicts the assumption of H having a disconnected spectrum. $\qquad\square$

Corollary 1. *Let H be hypertree with disconnected spectrum, with minimum possible number of vertices, and with minimum possible size. Then $\mathcal{D} = E(T)$.*

We therefore restrict our attention to mixed hypertrees with $\mathcal{D} = E(T)$, and hence use the notation $H = (T, \mathcal{C}, \mathcal{D} = E(T))$. Our goal is to show that all such hypertrees have continuous spetra. That would prove Theorem 1, since if there were a counterexample, then a minimum counterexample would satisfy $\mathcal{D} = E(T)$. In order to prove this, we introduce a seemingly more general problem of finding strict colorings of *precolored* hypertrees. The input of this problem consists of a mixed hypertree $H = (T, \mathcal{C}, E(T))$ and k disjoint independent sets $S_1, S_2, \ldots, S_k$ of vertices, and the task is to find a strict coloring which colors all vertices of S_i with color b_i, for each $i = 1, 2, \ldots, k$. Such a coloring will be called a *precoloring extension*. (Here and later on, we refer to a set of vertices as an independent set if it is independent in the underlying tree T.)

Lemma 3. *Let H be a hypertree and let $S_1, S_2, \ldots, S_k$ be disjoint nonempty independent sets of vertices of H precolored by $b_1, b_2, \ldots, b_k$, respectively. If H has a precoloring extension using $c \geq k + 2$ colors, then it has a precoloring extension which uses exactly $k + 2$ colors.*

Proof. Let the precoloring extension of H and $S_1, \ldots, S_k$ which uses c colors be f. For every $\mathcal{C}$-edge e, fix two different vertices $x_e, y_e \in e$ such that $f(x_e) = f(y_e)$ (every $\mathcal{C}$-edge contains at least one pair of such vertices). Replace every $\mathcal{C}$-edge e by the $\mathcal{C}$-edge e' consisting of all vertices of the x_e, y_e-path in T. For the resulting hypertree H', f is still a strict precoloring extension, and vice versa, every precoloring extension for H' (and $S_1, \ldots, S_k$) is also a strict precoloring extension for H (and $S_1, \ldots, S_k$). Note that in H', every $\mathcal{C}$-edge induces a path in the underlying tree, and therefore it makes sense talking about end-vertices of $\mathcal{C}$-edges.

Now for H', enlarge the sets S_i (if necessary) according to the following procedure: If there is a $\mathcal{C}$-edge e' which starts in a vertex $x_e \in S_i$ for some i, and ends in a vertex $y_e \notin S_i$, then add y_e into S_i. Repeat this step until no such edge exists. Note that the new S_i's are still disjoint after this procedure is finished, since we have only added vertices of color b_i into the set S_i.

For the sake of brevity write

$$S = \bigcup_{i=1}^{k} S_i.$$

Next, we define the set A of auxiliary two-element edges, which contains the pairs $x_e y_e$, $e \in \mathcal{C}$ such that $e \setminus \{x_e, y_e\} \subseteq S$.

Define the auxiliary graph G as the graph obtained from T by deleting S and then contracting all edges of A by collapsing their end-vertices. We claim that G has at least two vertices and is acyclic. Indeed, since the coloring f uses at least two colors other than $b_1 \dots b_k$, these two additional colors must remain in G, since vertices of different colors cannot be collapsed. Also, suppose G has a cycle. This cycle corresponds to a cycle in $T + A$ which contains at least three edges of T. Replacing every edge in A by the corresponding $(x_e y_e)$-path in T, we get a closed walk in T which traverses these (at least) three edges each only once, a contradiction with acyclicity of T.

Since G is acyclic, it can be colored by two colors, say colors b_{k+1} and b_{k+2}. And since G has at least two vertices, we can take such coloring which actually uses both colors. We claim that this coloring yields a strict precoloring extension of H' and $S_1, \dots, S_k$, and it obviously uses exactly $k + 2$ colors.

This coloring is strict on all $\mathcal{D}$-edges (i.e., edges of the tree T) by its definition. We will show that every $\mathcal{C}$-edge e' contains two vertices of the same color:

Case 1: e' has at least one end-vertex in S: Then e' has both end-vertices in S by the enlargment procedure, and these endpoints are in the same S_i, i.e., they are colored by the same color.

Case 2: e' has both endpoints in $T \setminus S$ and all other vertices in S: Then $x_e y_e \in A$, x_e and y_e are unified into one vertex of G, and so x_e and y_e get the same color.

Case 3: e' has at least 3 vertices in $T \setminus S$: Then these 3 vertices are colored by colors b_{k+1} and b_{k+2} and two of them must get the same color.   □

Lemma 4. *Let H be a mixed hypertree and let $S_1, S_2, \dots, S_k$ be disjoint nonempty independent sets of vertices of H. If H has a precoloring extension using $c \geq k + 2$ colors, then it has precoloring extensions using exactly t colors for every t, $k + 2 \leq t \leq c$.*

Proof. We will prove the statement by induction on the number of vertices in $T \setminus S$ (again $S = \bigcup_{i=1}^{k} S_i$). Note also that the statement holds trivially true due to Lemma 3 if $c = k + 2$, and so we assume that $c > k + 2$ further on.

Let the precoloring extension of H and $S_1, \dots, S_k$ which uses c colors be f. Replace every $\mathcal{C}$-edge e by e' as in the preceding proof (e' is a x_e, y_e-path in T and $f(x_e) = f(y_e)$). Enlarge the sets S_i as in the preceding proof (if $x_e \in S_i$ and $y_e \notin S$ then add y_e into S_i and repeat). Denote S_i' the enlarged set S_i, $i = 1, \dots, k$, and $S' = \bigcup_{i=1}^{k} S_i'$. Let H' be the hypertree obtained by replacing each e by e'. Again, f is a strict precoloring extension of H' (and S')

using exactly k colors, and every precoloring extension of H' (and S') is also a precoloring extension of H (and S).

H contains at least one vertex not belonging to S' (e.g., the vertex colored by a color different from $b_1, \ldots, b_k$). Choose a vertex not belonging to S', call it x, such that all other vertices in $V \setminus S'$ lie in the same connected component of $T - \{x\}$. Let T_2 be this connected component and let $T_1 = T \setminus T_2$ be the subtree rooted in x that is disjoint with T_2. Finally, let y be the neighbor of x in T_2.

Now let C be the set of endpoints of $\mathcal{C}$-edges e' that have x as the other endpoint, formally

$$ C = \{y_e : x_e = x, e \in \mathcal{C}\} \cup \{x_e : y_e = x, e \in \mathcal{C}\}. $$

We distinguish three cases:

Case 1: $f(x) = b_i$ for some i. Then redefine $S_i' := S_i' \cup C \cup \{x\}$. The hypertree H' has a strict precoloring extension for this new $S_1', \ldots, S_k'$ (namely f) and has less unprecolored vertices than H (with precoloring $S_1, \ldots, S_k$). Hence by induction hypothesis, H' has precoloring extensions using exactly t colors for every $k+2 \le t \le c$. Each of these extensions is also a strict precoloring extension for H and $S_1, \ldots, S_k$.

Case 2: $f(x) \ne b_i$ for all $i = 1, 2, \ldots, k$ and $C = \emptyset$. Consider H_2 as the hypertree on T_2 with all edges e' that lie entirely within T_2 as the $\mathcal{C}$-edges of H_2. All the $\mathcal{C}$-edges e' not entirely lying in T_2 will be colored properly, since their both end points are in S_i' for the same i and thus the following construction assign their ends the same color (b_i). Set $S_i'' = S_i' \cap T_2$ and say that k' of these are nonempty. The hypertree H_2 has less unprecolored vertices than H (at least x is missing) and we may use induction hypothesis. The coloring f restricted to H_2 is a strict precoloring extension and it uses at least $k' + c - k - 1$ colors (all the new colors used by f are used in T_2 except possibly for the color of x). But since $c - k \ge 3$, $k' + c - k - 1 \ge k' + 2$ as we need for Lemma 3. It follows that H_2 has a precoloring extension which uses exactly t colors for every t, $k' + 2 \le t \le k' + c - k - 1$. Each such coloring g' can be re-extended to a coloring g of H by giving x the color $g(x) = b_{k+2}$ if $g'(y) = b_{k+1}$ and setting $g(x) = b_{k+1}$ otherwise. Such a coloring g then uses $t + k - k'$ colors, and this number ranges from $k + 2$ (for $t = k' + 2$) to $c - 1$ (for $t = k' + c - k - 1$). This is what we needed, as a precoloring extension using c colors exists by the assumption itself.

Case 3: $f(x) \ne b_i$ for all $i = 1, 2, \ldots, k$ and $C \ne \emptyset$. Then $C \subset T_2$ and $C \cap S' = \emptyset$. Consider H_2 as the hypertree on T_2 with all edges e' that lie entirely within T_2 as the $\mathcal{C}$-edges of H_2; the other $\mathcal{C}$-edges will be colored properly for the same reason as in the previous case. Set $S_i'' = S_i' \cap T_2$, $S_{k+1}' = C \cup \{x\}$ and let k' of the sets S_i'', $i \le k$ be nonempty. The hypertree H_2 has less unprecolored vertices than H (at least x is missing), but now $k' + 1 \le k + 1$ is the number of colors used for the precoloring. By induction hypothesis, since H_2 has a strict precoloring extension (namely f restricted to H_2) which uses exactly $c - k + k' \ge k' + 3$ colors, H_2 has a precoloring extension which uses exactly t colors for every t, $k' + 3 \le t \le k' + c - k$. Each such coloring g' can be re-extended to a coloring g of H by giving x the color $g(x) = b_{k+2}$ if $g'(y) = b_{k+1}$ and setting $g(x) = b_{k+1}$

otherwise (note that in the former case all $\mathcal{C}$ edges starting in x and ending in C are well colored because they contain vertex $y \in T_2$ of the same color b_{k+1} as the endpoint in C). Such a coloring g then uses $t + k - k'$ colors, and this number ranges from $k + 3$ (for $t = k' + 3$) to c (for $t = k' + c - k$). This is what we needed, since a precoloring extension using $k + 2$ colors exists by Lemma 3.   □

For the proof of Theorem 1, use Lemma 4 with $k = 0$ and $c = \overline{\chi}(H)$.

3   Dynamic Programming Algorithm

One might expect that deciding strict k-colorability of mixed hypertrees should be solvable in polynomial time due to the structure of the underlying tree. We will see in the next section that this is not true (unless P=NP). In this section we study the more or less straightforward dynamic programming algorithm to decide the existence of a strict coloring of a mixed hypertree H. Henceforth, n will denote the number of vertices of H, k the number of colors, s the maximum size of a hyperedge and l_e (l_v) the maximum H-load of an edge (a vertex) of the underlying tree T, where the load of an edge (a vertex) is the number of hyperedges that include both of its end points (the vertex itself).

Consider a given mixed hypertree $H = (T, \mathcal{C}, \mathcal{D})$ on an underlying tree T rooted in a leaf $r \in V(T)$. Let $e = \{u, v\}$ be an edge of the underlying tree T with v being closer to r than u (in T). The hypergraph induced by the vertices of T in the same connected component of $T - \{v\}$ as u will be denoted by H_e. (This subhypertree contains exactly those hyperedges of H that have all vertices in H_e.) Let $e_1, e_2, \ldots, e_m$ be the hyperedges containing v and not fully contained in H_e. We maintain boolean arrays $\Phi^e(a_1, a_2, b_1, b_2, \ldots, b_m, c)$ indexed by $a_1, a_2 \in \{1, 2, \ldots, k\}$, $b_1, \ldots, b_m \in \{0, 1, \ldots, k\}$ and $c \in \{1, 2, \ldots, k\}$ such that

$$\Phi^e(a_1, a_2, b_1, b_2, \ldots, b_m, c) = \text{true}$$

if and only if H_e allows a strict coloring $\phi : \longrightarrow \{1, 2, \ldots, k\}$ such that

1. $\phi(v) = a_1, \phi(u) = a_2$,
2. for every $j = 1, 2, \ldots, m$, if $b_j > 0$ then $e_j \cap V(H_e)$ contains a vertex of color b_j, and if $b_j = 0$ then e_j is satisfied within $V(H_e)$, i.e., $e_j \cap V(H_e)$ contains two vertices of the same color if $e_j \in \mathcal{C}$ and $e_j \cap V(H_e)$ contains two vertices of different colors if $e_j \in \mathcal{D}$,
3. ϕ uses exactly c colors on $V(H_e)$.

The space needed to maintain these arrays is $O(nk^{3+l_e})$. It is also clear that H allows a strict coloring using exactly k colors if and only if

$$\Phi^{r_e}(a_1, a_2, k) = \text{true}$$

for some a_1 and a_2 where r_e is the tree-edge containing the root r (note that $m = 0$ in the case of the root since all hyperedges are fully contained in $H_{r_e} = H$).

The initialization of the arrays Φ for the leaves of T is obvious. We will only add hints how to update the arrays. Straightforward bottom-up strategy is used: consider for vertex x with children $y_1, y_2, \ldots, y_g$, a true entry of the arrays Φ^{xy_i} for each i and derive the information for Φ^{zx}, where z is the parent of x. This would be, however, exponential in g, the number of children of x, i.e., exponential in d. Our aim is to be a little more careful here. Let $e_1, \ldots, e_m$ be the hyperedges that contain x. We introduce another array Ψ_j^x, $1 \leq j \leq g$, such that

$$\Psi_j^x(a_1, a_2, b_1, \ldots, b_m, c) = \text{true}$$

if and only if the hypertree H_j^x induced by $\{z\} \cup \bigcup_{i=1}^j H_{xy_i}$ allows a strict coloring $\psi :\longrightarrow \{1, 2, \ldots, k\}$ such that

1. $\psi(z) = a_1, \phi(x) = a_2$,
2. for every $h = 1, 2, \ldots, m$, if $b_h > 0$ then $e_h \cap V(H_j^x)$ contains a vertex of color b_h, and if $b_h = 0$ then e_h is satisfied within $V(H_j^x)$,
3. ψ uses exactly c colors on $V(H_j^x)$.

The "horizontal" recursion step $j \to j+1$ consists of combining true entries of Ψ_j^x with a true entry of $\Phi^{xy_{j+1}}$, which takes $O(k^{3+l_e}).O(k^{2+l_v}).O(2^{l_v}).O(k) = O(2^{l_v}k^{6+l_e+l_v})$ time, since there are at most $O(k^{3+l_e})$ true entries $\Phi^{xy_{j+1}}$, combined with at most $O(k^{2+l_v})$ true entries Ψ_j^x (we only consider those that use the same color for x), and for this choice of two entries we check all hyperedges containing x. If a hyperedge is fully contained in H_j^x we check if it is satisfied in $\Phi^{xy_{j+1}}$ or in Ψ_j^x or by their combination. For hyperedges that are not fully contained in H_j^x, we choose which color will be propagated (henceforth the factor $O(2^{l_v})$). Finally, we compare the colors appearing in $\Phi^{xy_{j+1}}$ and in Ψ_j^x. Say $\Phi^{xy_{j+1}}$ explicitly names k_1 colors, Ψ_j^x names k_2 colors, and k_3 of these are in common. In total, $\Phi^{xy_{j+1}}$ uses c_1 colors and Ψ_j^x uses c_2 colors. The anonymous $c_1 - k_1$ colors may be those from $k_2 - k_3$ explicit colors for Ψ_j^x, but some or all of them may be completely new. This consideration gives an interval of possible numbers of colors used in Ψ_{j+1}^x, and the range of the interval can be determined in time linear in k. For all these values of parameters we set Ψ_{j+1}^x to true.

Skipping the tedious details, we have sketched the argument for the following proposition:

Theorem 2. *It can be decided if H allows a k-strict coloring in time $O(n \cdot 2^{l_v}k^{6+l_e+l_v}))$.*

A straightforward corollary reads that the upper chromatic number of a mixed hypertree can be determined in polynomial time if the maximum vertex load l_v is bounded.

4   Upper Chromatic Number is NP-Hard

Theorem 3. *It is NP-complete to decide if $\overline{\chi}(H) \geq \frac{2}{3}(n+2)$ even for hypertrees $H = (V, \mathcal{C}, E(T))$ with maximum edge load 4 and hyperedge size at most 7, where n is the number of the vertices of H.*

Proof. Given a CNF formula Φ with the set C of clauses over a set X of variables, define a tree T with $3m + 1$ vertices (here $m = |X|$) by

$$V(T) = \{r\} \cup X \cup \{x^+ : x \in X\} \cup \{x^- : x \in X\}$$

and edges

$$E(T) = \{x^+x, x^-x, xr : x \in X\}$$

and define a mixed hypergraph H on the same vertex set by

$$\mathcal{D} = E(T)$$

$$\mathcal{C} = \{\{x, x^+, x^-, r\} : x \in X\} \cup \{C_c : c \in C\}$$

where the set C_c is the smallest connected subtree of T containing the literals of the clause c (x^+ is the literal for positive occurrence of x and x^- is the literal for negated occurrence).

We claim that H allows a $(2m+1)$-strict coloring if and only if Φ is satisfiable. Suppose first that $f : V(T) \longrightarrow \{1, 2, \ldots, 2m + 1\}$ is a strict coloring. For every variable x, the $\mathcal{C}$ edge $\{x, x^+, x^-, r\}$ implies that at least one of the following three cases occurs:

$$f(x^+) = f(r)$$
$$f(x^-) = f(r)$$
$$f(x^+) = f(x^-)$$

and to have the total number of colors $2m + 1$, exactly one of these three equalities holds true and no other equalities among the colors hold whatsoever, in particular, for every variable x

$$f(x^+) = f(x^-) \rightarrow f(x^+) \neq f(r) \neq f(x^-).$$

Define a truth valuation ϕ by

$$\phi(x) = \begin{cases} \text{true} & \text{if } f(r) = f(x^+) \\ \text{false} & \text{if } f(r) = f(x^-) \end{cases}$$

(some variables may not be defined any value but none is set both true and false). The only possibility for an edge C_c to contain two vertices of the same color is that it contains a literal of the same color as r, i.e., every clause c contains a positive literal x such that $f(x^+) = f(r)$ (and hence $\phi(x) = \text{true}$) or it contains a negated literal $\neg y$ such that $f(y^-) = f(r)$ (and hence $\phi(y) = \text{false}$) and $\neg y$ satisfies c.

If Φ is satisfiable then a coloring using $2m+1$ colors can be constructed along the lines above.

Finally for the claim about edge loads, start with a formula such that every clause contains two or three literals and every veriable occurs in at most three clauses. Such restricted SAT problem is well known to be NP-complete.   □

The following theorem states that the hardness of the decision problem of the upper chromatic number for mixed hypertrees is caused by $\mathcal{C}$-edges.

Theorem 4. *It is NP-complete to decide if* $\overline{\chi}(H) \geq k$ *even for hypertrees* $H = (V, \mathcal{C}, \emptyset)$.

Proof. Given a graph G containing n vertices, define a tree T with $n+1$ vertices (let w be a new vertex not contained in $V(G)$) by

$$V(T) = V(G) \cup \{w\}$$

and edges

$$E(T) = \{vw : v \in V(G)\}$$

and define a mixed hypertree H on the same vertex set by

$$\mathcal{C} = \{\{uwv\} : uv \in E(G)\}$$

$$\mathcal{D} = \emptyset$$

We claim that $\overline{\chi}(H) = \alpha(G) + 1$ and thus this construction yields a reduction from the well known NP-complete problem INDEPENDENT SET.

Let f be a coloring of H using $k+1$ colors. Let $v_1, \ldots, v_k$ be vertices of H such that $f(v_i) \neq f(v_j)$ for all $i \neq j$ and $f(v_i) \neq f(w)$ for all $1 \leq i \leq k$. If there were an edge $v_i v_j$, then the edge $\{v_i w v_j\} \in \mathcal{C}$ would not be colored properly. Thus the vertices $v_1, \ldots, v_k$ create an independent set of G and $\alpha(G) \geq k$.

On the other hand, let $v_1, \ldots, v_k$ be an independent set of G and $b_1, \ldots, b_{k+1}$ be $k+1$ different colors. Define a coloring f of H by $f(v_i) = b_i$, $1 \leq i \leq k$, and $f(v) = b_{k+1}$ for all $v \in V(H) \setminus \{v_1, \ldots, v_k\}$. This coloring is strict, since for every edge $uwv \in \mathcal{C}$ either u or v is different from all v_i, $1 \leq i \leq k$ (otherwise $v_1, \ldots, v_k$ would not be an independent set), and thus its color is the same as the color of w. This implies that $\overline{\chi}(H) \geq k+1$. We conclude that $\alpha(G) + 1 = \overline{\chi}(H)$. □

5   Conclusion

Several questions concerning the computational complexity remain open. E.g., we have shown that strict colorability is polynomial if the loads of vertices are bounded, while the problem is NP-complete for unbounded vertex loads. The complexity of the case when the number of colors is fixed (but the load of vertices unbounded) is presently open.

Another most probably hard problem is to decide if the chromatic spectrum of a given mixed hypergraph is continuous.

References

1. R. Ahlswede, Coloring hypergraphs: a new approach to multi-user source coding, I and II, *Journal of Combinatorics, Information, and System Sciences* **4** (1979), 76–115, and **5** (1980), 220–268.
2. E. Bulgaru and V. Voloshin, Mixed interval hypergraphs, *Discrete Applied Math.* **77** (1997), 24–41.
3. T. Etzion and A. Hartman, Towards a large set of Steiner quadruple systems, *SIAM J. Discrete Math.* **4** (1991), 182–195.
4. T. Jiang, D. Mubayi, Zs. Tuza, V. Voloshin, D.B. West, The chromatic spectrum of mixed hypergraphs, submitted
5. H. Lefmann, V. Rödl, and R. Thomas, Monochromatic vs. multicolored paths, *Graphs Combin.* **8** (1992), 323–332.
6. L. Milazzo, On upper chromatic number for SQS(10) and SQS(16), *Le Matematiche* **L** (Catania, 1995), 179–193.
7. L. Milazzo and Zs. Tuza, Upper chromatic number of Steiner triple and quadruple systems, *Discrete Math.* **174** (1997), 247–259.
8. L. Milazzo and Zs. Tuza, Strict colorings for classes of Steiner triple systems, *Discrete Math.* **182** (1998), 233–243.
9. Zs. Tuza and V. Voloshin, Uncolorable mixed hypergraphs, *Discrete Applied Math.*, (to appear).
10. V. Voloshin, On the upper chromatic number of a hypergraph, *Australasian J. Comb.* **11** (1995), 25–45.
11. V. Voloshin, Mixed hypergraphs as models for real problems, (in preparation).

A Linear-Time Algorithm to Find Independent Spanning Trees in Maximal Planar Graphs

Sayaka Nagai and Shin-ichi Nakano

Department of Computer Science, Gunma University, Kiryu 376-8515, Japan.
nakano@cs.gunma-u.ac.jp

Abstract. Given a graph G, a designated vertex r and a natural number k, we wish to find k "independent" spanning trees of G rooted at r, that is, k spanning trees such that, for any vertex v, the k paths connecting r and v in the k trees are internally disjoint in G. In this paper we give a linear-time algorithm to find k independent spanning trees in a k-connected maximal planar graph rooted at any designated vertex.

1 Introduction

Given a graph $G = (V, E)$, a designated vertex $r \in V$ and a natural number k, we wish to find k spanning trees $T_1, T_2, \cdots, T_k$ of G such that, for any vertex v, the k paths connecting r and v in $T_1, T_2, \cdots, T_k$ are internally disjoint in G, that is, any two of them have no common intermediate vertices. Such k trees are called k *independent spanning trees of G rooted at r*. Five independent spanning trees are drawn in Fig. 1 by thick lines. Independent spanning trees have applications to fault-tolerant protocols in networks [1,7,11,15,17].

Given a graph $G = (V, E)$ of n vertices and m edges, and a designated vertex $r \in V$, one can find two independent spanning trees of G rooted at any vertex in linear time if G is biconnected [2,3,11], and find three independent spanning trees of G rooted at any vertex in $O(mn)$ and $O(n^2)$ time if G is triconnected [2,3,6]. It is conjectured that, for any $k \geq 1$, every k-connected graph has k independent spanning trees rooted at any vertex [14,20]. For general graphs with $k \geq 4$ the conjecture is still open, however, for planar graphs the conjecture is verified by Huck for $k = 4$ [9] and $k = 5$ [10] (i.e., for all planar graphs, since every planar graph has a vertex of degree at most 5 [19, p269] means there is no 6-connected planar graph). The proof in [10] yields an algorithm to actually find k independent spanning trees in a k-connected planar graph, but it takes time $O(n^3)$. On the other hand, for k-connected maximal planar graphs we can find k independent spanning trees in linear time for $k = 2$ [2,3,11], $k = 3$ [2,3, 18] and $k = 4$ [15].

In this paper we give a simple linear-time algorithm to find five independent spanning trees of a 5-connected maximal planar graph rooted at any designated vertex. Note that, since there is no 6-connected planar graph, our result, together with previous results [2,3,11,15,18], yields a linear-time algorithm to find k independent spanning trees in a k-connected maximal planar graph rooted

U. Brandes and D. Wagner (Eds.): WG 2000, LNCS 1928, pp. 290–301, 2000.
© Springer-Verlag Berlin Heidelberg 2000

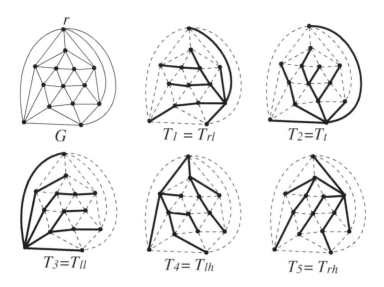

Fig. 1. Five independent spanning trees T_1, T_2, T_3, T_4 and T_5 of a graph G rooted at r.

at any designated vertex. Our algorithm is based on a "5-canonical decomposition" of a 5-connected maximal planar graph, which is a generalization of an st-numbering [8], a canonical ordering [12], a canonical decomposition [4,5], a canonical 4-ordering [13] and a 4-canonical decomposition [15,16].

The remainder of the paper is organized as follows. In Section 2 we introduce some definitions. In Section 3 we present our algorithm to find five independent spanning trees based on a 5-canonical decomposition. In Section 4 we give an algorithm to find a 5-canonical decomposition. Finally we put conclusion in Section 5.

2   Preliminaries

In this section we introduce some definitions.

Let $G = (V, E)$ be a connected graph with vertex set V and edge set E. Throughout the paper we denote by n the number of vertices in G, and we always assume that $n > 5$. An edge joining vertices u and v is denoted by (u, v). The *degree* of a vertex v in G, denoted by $d(v, G)$ or simply by $d(v)$, is the number of neighbors of v in G. The *connectivity* $\kappa(G)$ of a graph G is the minimum number of vertices whose removal results in a disconnected graph or a single-vertex graph K_1. A graph G is k-*connected* if $\kappa(G) \geq k$. A *path* in a graph is an ordered list of distinct vertices $(v_1, v_2, \cdots, v_l)$ such that (v_{i-1}, v_i) is an edge for all i, $2 \leq i \leq l$. We say that two paths are *internally disjoint* if their intermediate vertices are disjoint. We also say that a set of paths are *internally disjoint* if every pair of paths in the set are internally disjoint.

A graph is *planar* if it can be embedded in the plane so that no two edges intersect geometrically except at a vertex to which they are both incident. A planar graph G is *maximal* if all faces including the outer face are triangles in some planar embedding of G. Essentially each maximal planar graph has a unique planar embedding except for the choice of the outer face. A *plane graph* is a planar graph with a fixed planar embedding. The *contour* $C_o(G)$ of a biconnected plane graph G is the clockwise (simple) cycle on the outer face. We write $C_o(G) = (w_1, w_2, \cdots, w_h)$ if the vertices $w_1, w_2, \cdots, w_h$ on $C_o(G)$ appear in this order.

Let $P_1 = (v_1, v_2, \cdots, v_\ell)$ and $P_2 = (w_1, w_2, \cdots, w_{\ell'})$ be two paths in a plane graph sharing a (maximal) subpath $P_c = (v_a, v_{a+1}, \cdots, v_b)$, where P_c may be a vertex, v_{a-1} be the preceding vertex of P_c in P_1, and v_{b+1} be the succeeding vertex of P_c in P_1. If we cannot add an edge $e = (w_1, w_{\ell'})$ to P_2 so that (1) $\{v_{a-1}, v_{b+1}\}$ is located inside of the simple cycle consisting of P_2 and e, and (2) e does not intersect to P_1, then we say P_1 and P_2 *cross* each other on P_c.

3   Algorithm

In this section we give our algorithm to find five independent spanning trees of a 5-connected maximal planar graph rooted at any designated vertex.

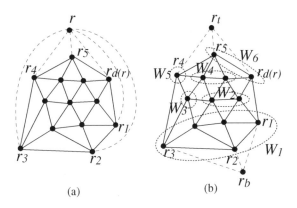

Fig. 2. (a) Five-connected plane graph G and (b) plane graph G''.

Given a 5-connected maximal planar graph $G = (V, E)$ and a designated vertex $r \in V$, we first find a planar embedding of G in which r is located on $C_o(G)$. Let $G' = G - \{r\}$ be the plane subgraph of G induced by $V - \{r\}$. In Fig. 2 (a) G is drawn by solid and dotted lines, and G' by solid lines. We may assume that all the neighbors $r_1, r_2, \cdots, r_{d(r)}$ of r in G appear on $C_o(G')$ clockwise in this order. Now $C_o(G') = (r_1, r_2, \cdots, r_{d(r)})$. Since G is 5-connected, $d(r) \geq 5$. We add to G' two new vertices r_b and r_t, join r_b with r_1, r_2 and r_3, and join r_t with $r_4, r_5, \cdots, r_{d(r)}$. Let G'' be the resulting plane graph where vertices r_1, r_b, r_3, r_4, r_t and $r_{d(r)}$ appear on $C_o(G'')$ clockwise in this order. Fig. 2 (b) illustrates G''.

Let $\Pi = (W_1, W_2, \cdots, W_m)$ be a partition of the vertex set $V - \{r\}$ of G'. We denote by G_k, $1 \le k \le m$, the plane subgraph of G'' induced by $\{r_b\} \bigcup W_1 \bigcup W_2 \bigcup \cdots \bigcup W_k$. We denote by $\overline{G_k}$, $0 \le k \le m-1$, the plane subgraph of G'' induced by $W_{k+1} \bigcup W_{k+2} \bigcup \cdots \bigcup W_m \bigcup \{r_t\}$. A partition $\Pi = (W_1, W_2, \cdots, W_m)$ of $V - \{r\}$ is called a *5-canonical decomposition* of G' if the following three conditions (co1)–(co3) are satisfied.

(co1) $W_1 = \{r_1, r_2, r_3\} \bigcup \{u_2, u_3, \cdots, u_{d(r_2)-2}\}$, where vertices $u_2, u_3, \cdots, u_{d(r_2)-2}$ are the neighbors of r_2 except r_1, r_3, r_b, and $W_m = \{r_{d(r)-1}, r_{d(r)}\}$

(co2) For each k, $1 \le k \le m$, G_k is triconnected, and for each k, $0 \le k \le m-1$, $\overline{G_k}$ is biconnected; and

(co3) For each k, $1 < k < m$, one of the following two conditions holds (See Fig. 3. The vertices in W_k are drawn in black dots.):

(a) $|W_k| \ge 2$, all vertices in W_k consecutively appear on $C_o(G_k)$ and each vertex $u \in W_k$ satisfies $d(u, G_k) = 3$ and $d(u, \overline{G_{k-1}}) \ge 3$; and

(b) $|W_k| = 1$, and the vertex $u \in W_k$ satisfies $d(u, G_k) \ge 3$ and $d(u, \overline{G_{k-1}}) \ge 2$.

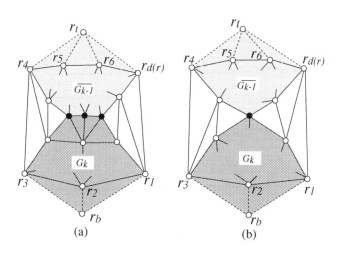

(a)   (b)

Fig. 3. Two conditions for (co3).

Fig. 2 (b) illustrates a 5-canonical decomposition of $G' = G - \{r\}$, where G' are drawn in solid lines and each set W_i is indicated by an oval drawn in a dotted line. A 5-canonical decomposition is a generalization of an "st-numbering" [8], a "canonical ordering" [12], a "canonical decomposition" [4,5], a "canonical 4-ordering" [13] and a "4-canonical decomposition" [15,16].

We have the following lemma. We will give a proof of Lemma 1 in Section 4.

Lemma 1. Let $G = (V, E)$ be a 5-connected maximal plane graph, and let r be a designated vertex on $C_o(G)$. Then $G' = G - \{r\}$ has a 5-canonical decomposition Π. Furthermore Π can be found in linear time.

We need a few more definitions to describe our algorithm. For a vertex $v \in V - \{r\}$ we write $N(v) = \{v_1, v_2, \cdots, v_{d(v)}\}$ if $v_1, v_2, \cdots, v_{d(v)}$ are the neighbors of vertex v in G'' and appear around v clockwise in this order. To each vertex $v \in V - \{r\}$ we assign five edges incident to v in G'' as *the right leg* $rl(v)$, *the tail* $t(v)$, *the left leg* $ll(v)$, *the left hand* $lh(v)$ and *the right hand* $rh(v)$ as follows. We will show later that such an assignment immediately yields five independent spanning trees of G. Let $v \in W_k$ for some k, $1 \le k \le m$, then there are the following four cases to consider.

Case 1: $k = 1$. (See Fig. 4(a).)

Now $W_1 = \{r_1, r_2, r_3\} \bigcup \{u_2, u_3, \cdots, u_{d(r_2)-2}\}$. We may assume that vertices $u_2, u_3, \cdots, u_{d(r_2)-2}$ consecutively appear on $C_o(G_1)$ clockwise in this order. Let $u_1 = r_3, u_0 = r_b, u_{d(r_2)-1} = r_1$ and $u_{d(r_2)} = r_b$. For each $u_i \in W_1 - \{r_2\}$ we define $rl(u_i) = (u_i, u_{i+1})$, $t(u_i) = (u_i, r_2)$, $ll(u_i) = (u_i, u_{i-1})$, $lh(u_i) = (u_i, v_1)$, and $rh(u_i) = (u_i, v_{d(u_i)-3})$ where we assume $N(u_i) = \{u_{i-1}, v_1, v_2, \cdots, v_{d(u_i)-3}, u_{i+1}, r_2\}$. For r_2 we define $rl(r_2) = (r_2, r_1)$, $t(r_2) = (r_2, r_b)$, $ll(r_2) = (r_2, r_3)$, $lh(r_2) = (r_2, u_2)$, and $rh(r_2) = (r_2, u_{d(r_2)-2})$.

Case 2: W_k satisfies Condition (a) of (co3). (See Fig. 4(b).)

Let $W_k = \{u_1, u_2, \cdots, u_l\}$ and $u_1, u_2, \cdots, u_l$ consecutively appear on $C_o(G_k)$ clockwise in this order. Since $d(u_i, G_k) = 3$ for each vertex u_i and G is maximal planar, vertices $u_1, u_2, \cdots, u_l$ have exactly one common neighbor, say v, in G_k. Let u_0 be the vertex on $C_o(G_k)$ preceding u_1, and let u_{l+1} be the vertex on $C_o(G_k)$ succeeding u_l. For each $u_i \in W_k$ we define $rl(u_i) = (u_i, u_{i+1})$, $t(u_i) = (u_i, v)$, $ll(u_i) = (u_i, u_{i-1})$, $lh(u_i) = (u_i, v_1)$, and $rh(u_i) = (u_i, v_{d(u_i)-3})$ where we assume $N(u_i) = \{u_{i-1}, v_1, v_2, \cdots, v_{d(u_i)-3}, u_{i+1}, v\}$.

Case 3: W_k satisfies Condition (b) of (co3). (See Fig. 4(c).)

Let $W_k = \{u\}$, let u' be the vertex on $C_o(G_k)$ preceding u, and let u'' be the vertex on $C_o(G_k)$ succeeding u. Let $N(u) = \{u', v_1, v_2, \cdots, v_{d(u)-1}\}$, and let $u'' = v_x$ for some x, $3 \le x \le d(u) - 2$. Then $rl(u) = (u, u'')$, $t(u) = (u, v_{d(u)-1})$, $ll(u) = (u, u')$, $lh(u) = (u, v_1)$, and $rh(u) = (u, v_{x-1})$.

Case 4: $k = m$. (See Fig. 4(d).)

Now $W_m = \{r_{d(r)-1}, r_{d(r)}\}$. Let $u_0 = r_t$, $u_1 = r_{d(r)-1}$, $u_2 = r_{d(r)}$ and $u_3 = r_t$. For each $u_i \in W_k$ we define $rl(u_i) = (u_i, v_1)$, $t(u_i) = (u_i, v_{d(u_i)-3})$, $ll(u_i) = (u_i, v_{d(u_i)-2})$, $lh(u_i) = (u_i, u_{i-1})$, and $rh(u_i) = (u_i, u_{i+1})$ where we assume $N(u_i) = \{u_{i+1}, v_1, v_2, \cdots, v_{d(u_i)-2}, u_{i-1}\}$.

We are now ready to give our algorithm.

Procedure FiveTrees(G, r)
begin
1   Find a planar embedding of G such that $r \in C_o(G)$;
2   Find a 5-canonical decomposition $\Pi = (W_1, W_2, \cdots, W_m)$ of $G - \{r\}$;

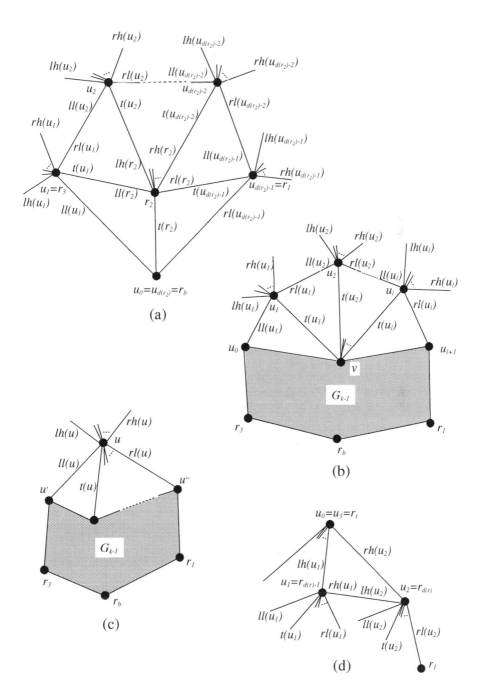

Fig. 4. Assignment.

3   For each vertex $v \in V - \{r\}$ find $rl(v), t(v), ll(v), lh(v)$ and $rh(v)$;

4   Let T_{rl} be a graph induced by the right legs of all vertices in $V - \{r\}$;

5   Let T_t be a graph induced by the tails of all vertices in $V - \{r\}$;

6   Let T_{ll} be a graph induced by the left legs of all vertices in $V - \{r\}$;

7   Let T_{lh} be a graph induced by the left hands of all vertices in $V - \{r\}$;

8   Let T_{rh} be a graph induced by the right hands of all vertices in $V - \{r\}$;

9   Regard vertex r_b in trees T_{rl}, T_t and T_{ll} as vertex r;

10   Regard vertex r_t in trees T_{lh} and T_{rh} as vertex r;

11   **return** $T_{rl}, T_t, T_{ll}, T_{lh}$ and T_{rh} as five independent spanning trees of G.

    end

We then verify the correctness of our algorithm. Assume that $G = (V, E)$ is a 5-connected maximal planar graph with a designated vertex $r \in V$, and that Algorithm FiveTrees finds a 5-canonical decomposition $\Pi = (W_1, W_2, \cdots, W_m)$ of $G - \{r\}$ and outputs $T_{rl}, T_t, T_{ll}, T_{lh}$ and T_{rh}. We first have the following lemma.

Lemma 2. Let $1 \leq k \leq m$, and let T_{rl}^k be a graph induced by the right legs of all vertices in $G_k - \{r_b\}$. Then T_{rl}^k is a spanning tree of G_k.

Proof. We can prove the claim by induction on k.   □

We then have the following lemma.

Lemma 3. $T_{rl}, T_t, T_{ll}, T_{lh}$ and T_{rh} are spanning trees of G.

Proof. By Lemma 2, T_{rl}^m is a spanning tree of G_m, and hence T_{rl} in which r_b is regarded as r is a spanning tree of G.

    Similarly T_t, T_{ll}, T_{lh} and T_{rh} are spanning trees of G.   □

Let v be any vertex in $V - \{r\}$, and let $P_{rl}, P_t, P_{ll}, P_{lh}$ and P_{rh} be the paths connecting r and v in $T_{rl}, T_t, T_{ll}, T_{lh}$ and T_{rh}, respectively. For any vertex u in $V - \{r\}$ we define $rank(u) = k$ if $u \in W_k$. For r, $rank(r)$ is undefined. If an edge (v, u) of G' is either a leg or a tail of vertex v, and (v, w) of G' is a hand of v, then $rank(u) \leq rank(v) \leq rank(w)$, and additionally if $v \neq r_2$ then $rank(u) < rank(w)$. Note that if (v, u) is in G' neither v nor u is r. Also note that only if $v = r_2$ then $rank(u) = rank(w)$. (See Fig. 4.) Now we have the following lemma.

Lemma 4. Every pair of paths $P_1 \in \{P_{rl}, P_t, P_{ll}\}$ and $P_2 \in \{P_{lh}, P_{rh}\}$ are internally disjoint.

Proof. We prove only that P_{rl} and P_{rh} are internally disjoint. Proofs for the other pairs are similar. If $v = r_1$, then $P_{rl} = (r_1, r)$ and P_{rl} is internally disjoint to any path, since P_{rl} has no intermediate vertices. If $v = r_{d(r)}$, then $P_{rh} = (r_{d(r)}, r)$ and similarly P_{rh} is internally disjoint to any path. If $v = r_2$, then $P_{rl} = (r_2, r_1, r)$, so only r_1 is the intermediate vertex on P_{rl}, $P_{rh} = (v, u_{d(r_2)-2}, \cdots)$, and, for each intermediate vertex v_i except $u_{d(r_2)-2}$ on P_{rh}, $rank(v_i) \geq 2$. Therefore P_{rl} and P_{rh} are internally disjoint if v is r_2. Thus we may assume that $v \neq r_1, r_2, r_{d(r)}$.

Let $P_{rl} = (v, v_1, v_2, \cdots, v_l, r)$, then $v_l = r_1$. Let $P_{rh} = (v, u_1, u_2, \cdots, u_{l'}, r)$, then $u_{l'} = r_{d(r)}$. The definition of a right leg implies that $rank(v) \geq rank(v_1) \geq rank(v_2) \geq \cdots \geq rank(v_l)$, and the definition of a right hand implies that $rank(v) \leq rank(u_1) \leq rank(u_2) \leq \cdots \leq rank(u_{l'})$. Thus $rank(v_l) \leq \cdots \leq rank(v_2) \leq rank(v_1) \leq rank(v) \leq rank(u_1) \leq rank(u_2) \leq \cdots \leq rank(u_{l'})$. We furthermore have $rank(v_1) < rank(u_1)$ since $v \neq r_2$. Therefore P_{rl} and P_{rh} are internally disjoint.   □

If $rl(v) = (v, u)$ then we say (v, u) is an *incoming right leg* of u. Similarly, if $t(v) = (v, u)$ then (v, u) is an *incoming tail* of u, and if $ll(v) = (v, u)$ then (v, u) is an *incoming left leg* of u.

We have the following lemma.

Lemma 5. Let $u \in V - \{r\}$. All incoming right legs of u appear consecutively around u, all incoming tails of u appear consecutively around u, and all incoming left legs of u appear consecutively around u. Furthermore $ll(u)$, the incoming right legs, incoming tails, incoming left legs and $rl(u)$ appear clockwise around u in this order.

Proof. (See Fig. 5.) Let $u \in V - \{r\}$, $ll(u) = (u, u')$, $rl(u) = (u, u'')$, and $N(u) = \{v_0, v_1, \cdots, v_{d(u)-1}\}$. One may assume that $u' = v_0$ and $u'' = v_z$ for some z, $3 \leq z \leq d(u) - 2$. Here it is possible that $rl(v_0) = (v_0, u)$ and/or $ll(v_z) = (v_z, u)$.

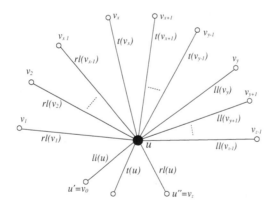

Fig. 5. Illustration for Lemma 5.

If $u = r_2$ then the claim is clearly holds. (In this case there is no incoming legs of u.) Thus we assume $u \neq r_2$.

If (u_i, u) is the tail of $u_i \in W_k$ then $u \in C_o(G_{k-1})$ and $u \notin C_o(G_k)$. (See Fig. 4.) Thus if $t(u_i) = (u_i, u)$ and $t(u_j) = (u_j, u)$ then $\{u_i, u_j\} \subset W_k$ for some k. Therefore all incoming tails of u appear consecutively around u. (See Fig. 4(b).)

If $1 \leq i \leq z-1$, $rl(v_i) = (v_i, u)$ and $v_i \in W_k$, then $(v_{i-1}, u) \notin C_o(G_k)$, and either $t(u) = (u, v_{i-1})$, $rl(v_{i-1}) = (v_{i-1}, u)$ or $ll(u) = (u, v_{i-1})$ hold. (If $rank(v_i) =$

$rank(u)$ then $t(u) = (u, v_{i-1})$. Otherwise $rank(v_i) > renk(u)$, and assume $rank(v_i) = k$. Now edge (v_{i-1}, u) is on $C_o(G_{k-1})$. If $rank(v_{i-1}) \le rank(u)$ then $ll(u) = (u, v_{i-1})$. If $rank(v_{i-1}) \ge rank(u)$ then $rl(v_{i-1}) = (v_{i-1}, u)$. See Fig. 4.) Thus if u has an incoming right leg e then the edge preceding e around u clockwise is either an incoming right leg of u, $t(u)$ or $ll(u)$. Since $t(u)$ and $ll(u)$ always appear consecutively around u, therefore all incoming right legs of u appear consecutively around u and $ll(u)$ precedes them. Similarly all incoming left legs of u appear consecutively around u and $rl(u)$ succeeds them. Thus the claim holds.   □

Lemma 5 immediately implies the following lemma.

Lemma 6. A pair of paths $P_1, P_2 \in \{P_{rl}, P_t, P_{ll}\}$ may cross at a vertex u, but do not share a vertex u without crossing at u.

From the definitions of a left leg , a tail and a right leg one can immediately have the following lemma.

Lemma 7. Let $1 \le k \le m$, $u \ne r_2$ and $u \in W_k$. Then u is on $C_o(G_k)$. Let u' be the succeeding vertex of u on $C_o(G_k)$. Assume that $N(u)$ starts with u'. Let $rl(u) = (u, v')$, $t(u) = (u, v'')$ and $ll(u) = (u, v''')$. Then v', v'', v''' appear in $N(u)$ in this order.

We then have the following lemma.

Lemma 8. A pair of paths $P_1, P_2 \in \{P_{rl}, P_t, P_{ll}\}$ are internally disjoint. Also P_{lh}, P_{rh} are internally disjoint.

Proof. We prove only that P_{rl} and P_{ll} are internally disjoint. Proofs for the other cases are similar. Suppose for a contradiction that P_{rl} and P_{ll} share an intermediate vertex. Let w be the intermediate vertex that is shared by P_{rl} and P_{ll} and appear last on the path P_{rl} going from r to v. Now $w \ne r_2$ because w is an intermediate vertex of P_{rl} and P_{ll}, and r_2 has degree one in both T_{rl} and T_{ll}. Then P_{rl} and P_{ll} cross at w by Lemma 6. However, the claim in Lemma 7 holds both for $k = rank(v)$ and $u = v$ and for $k = rank(w)$ and $u = w$, and hence P_{rl} and P_{ll} do not cross at w, a contradiction.   □

By Lemmas 4 and 8 we have the following lemma.

Lemma 9. $T_{rl}, T_t, T_{ll}, T_{lh}$ and T_{rh} are five independent spanning trees of G rooted at r.

Clearly the running time of Algorithm FiveTrees is $O(n)$. Thus we have the following theorem.

Theorem 1. Five independent spanning trees of any 5-connected maximal planar graph rooted at any designated vertex can be found in linear time.

4   Proof of Lemma 1

In this section we give an algorithm to find a 5-canonical decomposition. Then we show it runs in linear time. First we need some definitions.

Let $G = (V, E)$ be a 5-connected maximal plane graph, r be a designated vertex on $C_o(G)$. We construct G'' from G as in Section 3. Let H be a triconnected plane subgraph of G'' such that (1) H is a subgraph of G'' inside some cycle C of G'' (including C), and (2) $r_b, r_1, r_3 \in C_o(H) = C$. Let $C_o(H) = (r_b = w_1, w_2 = r_3, \cdots, w_l = r_1)$.

A set of edges $(v_1, u), (v_2, u), \cdots, (v_h, u)$ in H is called a *fan with center u* if (1) $u \notin C_o(H)$, (2) the neighbors of u on $C_o(H)$ are $v_1, v_2, \cdots, v_h$, called *leaves*, and they appear in $C_o(H)$ clockwise in this order, (but may not appear consecutively,) and (3) either (i) $h = 2$ and H does not have $edge(v_1, v_2)$, or (ii) $h \geq 3$. Assume a set of edges $(v_1, u), (v_2, u), \cdots, (v_h, u)$ is a fan F with center u. Now, for $1 \leq i \leq h - 1$, $v_i = w_a$ and $v_{i+1} = w_b$ hold for some a, b such that $1 \leq a < b \leq l$, and let C_i be the cycle consisting of the subpath $(w_a, w_{a+1}, \cdots, w_b)$ of $C_o(H)$ and two edges $(w_b, u), (u, w_a)$. Each plane subgraph F_i of H inside C_i (including C_i) is called a *piece* of F. F_i is called an *empty piece* if $a + 1 = b$. If F_i is an empty piece then C_i is a triangle face of H. (Since G is 5-connected, if $a + 1 = b$ then F_i has no vertex in the proper inside.) Note that by the definition if a fan has exactly two leaves then it has exactly one piece and the piece is not empty. Also note that F has exactly $h - 1$ pieces. If none of pieces of F contains a distinct fan, then F is a *minimal* fan.

A *cut-set* is a set of vertices whose removal results in a disconnected graph. Since G is 5-connected and maximal planar, every cut-set of H consisting of three vertices has (1) exactly one vertex not on $C_o(H)$ and (2) exactly two vertices on $C_o(H)$. Thus each cut-set of H consisting of three vertices corresponds to a center of a fan and its two leaves.

We have the following lemmas.

Lemma 10. If a vertex $v \in C_o(H)$ is on none of fans of H (Note that, however, v may be contained in a piece of a fan.), then $H - \{v\}$ is triconnected, where $H - \{v\}$ is the plane subgraph of H obtained from H by deleting v and all edges incident to v.

Lemma 11. If all pieces of a fan $F = (v_1, u), (v_2, u), \cdots, (v_h, u)$ of H is empty (Now $d(v_1) \geq 4$, $d(v_h) \geq 4$ and, for $j = 2, 3, \cdots, h - 1$, $d(v_j) = 3$.) and $u \neq r_2$, then $H - \{v_2, v_3, \cdots, v_{h-1}\}$ is triconnected, where $H - \{v_2, v_3, \cdots, v_{h-1}\}$ is a plane subgraph of H obtained from H by deleting $v_2, v_3, \cdots, v_{h-1}$ and all edges incident to them.

Now we give our algorithm to find a 5-canonical decomposition.

First, by Condition (co1) we can find W_m. Now $\overline{G_{m-1}}$ is biconnected since $\overline{G_{m-1}}$ is a triangle cycle. Since $G = (V, E)$ is 5-connected, the vertex set $V - \{r\}$ induces a 4-connected graph G'. And G_m is obtained from G' by adding a new vertex r_b adjacent three vertices of G'. Now G_m is triconnected since a graph obtained from a k-connected graph G by adding a new vertex adjacent k vertices of G is also k-connected [19, p145]. Also G_{m-1} is triconnected, since otherwise G_{m-1} has a cut-set S with two or less vertices and then $S \bigcup W_m$ is a cut-set of G with four or less vertices, a contradiction. Thus for $k = m - 1$ and m, G_k is triconnected, and for $k = m - 1$, $\overline{G_k}$ is biconnected. Clearly $r_1, r_2, r_3 \notin W_m$.

Then, inductively assume that we have chosen $W_m, W_{m-1}, \cdots, W_{i+1}$ such that for each $k = i, i + 1, \cdots, m$, G_k is triconnected, and for each $k = i, i + 1, \cdots, m - 1$, $\overline{G_k}$ is biconnected, $r_1, r_2, r_3 \notin W_m \bigcup W_{m-1} \bigcup \cdots \bigcup W_{i+1}$ and each W_k, $k = i + 1, i + 2, \cdots, m$, satisfies either (co1) or (co3). Now we can choose W_i as follows. We have two cases. If G_i has exactly one vertices in the proper inside of G_i then it is r_2 and we have done by setting all vertices in G_i except r_b as W_1. Otherwise we can find $W_i \subseteq V - W_m \bigcup W_{m-1} \bigcup \cdots \bigcup W_{i+1}$ such that (1) G_{i-1} is triconnected, (2) $\overline{G_{i-1}}$ is biconnected, (3) $r_1, r_2, r_3 \notin W_i$, and (4) W_i satisfies (co3), as follows.

Let $F = (v_1, u), (v_2, u), \cdots, (v_h, u)$ be a minimal fan of G_i. Note that G_i always has a fan $(r_b, r_2), (r_3, r_2), \cdots, (r_1, r_2)$ with center r_2 implies G_i always has a fan.

If every piece of F is empty then F has three or more leaves, and we can set $W_i = \{v_2, v_3, \cdots, v_{h-1}\}$. Now if $h \geq 4$ then W_i satisfies (a) of (co3) and G_{i-1} is triconnected by Lemma 11, and $\overline{G_{i-1}}$ is biconnected since each vertex in W_i has degree exactly three in G_i means each vertex in W_i has two or more neighbors in $\overline{G_i}$. Similarly if $h = 3$ then W_i satisfies (b) of (co3), and G_{i-1} is triconnected by Lemma 11, and $\overline{G_{i-1}}$ is biconnected as above.

Otherwise, let F' be a non-empty piece of F. Now F' has four or more vertices on $C_o(G_i)$ since otherwise G has a cut-set with four or less vertices, a contradiction. Now there exists at least one vertex of F' on $C_o(G_i)$ such that (1) it is not a leaf of F, and (2) it has two or more neighbors in $\overline{G_i}$. (Since otherwise each vertices of F' on $C_o(G_i)$ except the two leaves w_a, w_b of F has at most one neighbor in $\overline{G_i}$, and for G is maximal planar each neighbor in $\overline{G_i}$ is a common vertices, say x, and $\{u, w_a, w_b, x\}$ forms a cut-set, a contradiction.) Thus we can find W_i satisfying (b) of (co3). Now G_{i-1} is triconnected by Lemma 10, and $\overline{G_{i-1}}$ is biconnected.

Thus we can find a 5-canonical decomposition. By maintaining a data structure to keep fans and the number of neighbors in $\overline{G_i}$ for each vertex, the algorithm runs in linear time.

5   Conclusion

In this paper we give a linear-time algorithm to find k independent spanning trees of a k-connected maximal planar graph rooted at any designated vertex. It is remained as future work to find a linear-time algorithm for planar graphs, which are not always maximal planar.

References

1. F. Bao and Y. Igarashi, *Reliable broadcasting in product networks with Byzantine faults*, Proc. 26th Annual International Symposium on Fault-Tolelant Computing (FTCS'96) (1996) 262-271.
2. G. Di Battista, R. Tamassia and L. Vismara, *Output-sensitive reporting of disjoint paths*, Technical Report CS-96-25, Department of Computer Science, Brown University (1996).
3. G. Di Battista, R. Tamassia and L.Vismara, *Output-sensitive reporting of disjoint paths*, Algorithmica, 23 (1999) 302-340.
4. M. Chrobak and G. Kant, *Convex grid drawings of 3-connected planar graphs*, Technical Report RUU-CS-93-45, Department of Computer Science, Utrecht University (1993).
5. M. Chrobak and G. Kant, *Convex grid drawings of 3-connected planar graphs*, International Journal of Computational Geometry and Applications, 7 (1997) 211-223.
6. J. Cheriyan and S. N. Maheshwari, *Finding nonseparating induced cycles and independent spanning trees in 3-connected graphs*, J. Algorithms, 9 (1988) 507-537.
7. D. Dolev, J. Y. Halpern, B. Simons and R. Strong, *A new look at fault tolerant network routing*, Proc. 16th Annual ACM Symposium on Theory of Computing (1984) 526-535.
8. S. Even, *Graph Algorithms*, Computer Science Press, Potomac (1979).
9. A. Huck, *Independent trees in graphs*, Graphs and Combinatorics, 10 (1994) 29-45.
10. A. Huck, *Independent trees in planar graphs*, Graphs and Combinatorics, 15 (1999) 29-77.
11. A. Itai and M. Rodeh, *The multi-tree approach to reliability in distributed networks*, Information and Computation, 79 (1988) 43-59.
12. C. Kant, *Drawing planar graphs using the cononical ordering*, Algorithmica, 16 (1996) 4-32.
13. G. Kant and X. He, *Two algorithms for finding rectangular duals of planar graphs*, Proc. 19th Workshop on Graph-Theoretic Concepts in Computer Science (WG'93), Lect. Notes in Comp. Sci., 790, Springer (1994) 396-410.
14. S. Khuller and B. Schieber, *On independent spanning trees*, Information Processing Letters, 42 (1992) 321-323.
15. K. Miura, D. Takahashi, S. Nakano and T. Nishizeki, *A Linear-Time Algorithm to Find Four Independent Spanning Trees in Four-Connected Planar Graphs*, WG'98, Lect. Notes in Comp. Sci., 1517, Springer (1998) 310-323.
16. S. Nakano, M. S. Rahman and T. Nishizeki, *A linear time algorithm for four-partitioning four-connected planar graphs*, Information Processing Letters, 62 (1997) 315-322.
17. K. Obokata, Y. Iwasaki, F. Bao and Y. Igarashi, *Independent spanning trees of product graphs and their construction*, Proc. 22nd Workshop on Graph-Theoretic Concepts in Computer Science (WG'96), Lect. Notes in Comp. Sci., 1197 (1996) 338-351.
18. W. Schnyder, *Embedding planar graphs on the grid*, Proc. 1st Annual ACMSIAM Symp. on Discrete Algorithms, San Francisco (1990) 138-148
19. D. B. West, *Introduction to Graph Teory*, Prentice Hall (1996)
20. A. Zehavi and A. Itai, *Three tree-paths*, J. Graph Theory, 13 (1989) 175-188.

Optimal Fault-Tolerant Routings for k-Connected Graphs with Smaller Routing Tables

Koichi Wada and Wei Chen

Nagoya Institute of Technology, Gokiso-cho, Syowa-ku, Nagoya 466-8555, Japan.
{wada,chen}@elcom.nitech.ac.jp

Abstract. We study the problem of designing fault-tolerant routings with small routing tables for a k-connected network of n processors in the surviving route graph model. The surviving route graph $R(G, \rho)/F$ for a graph G, a routing ρ and a set of faults F is a directed graph consisting of nonfaulty nodes with a directed edge from a node x to a node y iff there are no faults on the route from x to y. The diameter of the surviving route graph could be one of the fault-tolerance measures for the graph G and the routing ρ and it is denoted by $D(R(G, \rho)/F)$. We want to reduce the total number of routes defined in the routing, and the maximum of the number of routes defined for a node (called route degree) as least as possible. In this paper, we show that we can construct a routing λ for every n-node k-connected graph such that $n \geq 2k^2$, in which the route degree is $O(k\sqrt{n})$, the total number of routes is $O(k^2 n)$ and $D(R(G, \lambda)/F) \leq 3$ for any fault set $F(|F| < k)$. We also show that we can construct a routing ρ_1 for every n-node biconnected graphs, in which the total number of routes is $O(n)$ and $D(R(G, \rho_1)/\{f\}) \leq 2$ for any fault f, and using ρ_1 a routing ρ_2 for every n-node biconnected graphs, in which the route degree is $O(\sqrt{n})$, the total number of routes is $O(n\sqrt{n})$ and $D(R(G, \rho_2)/\{f\}) \leq 2$ for any fault f.

1 Introduction

Consider a communication network or an undirected graph G in which a limited number of link and/or node faults F might occur. A *routing* ρ for a graph defines at most one path called *route* for each ordered pair of nodes. A routing is called *minimal-length* if any route from x to y is assigned to a shortest path from x to y. We assume that it must be chosen without knowing which components might be faulty.

Given a graph G, a routing ρ and a set of faults F, the *surviving route graph* $R(G, \rho)/F$ is defined to be a directed graph consisting of all nonfaulty nodes in G, with a directed edge from a node x to a node y iff the route from x to y is intact. The diameter of the surviving route graph (denoted by $D(R(G, \rho)/F)$) could be one of the fault-tolerance measures for the graph G and the routing ρ [2,4]. The routing ρ on G is called (d, f)-*tolerant* if $D(R(G, \rho)/F) \leq d$ for any set F with at most f faults.

U. Brandes and D. Wagner (Eds.): WG 2000, LNCS 1928, pp. 302–313, 2000.

When we consider the fault tolerance of ATM and/or optical networks, routings must satisfy several constraints such as the number of routes defined for a node (*route degree of the node*) and the total number of routes defined in the routing [12]. Since the size of the routing table is dominated by the route degree of the node and the edge-load of the routing (the maximum number of routes passing through the edge over all edges)is dependent of the total number of routes, the route degree and the total number of routes should be as least as possible[3].

Many results have been obtained for the diameter of the surviving route graph [6,10,11,13]. As long as we use minimal-length routings for general graphs, we can not expect good behavior for the diameter of the surviving route graph, say constant diameter [4]. It is also shown that the graph connectivity does not help to reduce the diamter of the surviving route graph if minimal-length routings are considered [4]. Therefore, we must consider non-minimal-length routings to obtain efficient fault-tolerant ones for k-connected graphs. For n-node k-connected graphs, a $(5, k - 1)$-tolerant routing and a $(3, k - 1)$-tolerant routing can be constructed if $n \geq k^2$ and $n \geq 2k^2$, respectively[9]. A $(2, k-1)$-tolerant routing can be constructed for every n-node k-connected graph such that $n \geq 7k^3 \lceil \log_2 n \rceil$ [15]. However, in these routings, the route degree of most nodes is $n - 1$ and it is undesirable. Stronger results have been known for n-node biconnected graphs[12]; We can construct a $(2, 1)$-tolerant routing with $O(n)$ routes and a $(3, 1)$-tolerant routing with $O(n)$ routes and route degree $O(\sqrt{n})$ [12]. It can be shown that the former routing is optimal in the sense that the diameter of the surviving route graph is minimum and the total number of routes in the routing is minimum, and the diameter of the surviving route graph for the latter routing is also optimal among routings with route degree $O(\sqrt{n})$ and the total number of route in the routing is minimum.

In this paper, we show the following results which improve the previous ones with respect to the route degree and the total number of routes.

1. For every n-node k-connected graph such that $n \geq 2k^2$, we can construct a $(3, k - 1)$-tolerant routing λ in which the route degree is $O(k\sqrt{n})$ and the total number of routes is $O(k^2n)$.
2. For every n-node biconnected graph, we can construct a $(2, 1)$-tolerant routing ρ_1 in which the total number of routes is $O(n)$ and a $(2, 1)$-tolerant routing ρ_2 in which the route degree is $O(\sqrt{n})$ and the total number of routes is $O(n\sqrt{n})$.

We improve the $(3, k - 1)$-tolerant routing shown in [9] to the routing λ so that the route degree of λ is reduced to $O(k\sqrt{n})$ with preserving the total number of routes. We also show that the diameter of the surviving route graph for λ is optimal among routings with the route degree $O(k\sqrt{n})$ if $n \geq 2k^2$ and $k = o(n^{1/6})$.

The routing ρ_1 does not improve the previous result. However, the idea to define ρ_1 is different from the previous ones and it induces the routing ρ_2 that is the first $(2, 1)$-tolerant routing with route degree $O(\sqrt{n})$ for biconnected graphs.

We also show that the total number of routes in the routing ρ_2 is minimum among $(2, 1)$-tolerant routings with route degree $O(\sqrt{n})$.

2   Preliminary

In this section, we give definitions and terminology. We refer readers to [8] for basic graph terminology.

Unless otherwise stated, we deal with an undirected graph $G = (V, E)$ that corresponds to a network. For a node v of G, $N_G(v) = \{u|(v, u) \in E\}$ and $deg_G(v) = |N_G(v)|$. $deg_G(v)$ is called *degree* of v and if G is apparent it is simply denoted by $deg(v)$. For a node set $U \subseteq V$, the subgraph induced by U is the maximal subgraph of G with the node set U and denoted by $G < U >$. A graph G is *k-connected* if there exist k node-disjoint paths between every pair of distinct nodes in G. For a node $v \in V$ and a node set $U \subseteq V - \{v\}$, *v-U fan* is a set of $|U|$ disjoint paths from v to all nodes of U. Usually 2-connected graphs are called *biconnected graphs*.

The *distance* between nodes x and y in G is the length of the shortest path between x and y and is denoted by $dis_G(x, y)$. The *diameter* of G is the maximum of $dis_G(x, y)$ over all pairs of nodes in G and is denoted by $D(G)$. Let $P(u, v)$ and $P(v, w)$ be a path from u to v and a path from v to w, respectively. In general, even if both $P(u, v)$ and $P(v, w)$ are simple, the concatenation of $P(u, v)$ and $P(v, w)$ is not always simple. Thus we consider two kinds of concatenation: one is a usual concatenation (denoted by $P(u, v) \cdot P(v, w)$) and the other is a special concatenation (denoted by $P(u, v) \odot P(v, w)$), which is defined as the shortest path from u to w in the graph $P(u, v) \cup P(v, w)$ to make the concatenated path simple.

Let $G = (V, E)$ be a graph and let x and y be nodes of G. Define $P_G(x, y)$ to be the set of all simple paths from the node x to the node y in G, and $P(G)$ to be the set of all simple paths in G. A *routing* is a partial function $\rho : V \times V \rightarrow P(G)$ such that $\rho(x, y) \in P_G(x, y)(x \neq y)$. The path specified to be $\rho(x, y)$ is called the *route from x to y*.

For a graph $G = (V, E)$, let $F \subseteq V \cup E$ be a set of nodes and edges called a set of *faults*. We call $F \cap V (= F_V)$ and $F \cap E(= F_E)$ the set of *node faults* and the set of *edge faults*, respectively. If an object such as a route or a node set does not contain any element of F, the object is said to be *fault free*.

For a graph $G = (V, E)$, a routing ρ on G and a set of faults $F(= F_V \cup F_E)$, the *surviving route graph*, $R(G, \rho)/F$, is a directed graph with node set $V - F_V$ and edge set $E(G, \rho, F) = \{< x, y > |\rho(x, y)$ *is defined and fault free*$\}$. In what follows, unless confusion arises we use notations for directed graphs as the same ones for undirected graphs.

In the surviving route graph $R(G, \rho)/F$, when $F = \phi$ the graph is called the route graph. In the route graph, the outdegree of a node v is called *the route degree of a node v* and the maximum of the route degree of all nodes is called *the route degree of the routing ρ*. The number of directed edges in the route graph corresponds to the total number of routes in the routing ρ. If the number of edges in the route graph is m, the routing ρ is called *m-route-routing* or simply *m-routing*.

For a graph $G = (V, E)$ and a routing ρ, if for any edge (x, y) in G such that $\rho(x, y)$ is defined, the route $\rho(x, y)$ is assigned to the edge, ρ is called *edge-routing*.

A routing ρ is *bidirectional* if $\rho(x, y) - \rho(y, x)$ for any node pair (x, y) in the domain of ρ. If a routing is not bidirectional, it is called *unidirectional*. Note that if the routing ρ is bidirectional, the surviving route graph $R(G, \rho)/F$ can be represented as an undirected graph.

Given a graph G and a routing property P, a routing ρ on G is *optimal* with respect to P if $max_{Fs.t.|F|\leq k} (D(R(G, \rho)/F))$ is minimum over all routings on G satisfying P. Note that from the definition of the optimality, if $D(R(G, \rho)/F)$ is 2 for any set of faults F such that $|F| \leq k$, the routing is obviously optimal with respect to any property. If the property P is known, we simply call the routing is optimal.

Lemma 1. [1] *Let $G = (V, E)$ be an n-node directed graph. If the maximum outdegree of G is d and the diameter of G is p, then $|E| = \Omega(n^2/d^{p-1})$.*

The next theorem can be derived from this lemma.

Theorem 1. *Let G be an n-node graph and let $f \geq 0$. There does not exist a $(2, f)$-tolerant $o(n\sqrt{n})$-routing with the route degree $O(\sqrt{n})$ on G.*

3 Optimal Routing for k-Connected Graphs

In this section, we show that for n-node k-connected graphs with $n \geq 2k^2$, we can construct a $(3, k - 1)$-tolerant bidirectional edge-routing λ such that the route degree is $O(k\sqrt{n})$ and the total number of routes in λ is $O(k^2n)$ and we show that the routing λ is optimal with respect to the route degree of $O(k\sqrt{n})$ if $n \geq 2k^2$ and $k = o(n^{1/6})$.

The routings for k-connected graphs are based on the following two properties [9,11,15].

Lemma 2. [8] *Let $G = (V, E)$ be a k-connected graph. Let U be any node set of V such that $|U| = k$ and let v be any node in $V - U$. Then there is a v-U fan.*

Let $u \in U$. The path from v to u in v-U fan is denoted by $P_{fan}(v, u; U)$. If there is an edge between v and u in U, the path $P_{fan}(v, u; U)$ in the v-U fan can be changed to this edge[4].

Lemma 3. [7] *Let $G = (V, E)$ be a k-connected graph. Let $v_1, v_2, \ldots, v_k$ be distinct nodes and $a_1, a_2, \ldots, a_k$ be positive integers such that $\Sigma_{i=1}^k a_i = |V|$. Then there is a partition $V_1, V_2, \ldots, V_k$ of V such that $v_i \in V_i$, $|V_i| = a_i$ and the induced subgraph $G < V_i >$ is connected for $i = 1, 2, \ldots, k$.*

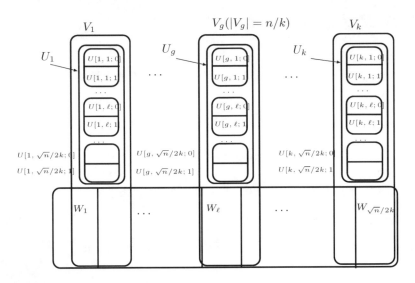

Fig. 1. The partition of nodes in G.

First we consider k-connected graphs with at least $4k^2$ nodes, and later we extend the result for k-connected graphs with at least $2k^2$ nodes.

Let $G = (V, E)$ be an n-node k-connected graph such that $n \geq 4k^2$. From lemma 3, there are k disjoint node subsets $V_1, V_2, \ldots, V_k$ such that $|V_g| = n/k$ [1] and the induced graph $G < V_g >$ is connected for $g = 1, 2, \ldots, k$. Let U_g be a subset of V_g such that $|U_g| = \sqrt{n}$ and let each U_g be partitioned into $\sqrt{n}/2k$ sets with each $2k$ nodes. These sets are denoted by $U[g, \ell](1 \leq \ell \leq \sqrt{n}/2k)$. Furthermore, each $U[g, \ell]$ is partitioned into two sets with cardinalities k. These sets are denoted by $U[g, \ell; 0]$ and $U[g, \ell; 1]$. Let $W = V - \cup_{g=1}^{k} U_g$ and let W be partitioned into $\sqrt{n}/2k$ sets. These sets are denoted by $W_1, W_2, \ldots W_{\sqrt{n}/2k}$. The partition of V is shown in Fig. 1.

A bidirectional routing λ is defined as follows:

1. For $x \in W_\ell$ and $y \in U[g, \ell; 0] (1 \leq g \leq k, 1 \leq \ell \leq \sqrt{n}/2k)$,
   $\lambda(x, y) = \lambda(y, x) = P_{fan}(x, y; U[g, \ell; 0])$.
2. For $x, y \in U_g (1 \leq g \leq k)$,
   $\lambda(x, y) = \lambda(y, x) =$ a shortest path between x and y in $G < V_g >$.
3. For $x \in U[g_1, \ell; i]$ and $y \in U[g_2, \ell; i]$
   $(1 \leq g_2 < g_1 \leq k, 1 \leq \ell \leq \sqrt{n}/2k, i = 0, 1)$
   $\lambda(x, y) = \lambda(y, x) = P_{fan}(x, y; U[g_2, \ell; i_2])$.
4. For $x \in U[g_1, \ell; i_1]$ and $y \in U[g_2, \ell; i_2]$
   $(1 \leq g_2 < g_1 \leq k, 1 \leq \ell \leq \sqrt{n}/2k, i_1 \neq i_2)$
   $\lambda(x, y) = \lambda(y, x) = P_{fan}(y, x; U[g_1, \ell; i_1])$.

---

[1] For simplicity, we assume that n/k and $\sqrt{n}/2k$ are integers.

Fig. 2 shows the routes in λ. Fig. 2 does not show z'-$U[g_1, \ell_1; 0]$ fan $(z' \in U[g_2, \ell_1; 1])$ and y'-$U[g_2, \ell_1; 1]$ fan$(y' \in U[g_1, \ell_1; 1])$ for the lack of space. Because of the property of v-U fan, λ is an edge-routing. It can be verified that the route degree of λ is $(2k+1)\sqrt{n} + 2k^2 - 2k - 1 = O(k\sqrt{n})$ and the total number of routes in λ is $(k^2 + 2k)n + (k^3 - 2k^2 - k)\sqrt{n} = O(k^2 n)$

From the definition of λ, the next lemma holds.

Lemma 4. *Let $G = (V, E)$ be a k-connected graph on which the routing λ is defined. For any node x and any node set U_g such that $x \notin U_g$, there are k node disjoint routes from x to k nodes in U_g.*

Theorem 2. *Let $G = (V, E)$ be an n-node k-connected graph such that $n \geq 4k^2$. The routing λ on G is (3,k-1)-tolerant.*

Proof. Let F be a faulty set with $|F| < k$ and let $R = R(G, \lambda)/F$. Since $|F| < k$ and the number of node sets V_g is k, there is a node set V_I such that $G < V_I >$ contains no elements of F. Let x and y be arbitrary non-faulty distinct nodes in $V - F$.

[Case:1] Suppose that $x, y \in U_I$. Since $U_I \subseteq V_I$ and the route $\lambda(x, y)$ is defined in $G < V_I >$, $\lambda(x, y)$ is fault free. Thus, $dis_R(x, y) = 1$.

[Case:2] Suppose that $x \in U_I$ and $y \notin U_I$. From Lemma 4, there are k node disjoint routes between y and k nodes in U_I. Since $|F| < k$, there is a fault-free

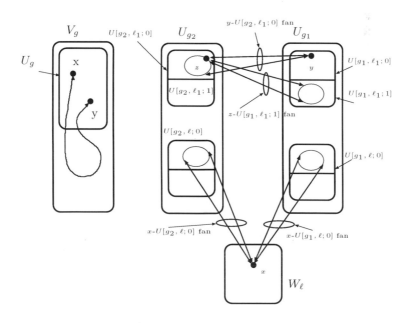

Fig. 2. The routing λ.

route between y and an node, say w, in U_I. Since $\lambda(x, w)$ does not contain any fault in F, $dis_R(x, y) = 2$.

The case that $x \notin U_I$ and $y \notin U_I$ can be proved similar to Case:2 by using Lemma 4.

For an n-node k-connected graph, if $2k^2 \leq n \leq 4k^2$, set $|U_g| = 2k$ instead of $|U_g| = \sqrt{n}$ in the definition. In this case the total number of routes in λ is $O(k^4) = O(k^2 n)$ and the route degree of λ is $O(k^2) = O(k\sqrt{n})$. Therefore, the following theorem holds.

Theorem 3. *Let $G = (V, E)$ be an n-node k-connected graph. If $n \geq 2k^2$, we can construct a $(3, k - 1)$-tolerant bidirectional edge-routing such that the total number of routes is $O(k^2 n)$ and the route degree is $O(k\sqrt{n})$. This routing is optimal with respect to the route degree of $O(k\sqrt{n})$ if $n \geq 2k^2$ and $k = o(n^{1/6})$.*

Proof. The optimality of λ can be shown as follows. If $k = o(n^{1/6})$ the least number of routes in $(2, k - 1)$-routings with route degree of $O(k\sqrt{n})$ is $\omega(n^{4/3})$ from Lemma 1. On the other hand, if $k = o(n^{1/6})$, the total number of routes in λ is $o(n^{4/3})$.

In the case that $k = 2$, we can construct a $(3, 1)$-tolerant bidirectional edge-routing such that the total number of routes is $O(n)$ and the route degree is $O(\sqrt{n})$ for any n-node biconnected graph with $n \geq 5$ [12].

4   Optimal Routings for Biconnected Graphs

As long as the authors have known, there do not exist $(2, k-1)$-tolerant routings for k-connected graphs such that their route degree are $O(k\sqrt{n})$. Although the case that $k \geq 3$ is still open, we show that we can construct a $(2, 1)$-tolerant edge-routing for biconnected graphs such that its route degree is $O(\sqrt{n})$ and the total number of routes is $O(n\sqrt{n})$. This routing is optimal because it is $(2, 1)$-tolerant. From Theorem 1, in order to define $(2, 1)$-tolerant routings with route degree $O(\sqrt{n})$, the total number of routes must be $\Omega(n\sqrt{n})$. Thus, the total number of routes in the routing shown here attains the lower bound.

4.1   Paths by Using s-t Numbering

Optimal routings for biconnected graphs are based on s-t numberings[5] which characterize biconnected graphs [12,14].

Given an edge (s, t) of a biconnected graph $G = (V, E)$, a bijective function $g : V \rightarrow \{0, 1, \ldots, |V| - 1 = n - 1\}$ is called an *s-t numbering* if the following conditions are satisfied:

1. $g(s) = 0$, $g(t) = n - 1$ and
2. Every node $v \in V - \{s, t\}$ has two adjacent nodes u and w such that $g(u) < g(v) < g(w)$.

In what follows, we assume that the node set of G is s-t numbered and it is denoted by $\{0, 1, \ldots, n-1\}$, where $s = 0$ and $t = n - 1$.

For a node v in G, we define two paths $P_I[v, t]$ and $P_D[v, s]$ as follows:

(1) $P_I[v, t] = (v_0(= v), v_1, \ldots, v_p(= t))$,

where $v_i = max\{u|u \in N_G(v_{i-1})\}(1 \leq i \leq p)$ and

(2) $P_D[v, s] = (v_0(= v), v_1, \ldots, v_q(= s))$,

where $v_i = min\{u|u \in N_G(v_{i-1})\}(1 \leq i \leq q)$.

Since we treat unidirectional routings, we consider directions for undirected paths. Therefore, for example, $P_I[v, t]$ denotes the path from v to t and the path from t to v of the same one is denoted by $P_I[t, v]$.

Note that if (v, s) and (v, t) are in E, $P_D[v, s] = (v, s)$ and $P_I[v, t] = (v, t)$ from the definition.

From the definition of the s-t numbering, two paths $P_I[v, t]$ and $P_D[v, s]$ are well defined and $P_I[x, t]$ and $P_D[x, s]$ are node-disjoint for any node $x(\neq s, t)$.

We define the following concatenated paths with P_Is and P_Ds. Let x and y be arbitrary distinct nodes. $P_s[x, y]$, $P_t[x, y]$ and $P_{st}[x, y]$ are defined as $P_D[x, s] \odot P_D[s, y]$, $P_I[x, t] \odot P_I[t, y]$ and $P_D[x, s] \cdot (s, t) \cdot P_I[t, y]$(if $x < y$) and $P_I[x, t] \cdot (t, s) \cdot P_D[s, y]$(if $x > y$), respectively, and they are called s-path, t-path and st-paths, respectively(Fig. 3).

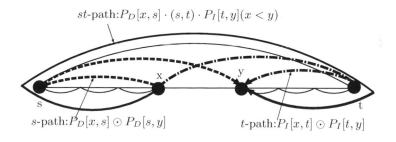

st-path:$P_D[x, s] \cdot (s, t) \cdot P_I[t, y](x < y)$

s-path:$P_D[x, s] \odot P_D[s, y]$ t-path:$P_I[x, t] \odot P_I[t, y]$

Fig. 3. Three kinds of paths, s-path, t-path and st-path.

4.2 Optimal Routing ρ

Let $G = (V, E)$ be a biconnected graph with n nodes. Assume that $n \geq 18$ and V is divided into $\lfloor n/18 \rfloor$ groups of 18 nodes each and the last group made up of the remaining (n mod 18) nodes. Each group except the last one is divided into two parts with 9 nodes each.

For a node $v \in V$, let $q = v$ div 18, $r = v$ mod 18, $g = r$ div 9 and $\ell = r$ mod 9. Each node v is represented as $[q; g, (i, j)]$, where $0 \leq q \leq \lfloor n/18 \rfloor$, $g = 0, 1$ and (i, j) is the ternary representation of ℓ ($0 \leq i, j \leq 2$).

$$\rho(x,y) = \begin{array}{|c|c|c|c|}
\hline
 & i_y = 0 & i_y = 1 & i_y = 2 \\
\hline
j_x = 0 & P_s[x,y] & P_{st}[x,y] & P_t[x,y] \\
\hline
j_x = 1 & P_t[x,y] & P_s[x,y] & P_{st}[x,y] \\
\hline
j_x = 2 & P_{st}[x,y] & P_t[x,y] & P_s[x,y] \\
\hline
\end{array}$$

where $x = [q_x; g_x, (i_x, j_x)]$ $y = [q_y; g_y, (i_y, j_y)]$.

Fig. 4. The routing ρ.

We define the routing ρ for G based on the ternary representation (i,j) of ℓ of each node as follows: Let x and y be represented as $[q_x; g_x, (i_x, j_x)]$ and $[q_y; g_y, (i_y, j_y)]$, respectively. The route $\rho(x,y)$ is defined as shown in Fig. 4. The route from x to y is determined based on j_x and i_y. For example, if $j_x = 0$ and $i_y = 2$ then the t-path from x to y is used to define $\rho(x,y)$ and $j_x = 2$ and $i_y = 0$ then the st-path from x to y is used to define $\rho(x,y)$ and so on. The routing ρ is well-defined. It is a unidirectional and $n(n-1)$-routing and its route-degree is $n-1$.

The following lemma is crucial to show that the routing ρ is $(2,1)$-tolerant. Let $I[q,g] = \{[q; g, (i,j)]|0 \le i \le j \le 2\}$, where $0 \le q \le \lfloor n/18 \rfloor - 1$ and $g = 0, 1$, and let $I[q] = I[q,0] \cup I[q,1]$.

Lemma 5. Let $G = (V, E)$ be a biconnected graph on which the routing ρ is defined. For arbitrary distinct nodes x and y and arbitrary two kinds of paths a-path and b-path$(a, b \in \{s, t, st\})$, there exists a node $z \in I[q,g]$ such that $\rho(x,z)$ is an a-path and $\rho(z,y)$ is b-path [2].

Proof. It can be shown by case analysis and it is omitted in this version.

Theorem 4. Let G be an n-node biconnected graph such that $n \ge 18$. The routing ρ on G is $(2,1)$-tolerant.

Proof. Let f be any fault and let $R = R(G, \rho)/\{f\}$. Let x and y be arbitrary distinct nonfaulty nodes in G. Except that $(x,y) \in E$ and $f = (x,y)$, we can assume that f is a node because if f is an edge we can consider that one of the endpoints of f is faulty. We write $f \in [a,b]$ if $f \in V$ and $a \le f \le b$.

Since there is one fault in G, either $I[0,0] = \{[0; 0, (i,j)]|0 \le i \le j \le 2\}$ or $I[0,1]$ does not contain f. Without loss of generality, we can assume that $I[0,1]$ is fault-free. Note that $I[0,1] = \{9, 10, \ldots, 17\}$. There are 12 cases according to the locations of x, y and $I[0,1]$ and they can be similarly proved by using Lemma 5. We show one case that $x < I[0,1] < y$(it means $x < 9$ and $17 < y$).

Suppose that $x < I[0,1] < y$. If $f = (x,y) \in E$, then from Lemma 5 there is a node $z \in I[0,1]$ such that $\rho(x,z)$ is an st-path and $\rho(z,y)$ is a t-path and

---
[2] It may be possible that $x = z$ or $y = z$. It can occur that $x, y \in I[q,g]$. In this case an empty path is considered to be $a(b)$-path, respectively

they do not contain f. If $f \in [s = 0, x - 1]$ then from Lemma 5 there is a node $z \in I[0, 1]$ such that both $\rho(x, z)$ and $\rho(z, y)$ are t-paths and their routes are fault-free. If $f \in [x + 1, 7]$, then from Lemma 5 there is a node $z \in I[0, 1]$ such that $\rho(x, z)$ is an s-path and $\rho(z, y)$ is a t-path and they are fault-free. The cases that $f \in [18, y - 1]$ and $f \in [y + 1, t = n - 1]$ can be proved similarly. Therefore $dis_R(x, y) \leq 2$.

The routing ρ is not an edge-routing because there is a case that $(x, y) \in E$ and $\rho(x, y)$ is defined as an st-path. However, it can be changed into an edge-routing as follows. Since both an s-path and a t-path from x to y become an edge if $(x, y) \in E$ from the definition, if $\rho(x, y)$ is defined by an st-path and $(x, y) \in E$, then $\rho(x, y)$ is defined as the edge (x, y). We can show that the modified ρ is $(2, 1)$-tolerant.

In the proof of Theorem 4, we only use the routes between nodes in $I[0]$, from nodes in $I[0]$ to other nodes and from nodes not in $I[0]$ to nodes in $I[0]$. Thus, we can obtain a $(2, 1)$-tolerant unidirectional edge-routing ρ_1 in which the total number of routes is $O(n)$ as follows.

routing ρ_1

Let x and y be represented as $[q_x; g_x, (i_x, j_x)]$ and $[q_y; g_y, (i_y, j_y)]$, respectively same as in ρ. $\rho(x, y)$ is defined as shown in Fig. 4 if $(q_x = q_y = 0)$, $(q_x \neq 0$ and $q_y = 0)$ or $(q_x = 0$ and $q_y \neq 0)$

Theorem 5. *Let G be an n-node biconnected graph such that $n \geq 18$. The routing ρ_1 on G is $(2, 1)$-tolerant edge-routing with $O(n)$ routes.*

4.3 Optimal Routing with Route-Degree o(n)

We construct a $(2, 1)$-tolerant routing ρ_2 with route degree $O(\sqrt{n})$ for n-node biconnected graphs by modifying the routing ρ.

Let $G = (V, E)$ be an n-node biconnected graph. We assume that each node x in G is denoted by $[q_x; g_x, (i_x, j_x)]$ same as in ρ. For simplicity, there is an integer ℓ such that $n/18 = \ell^2$. This condition can be easily removed and it will be shown in the final version. Thus, q_x can be represented by (q_x^L, q_x^R), where $1 \leq q_x^L, q_x^R \leq \ell = \sqrt{n/18}$.

routing ρ_2

Let x and y be represented as $[q_x = (q_x^L, q_x^R); g_x, (i_x, j_x)]$ and $q_y = [(q_y^L, q_y^R); g_y, (i_y, j_y)]$, respectively. $\rho_2(x, y)$ is defined as shown in Fig. 4 if $q_x = q_y$ or $q_x^R = q_y^L$.

In the routing ρ_2, the route $\rho_2(x, y)$ is defined if x and y are in the same interval $I[q]$ or the right part q_x^R of q_x and the left part q_y^L of q_y are equal. From the definition of ρ_2, we can verify that the route degree is $O(\sqrt{n})$. We can show that the routing ρ_2 is $(2, 1)$-tolerant by using Lemma 5 and the following lemma.

Lemma 6. *Let $x = [q_x = (q_x^L, q_x^R); g_x, (i_x, j_x)]$ and $y = [q_y = (q_y^L, q_y^R); g_y, (i_y, j_y)]$ be arbitrary distinct nodes of G on which ρ_2 is defined. Then, one of the following conditions holds.*

1. $q_x = q_y$, that is, x and y are in the same group $I[q_x = q_y]$.
2. $q_x^R = q_y^L$, that is, the routes from x to nodes in $I[q_y]$ are defined.
3. There is a group $I[q_z]$ such that $q_x^R = q_z^L$ and $q_z^R = q_y^L$, that is, the routes from x to nodes in $I[q_z]$ and from nodes in $I[q_z]$ to nodes in $I[q_y]$ are defined.

The total number of routes defined in ρ_2 is $O(n\sqrt{n})$, because the route degree of each node is $(\sqrt{n})$. From Lemma 1 the total number of routes is at least $\Omega(n\sqrt{n})$ to define $(2,1)$-tolerant routings with route degree $O(\sqrt{n})$ The routing ρ_2 attains the lower bound of the total number of routes.

Since in the case that $n < 18$ we can construct a $(2,1)$-tolerant edge-routing for n-node biconnected graphs[12], the following theorem holds.

Theorem 6. *Let G be an n-node biconnected graph. We can construct $(2,1)$-tolerant edge-routing on G with $O(n\sqrt{n})$ routes and route degree $O(\sqrt{n})$.*

5   Concluding Remarks

We have shown two optimal edge-routings with smaller routing tables. It is an interesting open question whether or not there exists an $(2, k-1)$-tolerant routing with route degree $O(k\sqrt{n})$ for n-node k-connected graphs($k \geq 3$). It is also an interesting open question whether or not there exists an $(2,1)$-tolerant bidirectional routing with route degree $O(\sqrt{n})$ for n-node biconnected graphs.

Acknowledgement.   The authors grateful to Mr. Yoriyuki Nagata and Mr. Takahiro Honda for their useful discussion. This research is supported in part by a Scientific Research Grant-In-Aid from the Ministry of Education, Science and Culture, Japan, under grant No. 10680352 and 10205209.

References

1. B.Bollobás: Extremal graph theory, *Academic Press*, 172(1978).
2. A.Broder, D.Dolev, M.Fischer and B.Simons: "Efficient fault tolerant routing in network," *Information and Computation*75,52–64(1987).
3. I.Cidon, O.Gerstel and S.Zaks: "A scalable approach to routing in ATM networks," *Proc. 8th International Workshop on Distributed Algorithms*, LNCS 859, 209–222 (1994).
4. D.Dolev, J.Halpern , B.Simons and H.Strong: "A new look at fault tolerant routing," *Information and Computation*72, 180–196(1987).
5. S. Evens: Graph algorithms, *Computer Science Press*, Potomac, Maryland(1979).
6. P.Feldman: "Fault tolerance of minimal path routing in a network,*in Proc. 17th ACM STOC*,327–334(1985).
7. E. Györi: "On division of connected subgraphs" in: Combinatorics(Proc. 5th Hungarian Combinational Coll.,1976, Keszthely) North-Holland, Amsterdam, 485–494(1978).
8. F.Harary, Graph theory, *Addison-Wesley*, Reading, MA(1969).

9. M.Imase and Y.Manabe: "Fault-Tolerant routings in a κ-connected networks," *Information Processing Letters*, 28, 171–175 (1988).
10. K.Kawaguchi and K.Wada: "New results in graph routing," *Information and Computation*, 106, 2, 203–233 (1993).
11. D.Peleg and B.Simons: "On fault tolerant routing in general graph," *Information and Computation*74,33–49(1987).
12. K.Wada, W.Chen Y.Luo and K.Kawaguchi: "Optimal Fault-Tolerant ATM-Routings for Biconnected Graphs," Proc. of the 23rd International Workshop on Graph-Theoretic Concepts in Computer Science(WG'97), Lecture Notes in Computer Science, 1335, 354–367(1997).
13. K.Wada and K.Kawaguchi: "Efficient fault-tolerant fixed routings on $(k + 1)$-connected digraphs," *Discrete Applied Mathematics*, 37/38, 539–552 (1992).
14. K.Wada, Y.Luo and K.Kawaguchi: "Optimal Fault-tolerant routings for Connected Graphs," *Information Processing Letters*, 41, 3, 169–174 (1992).
15. K.Wada, T.Shibuya, K.Kawaguchi and E.Syamoto: "A Linear Time (L,k)-Edge-Partition Algorithm for Connected Graphs and Fault-Tolerant Routings for k-Edge-Connected Graphs" Trans. of IEICE, J75-D-I, 11, 993–1004(1992).